ENERGY STORAGE IN ENERGY MARKETS

ENERGY STORAGE IN ENERGY MARKETS

UNCERTAINTIES, MODELLING, ANALYSIS AND OPTIMIZATION

Edited by

BEHNAM MOHAMMADI-IVATLOO
Faculty of Electrical and Computer Engineering, University of Tabriz, Tabriz, Iran

AMIN MOHAMMADPOUR SHOTORBANI
Faculty of Applied Science, University of British Columbia, Vancouver, BC, Canada

AMJAD ANVARI-MOGHADDAM
Department of Energy Technology, Aalborg University, Aalborg, Denmark

ELSEVIER

ACADEMIC PRESS
An imprint of Elsevier

Academic Press is an imprint of Elsevier
125 London Wall, London EC2Y 5AS, United Kingdom
525 B Street, Suite 1650, San Diego, CA 92101, United States
50 Hampshire Street, 5th Floor, Cambridge, MA 02139, United States
The Boulevard, Langford Lane, Kidlington, Oxford OX5 1GB, United Kingdom

Notices

Knowledge and best practice in this field are constantly changing. As new research and experience broaden
our understanding, changes in research methods, professional practices, or medical treatment may become
necessary.

Practitioners and researchers must always rely on their own experience and knowledge in evaluating and
using any information, methods, compounds, or experiments described herein. In using such information or
methods they should be mindful of their own safety and the safety of others, including parties for whom
they have a professional responsibility.

To the fullest extent of the law, neither the Publisher nor the authors, contributors, or editors, assume any
liability for any injury and/or damage to persons or property as a matter of products liability, negligence or
otherwise, or from any use or operation of any methods, products, instructions, or ideas contained in the
material herein.

Library of Congress Cataloging-in-Publication Data
A catalog record for this book is available from the Library of Congress

British Library Cataloguing-in-Publication Data
A catalogue record for this book is available from the British Library

ISBN: 978-0-12-820095-7

For information on all Academic Press publications visit our
website at https://www.elsevier.com/books-and-journals

Publisher: Brian Romer
Acquisitions Editor: Graham Nisbet
Editorial Project Manager: Chris Hockaday
Production Project Manager: Selvaraj Raviraj
Cover Designer: Miles Hitchen

Typeset by TNQ Technologies

Contents

14. Electric energy storage systems integration in energy markets and balancing services

Vahid Vahidinasab and Mahdi Habibi

15. Reliability modeling of renewable energy sources with energy storage devices

Vasundhara Mahajan, Soumya Mudgal, Atul Kumar Yadav, and Vijay Prajapati

16. Reliability and resiliency assessment in integrated gas and electricity systems in the presence of energy storage systems

Mohammad Taghi Ameli, Hossein Ameli, Goran Strbac, and Vahid Shahbazbegian

17. Reliability analysis and role of energy storage in resiliency of energy systems

Mohammad Taghi Ameli, Kamran Jalilpoor, Mohammad Mehdi Amiri, and Sasan Azad

18. Electric vehicles and electric storage systems participation in provision of flexible ramp service

Sajjad Fattaheian-Dehkordi, Ali Abbaspour, and Matti Lehtonen

Contributors

Ali Abbaspour Sharif University of Technology, Tehran, Iran

Masoud Agabalaye-Rahvar Faculty of Electrical and Computer Engineering, University of Tabriz, Tabriz, East Azerbaijan, Iran

Masoumeh Ahrabi Department of Electrical Engineering, Amirkabir University of Technology, Tehran, Iran

Alireza Akbari-Dibavar Faculty of Electrical and Computer Engineering, University of Tabriz, Tabriz, East Azerbaijan, Iran

Ameena Saad Al-Sumaiti Advanced Power and Energy Center, Department of Electrical Engineering and Computer Science, Khalifa University, Abu Dhabi, United Arab Emirates

Mohammad Taghi Ameli Department of Electrical Engineering, Shahid Beheshti University, Tehran, Iran

Ali Ameli Entrepreneurship Incubator, Shahid Beheshti University, Tehran, Iran

Hossein Ameli Department of Electrical and Electronic Engineering, Imperial College London, London, United Kingdom

Mohammad Amin Mirzaei Faculty of Electrical and Computer Engineering, University of Tabriz, Tabriz, East Azerbaijan, Iran

Mohammad Mehdi Amiri Department of Electrical Engineering, Shahid Beheshti University, Tehran, Iran

Meisam Ansari Electrical and Computer Engineering, Southern Illinois University, Carbondale, IL, United States

Jaber Fallah Ardashir Department of Electrical Engineering, Tabriz Branch, Islamic Azad University, Tabriz, East Azerbaijan, Iran

Somayeh Asadi Department of Architectural Engineering, Pennsylvania State University, State College, PA, United States

Mohammad Hosein Asgharinejad Keisami Department of Electrical Engineering, Shahid Beheshti University, Tehran, Iran

Sasan Azad Department of Electrical Engineering, Shahid Beheshti University, Tehran, Iran

Hamid Reza Baghee Department of Electrical Engineering, Amirkabir University of Technology, Tehran, Iran

Mahsa Bagheri Tookanlou Faculty of Engineering and Environment, Department of Maths, Physics and Electrical Engineering, Northumbria University Newcastle, Newcastle upon Tyne, United Kingdom

Sajjad Fattaheian-Dehkordi Aalto University, Espoo, Finland; Sharif University of Technology, Tehran, Iran

Hadi Vatankhah Ghadim Department of Electrical Engineering, Tabriz Branch, Islamic Azad University, Tabriz, East Azerbaijan, Iran

Gevork B. Gharehpetian Department of Electrical Engineering, Amirkabir University of Technology, Tehran, Iran

Mahdi Habibi Faculty of Electrical Engineering, Shahid Beheshti University, Tehran, Iran

Alireza Heidari School of Electrical Engineering and Telecommunications (EE&T), The University of New South Wales (UNSW), Sydney, NSW, Australia

Mohammad Hemmati Faculty of Electrical and Computer Engineering, University of Tabriz, Tabriz, East Azerbaijan, Iran

Vahid Hosseinnezhad School of Engineering, University College Cork (UCC), Cork, Ireland

Faezeh Jalilian Faculty of Electrical and Computer Engineering, University of Tabriz, Tabriz, East Azerbaijan, Iran

Kamran Jalilpoor Department of Electrical Engineering, Shahid Beheshti University, Tehran, Iran

Hossein Khounjahan Azarbijan Regional Electric Company, Tabriz, East Azarbaijan, Iran

Matti Lehtonen Aalto University, Espoo, Finland

Vasundhara Mahajan Department of Electrical Engineering, Sardar Vallabhbhai National Institute of Technology, Surat, Gujarat, India

Amin Mansour-Saatloo Faculty of Electrical and Computer Engineering, University of Tabriz, Tabriz, East Azarbaijan, Iran

Mousa Marzband Faculty of Engineering and Environment, Department of Maths, Physics and Electrical Engineering, Northumbria University Newcastle, Newcastle upon Tyne, United Kingdom; Center of Research Excellence in Renewable Energy and Power Systems, King Abdulaziz University, Jeddah, Saudi Arabia

Behnam Mohammadi-Ivatloo Faculty of Electrical and Computer Engineering, University of Tabriz, Tabriz, East Azarbaijan, Iran; Department of Energy Technology, Aalborg University, Aalborg, Denmark

Arash Moradzadeh Faculty of Electrical and Computer Engineering, University of Tabriz, Tabriz, East Azarbaijan, Iran

Fariba Mousavi Faculty of Electrical and Computer Engineering, University of Tabriz, Tabriz, East Azarbaijan, Iran

Soumya Mudgal Department of Electrical Engineering, Sardar Vallabhbhai National Institute of Technology, Surat, Gujarat, India

Mehrdad Setayesh Nazar Faculty of Electrical Engineering, Shahid Beheshti University, Tehran, Iran

Morteza Nazari-Heris Faculty of Electrical and Computer Engineering, University of Tabriz, Tabriz, East Azarbaijan, Iran; Department of Architectural Engineering, Pennsylvania State University, State College, PA, United States

Ali Parizad Electrical and Computer Engineering, Southern Illinois University, Carbondale, IL, United States

Roghayyeh Pourebrahim Faculty of Electrical and Computer Engineering, University of Tabriz, Tabriz, East Azarbaijan, Iran

Vijay Prajapati Department of Electrical Engineering, Sardar Vallabhbhai National Institute of Technology, Surat, Gujarat, India

Vahid Shahbazbegian Department of Electrical Engineering, Shahid Beheshti University, Tehran, Iran

Goran Strbac Department of Electrical and Electronic Engineering, Imperial College London, London, United Kingdom

Sajjad Tohidi Faculty of Electrical and Computer Engineering, University of Tabriz, Tabriz, East Azarbaijan, Iran

Vahid Vahidinasab School of Engineering, Newcastle University, Newcastle upon Tyne, United Kingdom

Atul Kumar Yadav Department of Electrical Engineering, Sardar Vallabhbhai National Institute of Technology, Surat, Gujarat, India

Kazem Zare Faculty of Electrical and Computer Engineering, University of Tabriz, Tabriz, East Azarbaijan, Iran

Introduction

The increasing penetration of intermittent renewable energy resources faces the energy systems with diverse challenges. Energy storage systems (ESSs) play a vital role in today's energy system to compensate for the intermittency and uncertainty in renewable energy generation. On the other hand, in a deregulated energy market, energy storage is utilized as a critical energy entity in response to market interactions and price variations.

Regarding the importance of ESSs in energy markets, this book aims to explore the modeling, optimization, and analyzing the impact of ESSs in the energy market, for an audience of energy, power, mechanical, chemical, process, and environmental engineers as well as the researchers. It is targeted at improving operational efficiency through new approaches involving utility systems and process integration.

New concepts of the energy market and innovative technologies of ESSs in power and energy systems have been developing significantly during the last decade. Different analyses and practices are investigated and improved by researchers and engineers. Lack of a comprehensive resource in integrating the various features of both energy markets and storages, including the technical and operational aspects, inspires a new resource in this area. This book recognizes and discusses the technology and operation of ESSs and the market issues for researchers and engineers.

Topics covered include energy storage technologies; different applications of ESSs; environmental impacts of energy storage; reliability and availability; design, integration, and operation of ESSs; optimal energy management systems; optimization of energy storage utilization; storage and market components; economic analysis of energy systems; scheduling methods; market analysis and response.

The ongoing penetration of renewable energy sources and the emergence of electric vehicles and ESSs have been growing worldwide and will propagate further since many researchers, power utility companies, and organizations in different countries are applying and administering new actions to control climate change by reducing greenhouse gas emissions. Nonetheless, this turnaround results in ever more challenges for the energy system operators to handle the system's energy needs due to the increasing difficulty in predicting demand.

Modern ESSs are extensively implemented as a new flexible solution to the abovementioned challenge by researchers and engineers. Storage systems provide flexibility to both utility and the customers, and thus are interesting to demand and supply sides. Although the technical aspect of energy storage is an active field of research and engineering, the assessment of storage applications requires more investigations from economic and investment points of view to fill the gaps.

Inspired by this background, the integration and technoeconomic evaluation of ESSs in the energy and ancillary multimarkets have become the *daily challenges* of the researchers, investors, utilities, and engineers.

In this context, this book addresses the challenges and defines the technoeconomic cases on the subject, describes the methodologies, and explains the solutions. Consequently, an evaluation framework is developed for energy markets with storage. The remarks, including the motivations and obstacles of deployment of ESSs in energy markets, are discussed. Besides, impacts of ESS on the energy market, as well as demand and price, are analyzed. In detail, the chapters are organized as the following.

Chapter 1 provides a comprehensive and updated review of the basics and operation models of the energy markets from the initial appearance to the current status. Dealings with the main challenges of the electrical energy markets and various types of electrical energy markets are discussed, and the fundamentals of their operation are introduced. Chapter 2 gives an overview of various types of ESSs. Moreover, the operation of the ESSs from different aspects of the electricity markets is studied, and the current status and the role of ESSs considering their impacts in electricity markets are reviewed. Chapter 3 discusses the use of various types of ESSs in wind farm applications. Chapter 4 focuses on the optimal planning of ESSs with renewable sources such as wind and solar, and dispatchable distributed generation to participate in the energy market. In this chapter, the impact of energy price uncertainty on the profitability of ESSs and related resources is also investigated. Chapter 5 addresses the effects of carbon capture power plants and compressed air ESSs on the power system performance in a carbon-constrained environment and in the presence of wind generation uncertainty using a power flow optimization framework. Chapter 6 highlights the various aspects of the economics of lead−acid, nickel−metal hydride, and Lithium-ion batteries and the market

dynamics, which affect the battery industry. The performance costs of the lead−acid, nickel−metal hydride, and lithium-ion batteries over the years and their impacts on the electric vehicles (EVs) industry are evaluated. This chapter indicates the market landscape and its growth prospects over the upcoming years. Chapter 7 develops a holistic day-ahead scheduling framework to guarantee economic and energy-efficient routing of EVs, where each EV and charging station (CS) finds optimal CSs and ESSs, respectively, for charging and discharging based on a cloud scheduling system. Chapter 8 studies the presence of massive functional entities like distributed energy resources (DERs) and EVs and their unpredictable actions that increase the risk of congestion in the system. It investigates the roles of new intermediary entities, like aggregators and microgrid operators on the fluency of the interactions using an adaptive framework in a deregulated active distribution network. Chapter 9 proposes a scheduling model, including a hydrogen storage system, power-to-hydrogen, and hydrogen-to-power technologies using a hydrogen fuel cell. The developed model can handle different uncertainties using stochastic programming. Chapter 10 presents a multilevel optimization framework for a resilient multienergy carrier distribution system in a day-ahead scheduling problem, in which the nonutility parking lots of plug-in hybrid EV aggregators' and electrical ESS aggregators' contribution strategies can change operational paradigms and sell their electricity to the distribution system. Chapter 11 focuses on solving the bidding strategy problem of EVs aggregators to optimally participate in the energy and flexible ramping market under the uncertainties of the system. Chapter 12 investigates the potential of applying EVs as ESSs in ancillary services markets, based on the vehicle-to-grid (V2G) concept and the characteristic of the EVs batteries, by offering various grid services, including load management, demand response, and grid regulation services like frequency control. Chapter 13 presents an optimal scheduling problem for integrated EV parking lots (EVPLs) in the day-ahead scheduling of wind-based power systems, which is formulated as a network-constrained unit commitment model considering the AC power flow model. Chapter 14 discusses the role of large-scale ESSs in joint energy and ancillary multimarkets. Moreover, the suitable technologies for large-scale storage facilities, their role in energy grids, and their services are explained. Chapter 15 analyzes the current progress in the integration of electric ESSs (EESSs) into the existing energy and balancing mechanisms by providing a detailed formulation of the problem. Chapter 16 proposes a multistate model using discrete Markov chains for reliability modeling of renewable energy sources and storage units to overcome the drawbacks of conventional methods such as Weibull and Beta Probability Density Functions. Chapter 17 presents the algorithms for enhancing the reliability and resiliency in the coordinated operation of gas and electricity networks by introducing a comprehensive framework, including a mathematical model for the operation of these networks in the presence of flexibility options, such as gas and electricity storage systems. Chapter 18 presents a comprehensive model for viewing the role of ESS in energy systems to improve network resiliency and reliability. The resiliency criterion for reducing the impact of severe accidents on the network is considered as a module of the objective function. Chapter 19 reviews the important concepts in flexible ramping products and state-of-the-art techniques that enable participation of EVs/ESSs in providing flexible ramping services in different operational management levels of restructured power systems.

Behnam Mohammadi-Ivatloo
Amin Mohammadpour Shotorbani
Amjad Anvari-Moghaddam

Energy market fundamentals and overview

Fariba Mousavi[1], *Morteza Nazari-Heris*[1,3],
Behnam Mohammadi-Ivatloo[1,2], *Somayeh Asadi*[3]

[1]Faculty of Electrical and Computer Engineering, University of Tabriz, Tabriz, East Azerbaijan, Iran; [2]Department of Energy Technology, Aalborg University, Aalborg, Denmark; [3]Department of Architectural Engineering, Pennsylvania State University, State College, PA, United States

1. History of electricity markets

The first modern power market and the first steps in line with the privatization and restructuring of power systems are related to Chile in the late 1970s. The model used in Chile in its entirety was successful in the entry of rationality and transparency to the power pricing system, but at the same time, it was accompanied by a number of big companies that threaten efficiency and competitiveness in the marketplace. Following that, Argentina renewed its restructuring with a view to Chile's experience by imposing restrictions on the market's focus and raising the reliability of its power system. One of the main objectives of the power market launch, which the Argentinean government pursued, was the privatization of power generating systems due to the state monopoly of undesirable technical conditions, which accordingly encountered continuous disturbances [1]. In fact, one of the main reasons for the restructuring of the electricity industry is the failure of the vertical monopoly of the government on the industry, which separates the integrated vertical and interconnected supply chain. It also separates the industry into competitive and noncompetitive sectors. The issue of vertical integration, which is interpreted as joint ownership of manufacturing and retail businesses in the supply chain (Fig. 1.1), is an obstacle to competition in the electricity market. At the same time, the government was seeking to attract the funds needed to restructure the assets and expand the existing system. During the 1990s, several restructuring projects in other Latin American countries, including Peru, Brazil, and Colombia, were implemented by the World Bank, but they were not successful [1].

Energy Storage in Energy Markets
https://doi.org/10.1016/B978-0-12-820095-7.00005-4

1

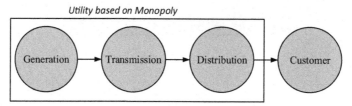

FIGURE 1.1   Monopoly model of the power market.

The most successful move in the power industry occurred in 1990 when Margaret Thatcher was the prime minister in the United Kingdom with the privatization of the British and Wales Electric Industry. The model employed in the United Kingdom has been used as a standard for liberalization and deregulation in other commonwealth of independent states (CIS) countries, notably Australia and New Zealand, and regional markets such as Alberta and Canada. Of course, it should be noted that in many of these cases, market liberalization was carried out without large-scale privatizations, such as the United Kingdom. Experiences of the energy market foundation in different countries in terms of newly created entities and market designs are different, but their underlying principles are the same. In 1998, the local market was established in California, Pennsylvania, New Jersey, and Maryland and served as a global market model. However, the crisis of 2000 and 2001 in California slowed down the pace of deregulation to some extent. Later, the New England region and the state of New York followed a similar market. In 2001, Britain finished the second market reform on contract-based free trade [2]. Thus, the process of market reform around the world has been followed and is currently underway in parts of Asia and Africa. The electricity industry in Southeast Asian countries such as Japan, South Korea, India, Singapore is in different stages of restructuring, and in many other countries, including Iran, this process is implemented with indigenous features and characteristics.

1.1 Foundation of the electricity markets

From the beginning, government or quasi-government agencies were responsible for the management and control of the power industry. In this type of production structure, transmission and distribution of electric energy are carried out exclusively by an entity, which is the result of the so-called monopoly in the industry. In the traditional structure of the electricity market, the electricity supply is a public service with an exclusive feature, in which only the government and government agencies monitor the market and regulate the rate of return and the cost structure. In exclusive markets, local electricity companies have a monopoly on macro/microsales of electrical energy in certain geographical areas. The basis of this monopoly is granting of all privileges as well as all restrictions on public services to electricity companies. Under these conditions, the power company is obliged to supply electricity at a certain level of reliability with a fully controlled price. The unfavorable technical condition caused by state monopoly in the electricity industry production sector led some countries to privatization by transitioning from a monopoly to the competitive industry. However, the transmission sector has still remained a monopoly all over the world [3,4].

The power system operation between supply and demand must be constantly balanced. Inadequate supply will lead to the allocation of new production capacity and will lead to high prices. The automatic production control (AGC) is widely used to maintain the safety and reliability of the system, in particular for regulation and control of the frequency deviation and support of commercial transactions. In deregulated power markets, one imprortant design issue is a suitable and desirable mechanism for supplying and pricing AGC services. In a competitive power market, the AGC services are made through either negotiations between the market operator and the concerned production companies or a special form of auction. It is evident that the need for AGC capacity has a considerable effect on the result of the AGC market. If there is a mismatch between the planned production output and the system load, the generation units in the AGC mode automatically either increase or decrease production output to lead to an equilibrium [5].

1.2 Deregulation of the electricity market

The liberation of the energy market began around 1996 in the European Union by Norway and Britain. At the same time, talks were held to issue an order to release electricity from the European Union. The result of these talks was signing an agreement on the liberalization of the energy market by the Council of Ministers. The goal of liberalizing the energy market was to achieve greater efficiency and lower prices for consumers by increasing market competition. This has been partially but sometimes not on an expected scale explored in some countries [5].

Based on economic theory, the price of a commodity should be equal to its production cost to maximize the social welfare and public satisfaction. Of course, the practical implementation of this proven economic principle is not so simple to achieve this result, and one can choose two trends:

- Establishing organized structure and forming controlled producers who sell goods at a price equal/near to the border cost.
- Relying on a default that the producers will select the same price. In other words, the trend is the creation of a natural process for selecting the same price by the manufacturer.

The conventional structure of the power industry is based on the first assumption. For this reason, only one producer is considered in the traditional structure for this product. Of course, if this exclusive producer considers the prices equal to the cost of its production, it cannot attain maximum benefit. After avoiding the dominance of this proprietary manufacturer in the market, it is possible to have a regulator that can handle this situation and is able to price it within specific bounds. But this structure is not efficient to guarantee the profit. The concept of free selection, which is important in many countries, is not included in such structure. Besides, the bargaining power of interest groups is detrimental to the regulation of market regulations to achieve greater profits in many countries. This trend has led to the formation of an exclusive structure that only benefits a particular group.

The ultimate goal of liberating electricity markets was to have an economic incentive to have a natural way to achieve a price at the limit of border expenses and, of course, an increase in efficiency (i.e., maximum social welfare). The efficiency in our discussion is only a specific meaning of a condition with significant diversity. In general, economic view and in terms of efficiency, especially in the field of liberalization of electricity markets, efficiency is not only meant for the maximization of the social welfare in a static situation, but it also involves the dynamic process of fixing production costs. In other words, competition is expected not only to reduce the unit prices of the border costs but also to attain efficient exploitation of resources. In short, the economic incentives of competition can be sought to achieve two main goals, including pricing on border costs and minimization of production costs. By adopting this economic idea, there is no choice but to remove the monopoly structure of the electricity industry, as otherwise it will be charged from social assets to compensate the costs of a monopoly system under regulation [6].

From the beginning, electricity was sold as a product at the market price, which became a basic product after a time with the development of the electricity industry with strict regulation. From the post–World War II period until the late 1970s, the European governments supported the formation of a vertical monopoly to exploit transmission networks or distribution networks. Deregulation began in the early 1960s, and since then, many developed countries have supplied electricity with market strategy [2,7].

The liberalization of the energy market aimed at improving efficiency and reducing the cost of consumers by increasing trade competition. Based on economic theory for social welfare maximization and public satisfaction, the price of a commodity should be equal to its production cost. The implementation of this principle requires a set of structures and the formation of controlled manufacturers who sell goods at a price equal to the cost of the border [8]. In fact, the conventional structure of the power industry is based on this assumption, which leads to the consideration of only one producer for this product in the traditional structure. It is obvious that if the sole producer price of goods is equal to the frontier cost, it cannot achieve maximum profit. So, to avoid this monopoly producer dominating the market, there is usually a regulatory body that controls this monopoly situation and ensures that prices do not exceed by a socially acceptable standard. However, this structure is not efficient for guaranteed profit. The concept of free selection, which is considered as an important factor in many countries, is not included in it, and the bargaining power of interest groups is detrimental to the market regulations in order to achieve greater profits in many countries. In fact, the monopoly structure has operated based on rules set up for profit by a particular group. This is where the importance of deregulation and liberalization of electricity markets is clear. The ultimate goal of the liberalization of electricity markets is to reach a price at the limit of border expenses and to maximize the social well-being and efficiency. In the general economic view of efficiency, especially in the field of liberalization of electricity markets, efficiency is not only meant to maximize the social welfare in a static state, but also the dynamic process of fixed modification of production costs. In other words, competition is expected not only to reduce the unit prices of the border costs but also to attain efficient exploitation of resources. In summary, economic incentives to create competition can be sought to achieve two major objectives: price-based cost definition and minimization of production cost. By adopting this economic idea, there is no choice but to remove the monopoly structure of the electricity industry, as otherwise, it would be used to compensate for the cost of a monopoly system under regulation. With the process of decentralization of the electricity industry in the world, the importance of optimal energy purchasing and formulation of a regular decision process cannot be achieved. Electricity supply industries have been restructured in steps leading to the creation of competitive energy markets in the industry. Generally, these arrangements require a kind of central bid to provide electricity in short periods of time. Given that electricity delivered directly, it is obvious that the day-ahead market will require a balanced service with a real-time market. The institution that is able to meet these conditions is the system's sovereign operator. The independent system operator (ISO) is responsible for controlling the network, setting transfer tariffs, maintaining system security, scheduling periodic maintenance, and contributing to long-term decisions. The real-time exploitation of the market is a fundamental responsibility of the ISO so that the energy supply bid and the procedures for submitting it are determined by the operator. It is also necessary to ensure that there is enough quantity of automatic production control and rapid production capacity meeting the standards of mandatory reliability. One way to satisfy these needs is to create a competitive market for ancillary services by the ISO. Providing operating reserves and automatic generation control (AGC) services is one of the other important tasks of this operator. These conditions are the result of deregulation and liberalization of the power industry, the creation of a competitive, safe, and transparent energy market, the abolition of tariffs, and the end of state monopoly by reducing and simplifying regulations [9,10].

1.3 Electric energy restructuring and integration

By the early 1980s, the politic, environment, and technology had significant pressure to reform the electric energy sector in Europe. It is while the energy market in the world has turned to privatization, deregulation, and liberalization with the aim of the withdrawal of the electricity industry from the monopoly of government bodies and government. However, there are two main reasons for the power of these entities and delay in the liberalization of the electrical energy sector, which are the security of supply and complexity of goods. These issues are two fundamental challenges in the restructuring of the electric grid [7].

The diversity of different sectors of the power industry makes unified analysis difficult and complicated. However, it can be shown that even if the reasons for initiating the restructuring process differ between the two countries (i.e., supply crisis, the inefficiency of management, and lack of resources for investment), most of them have successfully achieved the goals proposed at the beginning of the reform. The reform processes are indeed based on a fundamental shift in the paradigm for e-commerce. The paradigm of a business sense has evolved vertically integrated by the government, in which the economic characteristics are substantially recognized at the stages of production, transmission, distribution, and supply. The main objectives of the reform process are to maximize social welfare and create effective economic conditions in the industry, enabling competitive markets in all possible sectors and efficient regulation. The restructuring has led to the development of low-cost energy supply, with levels of economic reliability and service quality, increasing the use of energy and reducing losses. On the other hand, governments have greatly encouraged private investment, and the role of traditional entrepreneurship has changed the government to the role of regulator. Indeed, through the introduction of competition, privatization of state industries, and the deregulation of key industrial sectors, a strong alternative was proposed instead of what was generally considered as a centralized and inefficient electrical industry. Going toward the privatization of the electricity market has resulted in a substantial change in the performance and organization of power systems, which are large and complex integrated engineering systems that require a certain level of makeup. The important achievements mentioned are the proceeds of these reforms, although this differs from one country to another. Inadequate investment in the new production, transmission capacity, and distribution are given to private investors vying for the development of new infrastructure. The inefficient management of the government in the field of electric installations, with low labor productivity, provides a way for efficient private companies that increase energy production, looking for new opportunities to cut costs and increase income [11].

While the basic model for structural and regulatory reform in electricity is relatively simple, the needed details of institutional reform to improve the performance of the current system are complex. Structural and regulatory reform of electricity sectors in developing countries and other countries is looking for a model that has already been applied to network industries such as telephone and natural gas. Potential competitive sectors (i.e., power generation) are structurally or functionally disconnected from the natural monopolies (i.e., physical transmission and power distribution). The prices were set for entry and exit from the deregulated competitive segments, and it gave consumers the opportunity to choose from competitors. However, there is interesting public attention to structural and regulatory reforms, which lead to increased competition, if appropriate use of institutional arrangements could result in real cost savings. Due to the crucial role of these reforms in the economy, there are deep public interests to ensure that such reforms are improved instead of reducing the performance of the electricity sector in the long run. In fact, it can be stated that the restructuring of the electricity industry is done in order to reinforce competition in the manufacturing sector and to reform the regulation of the transfer and distribution functions [12].

2. Challenges of energy markets

One of the main challenges of liberalization of the energy market is that there is no precise analysis to examine the broader social and environmental consequences of liberalization. There are problems such as environmental consequences, supply security, sustainable long-term planning, resource exhaustion, improving renewable energy systems, and effective investment in sustainable energy systems. To solve these problems, long-term social-term planning is needed to contradict the short-term time horizon for the benefit of a deregulated business market. It is often said that the industry transition requires a free market for new investments that include new transaction costs ending with customers. Therefore, the transition to a free power market requires establishing large updated databases in order to control the transit of customers between suppliers [6]. Controlling air pollution, reduction of greenhouse gas emissions, integration of renewable energy, and supply of modern energy services are some of the energy challenges that have always been discussed. Because the supply of electricity must continuously meet demand, and storage is not cheap, energy suppliers in the market will be able to exercise market power and other issues in the market. Governments and quasi-government organizations pose other challenges to policymakers and electricity legislators to reduce greenhouse gas emissions and increase the use of renewable energy. In the following, we will explain more about these challenges.

To change resources from fossil fuels to renewables, electricity markets are changing. Wind and sun can be considered as the main renewable energies that bring challenges. The main challenge for renewable sources is the availability of alternating access and the variability of these resources, which have zero final cost and no inertia. These challenges can be improved by the application of battery storage responding to better demand, which is currently being explored in several studies. On the other hand, following the reduction of air pollutants and the issue of decarbonization, as the most important environmental challenges, the advancement of electrical systems, electrical modifications, and facilitating the transfer of the current system to clean energy systems have been considered by various countries. However, there are obstacles that make market reform difficult. For example, one of the most important barriers to China's current clean energy system is the market reform of coal-fired power generation, which currently accounts for most of China's electricity generation. China ranks first in the world in coal production, accounting for more than 80% of its heat production through coal. The challenges that China faces in transferring the system are two separate issues that may have a common solution. The first is the issue of transfer: China has the capacity to produce coal compared to the expectation supported by the market price in the short term. The second issue of the transfer is that coal producers in China who are unable to pay their fixed costs will be forced to retire. These issues overshadow China economically. Therefore, it is important for Chinese policymakers to anticipate and address the impact of reforms in the electricity market on existing coal producers and to provide sustainable political solutions. Electricity markets can play an important role in China's transition to a low-carbon energy system. However, if markets are placed on transfer issues, their potential may not be realized. Therefore, in many countries, there may be obstacles and problems for decarbonization, which will challenge the transition of the current transmission system to clean energy systems. Policies aimed at decarbonization of the electricity sector through the expansion and invasion of wind energy penetration and solar energy production have significant implications for the performance of electricity wholesale markets [13].

In addition, a transition from a system in which fossil fuels have a high share of energy to a system in which renewable energy has been substantially replaced requires newer mechanisms with real-time performance and the necessary infrastructure. Lack of a coordinated policy on renewable energy could lead to more centralized market models. With the increase in production through renewable energy production, the need for government funding for more vulnerable areas to establish renewable energy plants is of importance. However, this is not the main challenge, and there is another challenge to create congestion in tender areas. Coupling is achieved by calculating the capacity and implicating auctions using a single optimization algorithm implemented by the power exchange or other institutions. These assumptions are derived from historical experience. However, with the increase in network flow, even the old control zones may not be able to withstand future flows. Tender areas are complex control areas. Another important issue facing electrical systems is the deregulation of electricity markets. In the deregulated electricity market, participants do not necessarily act as price takers. Participants in the transmission network can manipulate electricity prices and increase their economic surplus beyond competitive levels through strategic proposals. This effect leads to higher price levels and loss of social welfare. This is where the issue of market surveillance by a neutral organization becomes even more important in order to ensure social welfare, including enough profits for producers and ultimate consumer satisfaction. Therefore, the new system requires sufficient legislation and oversight to prevent potential problems [14,15].

Another issue that most European countries are pursuing is the development of national energy markets to integrated border wholesale markets, which require the establishment of coordinated executive laws and regulations. This has been the case for progress toward a greater focus or better cooperation at the European level. In order to provide an efficient market, there is a need for rules and regulations that lead to optimal regional and national decisions and solutions to the general interests and optimal utilization. Inefficient planning and decisions will exacerbate and increase volatility in the system. The challenges ahead may call into question this plan [13]. An unresolved issue is the management of the interaction between transfer investment and investment in production. Optimal production investments require knowledge of long-term transmission programs, but production investments ultimately affect transmission programs. There is no clear answer to this coordination problem. Finally, in general, the unpreparedness of society in terms of technology and economics, and on the other hand, the complex constraints on the system's transition network from a traditional structure to a restructured structure are challenged. Given the solid foundation that has been built over the past two decades, it is expected that the electricity market plans will continue to advance and meet future challenges. The key to effective management is careful planning and focus on the basic principles of the market in order to achieve market goals.

3. Different energy market services

The type of service provided by the electricity industry, as the beating heart of the industry, has undergone many changes over the years. This industry is one of the most important factors influencing economic, political, social, cultural, and welfare conditions, which has been considered as a public service for many years. As a result, the electricity industry has been monopolized for years. But that is no longer the case, and the process has changed. The electricity industry around the world has seen significant structural changes over the past two decades. These structural changes, which were made to create a competitive and nonexclusive environment to ensure social welfare and increase the reliability of the power system, also greatly affected the type of market service. In this section, we will review the type of market service and provide services to consumers in the way it is currently managed.

3.1 Competitive electricity service

One of the most important competitive markets is the inability of market operators to exploit demand fluctuations. One of the main concerns of market regulators is the increase and fluctuations in the exchange rate of electricity, which is rooted in the abuse of market power by actors. In a competitive market, the legislature checks all prices before introducing them to the market to avoid high prices. This price monitoring is aimed at ensuring a reasonable price for all products, without distorting the price formation process between electricity suppliers. In Ibn Bazaar, the innovators create competition in the market by innovating in providing a strong product, especially in the case of green energy, contracts, additional services, and sustainability. Although the overall changes in many markets are similar, there are differences in the rules and structure of the primary market that affect market outcomes and responses. In recent years, it has intensified to improve competition and eliminate market failures, regulating the market and monitoring the performance of market participants. These failures may be due to a lack of transparency and negative experiences with electricity providers such as bankruptcy, fraudulent billing, and poor customer support, which can lead to loss of trust and customer satisfaction. To effectively monitor the market, it is necessary to consider the market structure, product innovation, pricing strategies, gross profit of electricity suppliers, as well as consumer behavior. The gross profit of electricity suppliers has fallen slightly, but the price difference between electricity suppliers remains high, and most consumers have never changed. These differences exist despite price monitoring by the legislature. In fact, the idea of a competitive market may be because of the intense competition between electricity suppliers. However, even if customers have a serious preference for minimizing costs, most customers who buy electricity are poor and need objective training and information to identify the best deal. In general, it can be said that the transition to a competitive and efficient market depends on the awareness and desire of families for individual well-being, which will lead to active search and selection of appointments tailored to their needs [16].

3.2 Wholesale electricity market

No central organization has the responsibility of supplying power in the wholesale electricity market, as demonstrated in Fig. 1.2. At the forefront of a restructured electricity market is a wholesale market where producers compete to withstand loads [1,17]. The price of energy offered in the retail market is heavily dependent on wholesale markets. By understanding how energy pricing works in the wholesale market, you will gain a clear understanding of

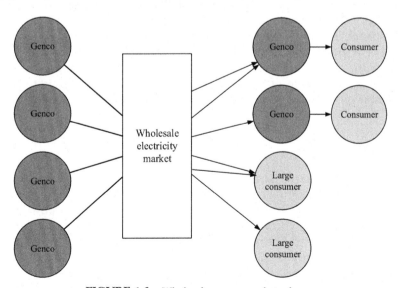

FIGURE 1.2 Wholesale power market scheme.

how short-term and long-term prices will move and optimize your energy management strategy. In many markets, there are four main components to determine the wholesale market price of electricity in a competitive market:

- Energy, which is the real commodity market for consumer consumption.
- Capacity or reliability, which provides available source services for deployment as needed.
- Auxiliary services, which are additional services to ensure the correct operation of the network.
- The density of transmission lines and losses in the line, where the cost of both power transmission in transmission lines with enough capacity and the probability of overcrowding losses due to remote transmission are considered.

These four elements are the main pricing components in the wholesale market. In this way, energy constitutes most of the market price, and then the capacity will be the main determinant of the price, but other components can add high costs in certain situations.

The final price of a unit of electricity (MWh) is usually referred to as the locational marginal price (LMP). This indicates that the price is obtained by the cost of transferring electricity from the source of production to the place of consumption (i.e., node). Marginal points to the fact that the cost of all power consumed at a given time and place is obtained by the most expensive marginal source in that compound [5]. The auction process is designed to respond to load at the lowest possible price under ISO supervision. After forecasting hourly demand by ISO, each generator will launch a specific production capacity (i.e., supply) at its proposed price, which is based on the launch of the facility. After submitting bids, ISO sorts them in ascending for obtaining the supply at different price points. Then, the winning bidder selects the combination of bids that have offered the lowest price to meet the required demand. After announcing the composition selected by ISO at the time set for the auction, the announced composition will supply energy to the market. The final price is adjusted based on the final cost of production required to satisfy demand.

Finally, in general, in the wholesale market of the energy exchange, through the participation of producers in the auction process and the announcement of the auction winners at the time specified by the market supervisor, the product is sold to market retailers and provided to consumers. In fact, the role of the stock market wholesale market is to allow traders, retailers, and other intermediaries to trade for short-term electricity supply or to deliver it in the future [18].

3.3 Retail electricity market

The retail electricity market is part of the electricity market and the energy exchange market, where retailers act as intermediaries and they buy/sell the electricity needed by small consumers from the wholesale market [16]. This market increases the power of customers allowing them to consciously choose their power supplier. Since the liberalization of energy markets, there has been a need to compete for retail and its potential benefits for consumers. Retailers in the electricity market compete as retailers for retail customers. In fact, a retailer is an intermediary entity that buys electricity from the wholesale market, as well as distributed products and sells them to subscribers who are not licensed to buy electricity directly.

These companies try to attract more customers and increase their sustainable profits by applying effective strategies. Competition in the retail market of any product is mainly affected by the innovation, quality, and price of the product. Competition in the downstream, the retail market, is an essential component of the upstream competition, the wholesale market.

As mentioned in the previous section, one of the important features of competition in the retail market is the presence of the bulk of inactive consumers, i.e., consumers who have not changed their electricity supply or contract. Inactive consumers in the electricity industry are more visible because as the market matures into a regulated and competitive market, many customers do not choose a particular contract and continue with their current retailers by default. This behavior may be caused by two different factors. The first factor is the lack of awareness and knowledge of consumers about contracts and the necessary conditions to identify the best contract that requires training and information. The second factor is that if they think that the costs associated with choosing (i.e., searching and trading) outweigh the expected benefits of change, the decision may not be logical. However, it is important to know that consumers have different assessments of the expected costs and benefits of becoming an active consumer in the retail market. Although some consumers may lose some financial benefits by nonchanging electricity supply or contract, and nonfinancial costs may outweigh these benefits, and their behavior may be quite reasonable.

In general, the performance of the retail market appears to have improved with a sharp decline in the number of consumer complaints. Innovation in product delivery, especially in the field of green energy, can be very popular. In this regard, consumers are willing to pay premiums for green electricity. Retailers have improved information presentation and comparison websites to explain product features. Retailers are trying to innovate in new products in a way that suits the tastes of consumers. Therefore, innovation seems to have positive welfare effects for a significant portion of active families. It is also not possible to deny harm to passive families who have not fully experienced the benefits of competition. Retail gross profit has declined slightly, although this financial profit is estimated at only a few euros per household per year.

Requirements in the retail market include network structure measures, contract restrictions, information rules, price monitoring, and market surveillance. Retailers offer long-term contracts (i.e., 1, 2, 3, 6, and 12 months) at a price that is subject to the limitation of the price ceiling announced by the retail market regulator, with customers selling electricity and mutual services to each other. The duties and responsibilities of retailers can be summarized as follows:

- Performing all subscriber services, including branch sales, applying branch changes, following up on technical matters from the distribution company on behalf of customers.
- Predicting customer demand coverage.
- Reading and preparing invoices and collecting materials
- Protection of customers' rights against noncontinuity and the level of quality expected by the distribution company
- Customer activation to participate in consumption management programs
- Identification of illegal branches and legal barriers against them in coordination with the distribution company

- Periodic testing of meters
- Coordinate to connect and disconnect the branch

In the retail electricity market, it is not possible to accurately predict consumption and production. For example, it is not possible to accurately predict the exact amount of consumption in the next few days, or it is not possible to predict the lack of fuel due to possible consumption or technical problems. For this reason, price fluctuations may be seen due to rising and falling supply and demand. In order to avoid possible problems caused by the phenomenon called volume risk, contracts are concluded between retailers and wholesalers, the implementation of which depends on the regulations of the regional market and the structure of the electricity market [16,18].

3.4 Electricity capacity markets

Product adequacy and capacity mechanisms are the most important issues in the energy market. The requirement of electricity to a separate power market is questionable. Energy markets alone are unlikely to generate adequate production capacity, especially if they are covered by accurate prices. Some believe that there is no need for capacity markets in the open market. But, others believe that the existence of a capacity market, followed by a change in the pricing mechanism required and, of course, a positive effect on electricity sales, will be competitive. In the energy market, there may always be shortcomings that lead to an increase in marginal prices and the exercise of market power. With careful design, the capacity market will be able to reduce market power and strategic offerings to ensure adequate production at a specific time and place. Other goals of the system include anticipating more revenue for generators, cheaper investment, and ultimately lower prices for consumers.

More precisely, the capacity market has improved the reliability and retained the attractiveness of the product development business by providing sufficient resources and compensating the investors' lost money in the long run. But considering a plan for consumption management, the first option in energy policies to achieve sustainable and reliable systems is consumption management activities, which resulted in the discussion of the presence of cargo in the capacity market. The authors in Ref. [5] state that the most efficient way to use resource management resources in the capacity market is to employ them by retailers. Also, in order to increase the efficiency of retailers in providing capacity goods, they must participate in an auction along with generators [19].

3.5 Day-ahead market

The day-ahead market is known as a trading environment where the electrical energy suppliers and the consumers transact energy, which is an auction accomplishing the day. Considering technical characteristics of electrical energy networks, most of the power trading process is performed 1 day before delivery. Thereafter, transmission system operator schedules the market position that is able to perform the required security checks and operation of the system in real-time condition. The day-ahead market is an environment for trading energy in the pool, which is mostly negotiated, while adjustment markets are generally based on adjustment of cleared energy in the day-ahead environment. In the day-ahead market, the power generators propose energy blocks and associated lower bound of selling prices for every hour of the market time interval and every generation plant. Simultaneously, consumers and retailers propose energy blocks and the associated upper bound of buying prices for every hour of the market time interval. The operator of the market gathers the bids of purchase and offers of sale and clears the market by applying a market-clearing method. The result of a market-clearing process is prices and scheduling of electrical energy generation and consumption. The market-clearing price will be identical for all market agents when the power transmission system is not taken into account in the process. However, a locational marginal price (LMP) will be considered for each bus of the network when the power transmission system is taken into account in the process. Because of losses and congestion of power transmission lines, LMPs are different across system buses. Considering congestion of power transmission lines, more expensive product is required for dispatch on the downstream side of the congested line, which yields an increment in the prices in buses located on the downstream side of this line [20].

3.5.1 Intraday market

In European electrical energy networks, market players are capable of adapting positions intraday considering expectations of the updated market. Such capability gives the chance of avoiding the price risk of balancing a market that is specifically beneficial for variable renewable energy resources based on higher prediction accuracy closer to real-time conditions. The intraday market is well demonstrated in the European market, where the bids and offers are matched based on a continuous scheme. Generally, the economic principles of intraday and day-ahead markets are the same. However, the intraday market mat has lower liquidity and more volatile prices, which can be justified considering the technical limits of production plants in changing power injections in real-time conditions. It is expected that the intraday prices are related to price expectations in the real-time balancing market. A low-price expectation for future results from a forecasted extra power production, which results in lower price. In other words, low price in the balance market is a result of expected extra power, which inclines producers with extra energy to enroll in intraday market. Accordingly, they will be avoided from the risk of low electrical energy prices in the balance market when expecting rigid real-time status. However, intraday prices will be down considering such action. Accordingly, the balancing price drivers, the expected forecasted error, and flexibility have an indirect influence on the intraday price [20,21].

3.6 Real-time market

The real-time pricing, along with saving for subscribers, allows them to be flexible in their choice of power consumption time depending on their preferences. The program is designed for commercial, agricultural, and pumping consumers who can shift or reduce electricity consumption when electricity prices are high. In addition to their effectiveness in improving the performance of the power system, consumers can benefit from hourly changes in electricity prices and, consequently, lower electricity bills. In the real-time pricing, customers can select a demand response program for optimal adjustment of energy usage within cooperating in the program to minimize the cost of power utilization. Real-time mutual communication and control between the load and market can be performed using smart meters that allow consumers to obtain real-time electrical energy prices. Also, the implementation of this mutual communication by an observant consumer will help in the optimization of its energy utilization pattern to minimize energy cost [22,23].

3.7 Regulation market

A regulation market is an exchange environment where government organizations or, less generally, industry or labor organizations, apply a level of overlooking and control. Market regulation is usually controlled by the government that includes deciding who can join the market and selecting the purchasing prices. The main role of the government in a market economy is to monitor and control the economic and financial systems. There are different types of regulations in a regulated market, which include control, supervision, antidiscrimination, environmental protection, taxes, and labor laws. Regulation reduces the freedom of market members or gives them appropriate privileges. Regulations involve laws about how shipment and services can be marketed; what services customers have to load payments or replacements; safety standards for goods, workplaces, food and medicines; mitigation of environmental and social impressions; and the level of the handle a provided member is permitted to assume covering a market. Followers of a presented regulation or regulatory governments usually tend to indicate benefits to the wider society [24].

Regulations are not always completely useful, though, nor are their bases always completely altruistic. Additionally, well-intentioned regulations can bring unintended results. Local-content conditions are oftentimes imposed to profit the domestic industry. These regulations do not certainly benefit nurturing local manufacturing, but often manage to letter-of-the-law workarounds or black markets. Some defendants of free markets ratiocinate that anything above the most basic regulations is ineffective, expensive, and maybe unfair. Some ratiocinate that also modest minimum wages increase unemployment by making a barrier to entry for low-skilled workers, for example. Defendants of the minimum wage tell historical cases in which very profitable organizations paid wages that did not give employees with even a basic standard of living, arguing that setting wages decreases the exploitation of weak workers.

3.8 Reserve market

Product adequacy and capacity mechanisms are one of the most important issues in the energy market. An important question is whether separate reserve markets are needed for supplying electricity load. Energy markets alone are unlikely to generate adequate production capacity, especially if they are covered by accurate prices. Some believe that there is no need for reserve markets in the open market. On the other side, others believe that the existence of a reserve market, followed by a change in the pricing mechanism required and, of course, a positive effect on the sale of electricity, will be competitive. In the energy market, there may always be shortcomings that lead to an increase in marginal prices and the exercise of market power. With careful design, the reserve market will be able to reduce power and propose strategic offerings to ensure adequate production at a specific time and place. Other goals of the system include anticipating more revenue for generators, cheaper investment, and ultimately lower prices for consumers [19].

More precisely, the reserve market has improved the reliability and retained the attractiveness of production development trade by providing sufficient resources and compensating for the lost money of investors in the long run. But since the strategic plan for consumption management of the IAEA states that the first option in all energy policies to achieve sustainable and reliable systems is consumption management activities, the discussion of the presence of cargo in the reserve market was raised. The most efficient way to use resource management resources in the reserve market is to employ them by retailers. Also, to increase the efficiency of retailers in providing capacity goods, they must participate and compete in an auction with generators.

3.9 Balancing market

In electricity markets, a balanced market is arranged to balance electricity supply and demand. The equilibrium market plays an essential role in the operation of the power system due to the requirement of equality of production and consumption. Also, this service plays a role as the last step in the trading market. Due to the impossibility of storing electrical energy on a large scale, the importance of market equilibrium becomes clear. Balance markets are usually single-period markets, meaning a separate session for each trading session. In addition to providing ancillary services such as voltage control to maintain the stability of the power system, this market also allows traders to participate in the market. Manufacturers in the equilibrium market participate in the equilibrium market to regulate the power in the electricity market through upward and downward production when necessary. Upward production means an increase in production if there is a shortage of power in the market, and downward production means a decrease in production in case of excess production in the market. In fact, stochastic manufacturers compensate for deviations from contract production by accessing the equilibrium phase. The pricing of these deviations is done differently according to the system imbalance. A single-price imbalance system can be distinguished from a two-price imbalance system. In the single-price imbalance system, deviations are settled through market prices regardless of excess production [25].

Generally, the equilibrium price can be higher/lower than the current market price. The equilibrium price is higher than the current market price in the event of a shortage of power, followed by an increase in production in the equilibrium market. Also, it is lower than the current price in the event of a surplus of power in the market, followed by a decrease in production in the equilibrium market. This settlement and price agreement provide arbitrage and profitability opportunities for electricity producers. If the producer and the system deviate in a contradictory way in the production, the producer will benefit from this deviation. For example, the manufacturer reduces the overall imbalance in the market by deviating from its production and thus receives its reward. But if this deviation is the same for the manufacturer and the system, the manufacturer will be punished. For example, if the market faces a shortage of power, the producer will be punished if the producer also suffers from a shortage of production. In the two-price imbalance system, the equilibrium price and deviations from the production plan are traded at different prices. When the producer deviates from the production plan in the opposite direction of the system deviation from the production plan, the price of the deviation is presented in the market price of the day-to-day market and there is no excess profit for the producer. In other words, in this situation, the producer helps the system imbalance. Conversely, when two imbalances occur in one direction, the deviation of the manufacturer is priced at a balanced market price that means it is penalized [25].

3.10 Novel energy markets including ramp markets

At present, new energies have a special place in the energy market. Following the environmental damage caused by fossil fuels, the world has sought to find alternatives to traditional fuels that minimize the devastating effects. In other words, the goal is to use cheap and clean energy in the industry, which has led to large investments and the use of the latest technologies in this field. Fossil fuels that have been in the spotlight for a long time have produced millions of tons of carbon dioxide a year, and this amount of carbon dioxide has warmed the Earth's temperature by several degrees. In recent years, there is more investment in new energy than fossil fuels. Much of this investment has been in the solar and wind sectors.

The integration of renewable energy resources such as wind power and the penetration of uncertain load demands to electrical energy networks such as electric vehicles, several issues have appeared for the system operators to deal with uncertainties of such elements. Flexible ramping product (FRP) is known as a means for handling uncertainty associated with energy demands and forecasted power output of renewable energy resources such as wind turbines. FRP is a product supplying additional upward and downward flexible ramping. FRP is applied in several power markets such as California independent system operator (CAISO) and Midwest ISO (MISO) [26,27]. Several differences exist between FRP and existing services in the power market, such as spinning/nonspinning reserve and frequency control. All these services handle specified system contingencies, and spinning/nonspinning reserves perform up-ramping flexibility. On the other side, FRP deals with both up-ramping and down-ramping support with 5 min checking time intervals. The checking time of frequency regulation is several seconds. FRP has been of great importance in modeling power system operation, market clearing of coupled FRP and energy for integrated power and gas system [28], and investigating the influence of such flexible resources such as electric vehicles, demand-side management and bulk storage technologies on the operation, spinning reserve and FRP of plants, and operation cost of the system [29].

4. Fundamental features of the energy market

The product features on the market are determined by the fundamental principles of the relevant market. Electrical characteristics that can be attributed to the continuity of production and demand processes, the impossibility of storage and accumulation, the importance of product reliability, and the standard range of the product determine the basic characteristics of the electricity market. These basic features are mainly related to the power and capacity of the electricity market. The purpose of creating an electricity market in the electricity industry is to create a platform for competition and the sale and purchase of electricity and to increase the efficiency and effectiveness of the system. The formation of day-ahead and real-time markets is planned and implemented with the formation of the electricity market, respectively.

4.1 Electricity spot prices

Given the nature of wholesale electricity markets, the dynamics of point prices can be partially understood, and the prediction process of such variables is a major challenge for market participants and system controllers even on a day-ahead time horizon. Among the reasons for the complexity of the problem and the challenges in predicting price are: (a) the momentary nature of the goods; (b) the shape of the supply function, which is inherently steep, discontinuous, and convex in the presence of a variety of technologies; (c) the exercise of market power by oligopolistic market structures; (d) complex market designs; and (e) frequent regulatory interventions and changes in market structure. These issues indicate that spot price forecasting is important. First, forward prices give a signal of limited information about spot movements. Due to the lack of electricity storage, the facility return ratio, which relates the forward curve to the spot price, is not established for electricity. Second, while stylized stochastic schemes of spot prices replicate the features of the statistical distribution well, and permit derivatives pricing based on analytical or numerical investigations, their short-term prediction performance is not enough, partially because of abrupt occurrence, fast-reverting spikes. Finally, a subset of market participants may have access to fundamental data on the future market, which makes forecasting hard [30].

Therefore, in a fully statistical model where market principles and factor behavior are not implemented, it may not be sufficient as a basis for forecasting. However, there are limitations in such basic models for predicting electricity prices. First, they are mainly limited to the effects of autoregression and price response to demand, fuel prices, or weather conditions, such as temperature, rainfall, and wind. However, the application of these restrictions is not sufficient, especially for emerging markets. Other factors are (i) aspects of plant dynamics; (ii) risk measures; (iii) market design effects (e.g., Ref. [31]); (iv) agent learning; and (v) strategic behavior. Second, price formation depends on the design and structure of the market. Although the application of market power has been widely reported, as well as competitive models in market valuations, the relevant indicators do not appear in economical pricing models. In order to obtain accurate market inferences and price forecasts, these characteristics must be reflected in the price models. Third, most features refer to average daily prices that hide patterns within the day. However, each day of daily trading shows relatively specific price specifications, which reflect the dynamics of demand, supply, and operating constraints. However, high-frequency studies have only recently emerged [32].

4.2 Fundamental features of the PJM day-ahead energy market

Ideally, in order to maximize social well-being and provide optimal conditions, management mechanisms need to be carefully designed and competition between participants should be strong enough. Yet the emerging electricity market structure is more like an oligopoly than the full competition in the market, which means that there is still domination in the market by some manufacturers and it reduces competition and eliminates full competition in the market. This is due to the special characteristics of electricity supply industry such as limitation of the number of producers, the availability of resources as input, transmission constraints that impede the effective availability of consumers to many of the generators, and the losses caused by the transmission process that consumers buy power from distant suppliers.

In today's market, market participants offer suggestions on the amount of energy that can be supplied and stored along with the price for each hour of the next day. As a result, the supply day schedule is determined to meet the projected demand for each hour and place. The leading day market allows participants to coordinate production at the same time as planning for the day and to protect more volatile prices in real time. Also, in the upcoming market, for the next few days and even next month, by collecting information in different ways, despite the appropriate time interval and load process in similar periods, contracts will be planned and contracted. Market inputs will be system parameters, resource parameters, shutdown information, bid information, ISO forecast of demand, constraints, and density of transmission lines. The requirements of the market ahead are reserves, residual unit participation, and energy to meet demand.

The premature market means an exchange of energy for a future time. Almost all cash in the world, including the Iranian market, is an early market and a day before. In the market the day before, assuming the physical delivery date of energy or ancillary services is 20, by May 19, all participants have the opportunity to make offers to buy or sell that can be about an hour's delivery or a few hours. The day before delivery, the auction ends at a certain time and the calculations are started by ISO or PX to determine the market settlement price and the program of production and consumption of the participants and announce it. The program is usually announced in the afternoon of the previous day, and from 0:00 on the day of delivery, the participants must receive or deliver energy from the network according to the schedule. The settlement of these markets is usually daily or weekly.

In fact, the planning of today's market is to create optimal auction strategies among competitive suppliers in an energy market. To ensure the adequacy of the resources of the system operators and the reliability, the suggestions are structured and their impact on the market of the next day is determined. Another important goal of creating a suitable environment for strategic bidding is to prevent the abuse of potential market power through existing weaknesses. It is clear that the development of tender strategies should be based on market models and rules of activity, especially auction laws and tender protocols [33,34].

4.3 Fundamental features of the real-time energy market

In the real-time market, all estimation, planning, and action operations are performed for a maximum of 5 min ahead and based on information 5 min before. The real-time market is a tender and security-based economic deployment. In this regard, the role of IT in information exchange, information security, and accuracy of exchanged information is considered as a basis and market requirement. The role and position of information technology and the need for a deep approach to its inefficient physical deployment and the real-time market formation and hardware and software requirements have always been considered. In fact, the emphasis is more on information technology and its requirements [35,36].

In fact, in order to maintain the frequency and reliable performance of the power system, production and consumption must be balanced at all times. Real consumption or production may be different from what is predetermined in the market; therefore, it is created simultaneously to ensure the equilibrium power of the market. This market is also called the equilibrium market. Simultaneous markets in most countries are managed by ISO. Participants who want to play a role in the market at the same time usually submit their bids to increase production (or decrease consumption) or decrease production (or increase consumption) 1 day after the end of the market. An increase in production or a decrease in consumption also called upward regulation is for a time when the supply of real time is less than the demand for real time. Decreased production or increase in consumption, also called downward regulation, is when the supply of real time exceeds the demand for real time. Recommended prices for increasing production or decreasing consumption are usually arranged in ascending order, and prices for decreasing production or increasing consumption are arranged in descending order. To address supply shortages, ISO starts at the lowest price on the side of those who have offered to increase production, and to address the supply surplus, ISO starts at the highest price on the side of those who have offered to reduce production. Participants selected at the right time should be able to follow ISO instructions to reduce or increase production or consumption [37].

5. Conclusions

The basic principles of the energy market and management mechanisms are developed with the aim of creating the conditions for the performance of a competitive industrial market. This is made possible by studying the positive and negative experiences of other countries, evaluating the performance of different market models and economic consequences, market participants, and consumers. The competitive environment of the electricity market is largely determined by the structure of production capacity. In addition, the state of the industry needs to change its capacity to generate electricity. To develop production and networking, it is essential to establish a reciprocal link between manufacturing and networking companies to connect a new manufacturing center to networks. As looked ahead in this chapter, service providers are also competing with programs and technologies that help customers minimize the cost of electricity according to their preferences. The future smart environment on the consumer side will make it possible to both change and reduce demand for the benefit of customers. Innovative service providers that do their best to maximize customer value will thrive. With the development and adoption of smart home technologies, the retail competition will become more important. The final element in the distribution market model is the low voltage lines that bring electricity to our homes and businesses. The distributor company is an exclusive tool in the reconstructed market. But even here, with the introduction of various forms of distribution production and storage, this landscape is changing. Currently, the electricity industry is growing significantly and is achieving its goals day by day. It is expected in the near future that fully competitive markets around the world will be able to maximize social welfare and address many of the challenges and problems facing the advancement of power systems.

References

[1] W.-J. Ding, B.-C. Wang, Q.-G. Chen, Y.-H. Xing, F. Zhang, F. Liu, Chile's electricity market construction and its enlightenment, 2019 4th International Conference on Automation, Mechanical and Electrical Engineering (AMEE) (2019).

[2] D. Gan, D. Feng, J. Xie, Electricity Markets and Power System Economics, CRC Press, 2013.

[3] M. Pollitt, Evaluating the evidence on electricity reform: lessons for the South East Europe (SEE) market, Util. Pol. 17 (1) (2009) 13–23.

[4] W.W. Hogan, Electricity market restructuring: reforms of reforms, J. Regul. Econ. 21 (1) (2002) 103–132.

[5] X. Zhao, F. Wen, D. Gan, M. Huang, C. Yu, C. Chung, Determination of AGC capacity requirement and dispatch considering performance penalties, Elec. Power Syst. Res. 70 (2) (2004) 93–98.

[6] N.I. Meyer, Distributed generation and the problematic deregulation of energy markets in Europe, Int. J. Sustain. Energy 23 (4) (2003) 217–221.

[7] R.J. Serrallés, Electric energy restructuring in the European Union: integration, subsidiarity and the challenge of harmonization, Energy Policy 34 (16) (2006) 2542–2551.

[8] M.N. Dudin, E.E. Frolova, V.N. Sidorenko, E.A. Pogrebinskaya, I.V. Nikishina, Energy policy of the European Union: challenges and possible development paths, Int. J. Energy Econ. Pol. 7 (3) (2017) 294–299.

[9] H. Yan, H. Yan, Optimal energy purchases in deregulated California energy markets, Conference Optimal Energy Purchases in Deregulated California Energy Markets, n.d. vol. 2, IEEE, 1249–1254.

[10] H. Singh, A. Papalexopoulos, Competitive procurement of ancillary services by an independent system operator, IEEE Trans. Power Syst. 14 (2) (1999) 498–504.

[11] H. Rudnick, J. Zolezzi, Electric sector deregulation and restructuring in Latin America: lessons to be learnt and possible ways forward, IEE Proc. Generat. Transm. Distrib. 148 (2) (2001) 180–184.

[12] P.L. Joskow, Restructuring, competition and regulatory reform in the US electricity sector, J. Econ. Perspect. 11 (3) (1997) 119–138.

[13] W. Boltz, The Challenges of Electricity Market Regulation in the European Union. Evolution of Global Electricity Markets, Elsevier, 2013, pp. 199–224.

[14] S. Borenstein, J. Bushnell, F. Wolak, Diagnosing Market Power in California's Restructured Wholesale Electricity Market, National Bureau of Economic Research, 2000.

[15] H. Dagdougui, R. Minciardi, A. Ouammi, M. Robba, R. Sacile, A dynamic decision model for the real-time control of hybrid renewable energy production systems, IEEE Syst. J. 4 (3) (2010) 323–333.

[16] M. Mulder, B. Willems, The dutch retail electricity market, Energy Policy 127 (2019) 228–239.

[17] A. Sumper, Micro and Local Power Markets, Wiley Online Library, 2019.

[18] P. Cramton, Electricity market design, Oxf. Rev. Econ. Pol. 33 (4) (2017) 589–612.

[19] B.F. Hobbs, J.G. Inon, M.-C. Hu, S.E. Stoft, Capacity markets: review and a dynamic assessment of demand-curve approaches, Conference Capacity Markets: Review and a Dynamic Assessment of Demand-Curve Approaches, n.d., IEEE, 514–522.

[20] K. De Vos, Negative wholesale electricity prices in the German, French and Belgian day-ahead, intra-day and real-time markets, Electr. J. 28 (4) (2015) 36–50.

[21] C. Weber, Adequate intraday market design to enable the integration of wind energy into the European power systems, Energy Policy 38 (7) (2010) 3155–3163.

[22] C. Gérard, A. Papavasiliou, A comparison of priority service versus real-time pricing for enabling residential demand response, Conference a Comparison of Priority Service versus Real-Time Pricing for Enabling Residential Demand Response, n.d.

[23] Y. Dai, L. Li, P. Zhao, J. Duan, Real-time pricing in smart community with constraint from the perspective of advertising game, Int. Trans. Electr. Energy Syst. 29 (9) (2019) e12043.

[24] D.J. Shiltz, S. Baros, M. Cvetković, A.M. Annaswamy, Integration of automatic generation control and demand response via a dynamic regulation market mechanism, IEEE Trans. Contr. Syst. Technol. 27 (2) (2017) 631–646.

[25] N. Mazzi, P. Pinson, Wind power in electricity markets and the value of forecasting. Renew. Energy Forecast. (2017) 259–278. Elsevier.

[26] B. Wang, B.F. Hobbs, Real-time markets for flexiramp: a stochastic unit commitment-based analysis, IEEE Trans. Power Syst. 31 (2) (2016) 846–860.

[27] N. Navid, G. Rosenwald, D. Chatterjee, Ramp capability for load following in the MISO markets, Midwest Indep. Syst. Oper. 20 (2011).

[28] X. Zhang, L. Che, M. Shahidehpour, A. Alabdulwahab, A. Abusorrah, Electricity-natural gas operation planning with hourly demand response for deployment of flexible ramp, IEEE Trans. Sustain. Energy 7 (3) (2016) 996−1004.

[29] E. Heydarian-Forushani, M.E.H. Golshan, M. Shafie-khah, P. Siano, Optimal operation of emerging flexible resources considering sub-hourly flexible ramp product, IEEE Trans. Sustain. Energy 9 (2) (2018) 916−929.

[30] N.V. Karakatsani, D.W. Bunn, Forecasting electricity prices: the impact of fundamentals and time-varying coefficients, Int. J. Forecast. 24 (4) (2008) 764−785.

[31] F.A. Wolak, An empirical analysis of the impact of hedge contracts on bidding behavior in a competitive electricity market, Int. Econ. J. 14 (2) (2000) 1−39.

[32] G. Guthrie, S. Videbeck, Electricity spot price dynamics: beyond financial models, Energy Policy 35 (11) (2007) 5614−5621.

[33] F.A. Rahimi, A. Vojdani, Meeting the emerging transmission market segments, IEEE Comput. Appl. Power 12 (1) (1999) 26−32.

[34] F. Wen, A. David, Strategic bidding for electricity supply in a day-ahead energy market, Elec. Power Syst. Res. 59 (3) (2001) 197−206.

[35] B. Hua, D.A. Schiro, T. Zheng, R. Baldick, E. Litvinov, Pricing in multi-interval real-time markets, IEEE Trans. Power Syst. 34 (4) (2019) 2696−2705.

[36] M. Alipour, K. Zare, H. Seyedi, M. Jalali, Real-time price-based demand response model for combined heat and power systems, Energy 168 (2019) 1119−1127.

[37] S. Acha, G. Bustos-Turu, N. Shah, Modelling real-time pricing of electricity for energy conservation measures in the UK commercial sector, Conference Modelling Real-Time Pricing of Electricity for Energy Conservation Measures in the UK Commercial Sector, n.d., IEEE, 1−6.

Energy storage fundamentals and components

Arash Moradzadeh[1], Morteza Nazari-Heris[1,3],
Behnam Mohammadi-Ivatloo[1,2], Somayeh Asadi[3]

[1]Faculty of Electrical and Computer Engineering, University of Tabriz, Tabriz, East Azerbaijan, Iran; [2]Department of Energy Technology, Aalborg University, Aalborg, Denmark; [3]Department of Architectural Engineering, Pennsylvania State University, State College, PA, United States

1. Introduction

Energy is an important element of both industrial and economic development all over the globe. The effective operation of energy systems and energy efficiency are essential factors for sustainable development. Today, the efforts of energy policies to decrease CO_2 emissions from energy sources are of great importance for experts in environmental science. On the other side, the energy demand is increasing exponentially and is predicted to be doubled by 2030 [1,2]. Electrical energy storage (EES) systems are known as an important element of energy systems as a solution for dealing with power supply stabilization and minimizing the peak load condition of the power systems. EES is beneficial for power systems in restraining power fluctuations made in the system considering the stochastic nature of the power output of renewable sources such as wind turbine and photovoltaic cells [3]. Also, power systems with installed EES systems take advantage of system imbalances reductions [4], demand shifting [5], and minimizing the operation cost of the system [6]. Such benefits are the reason that the US Department of Energy (DOE) has defined EES as an appropriate solution to power system stability [7]. Researchers have forecasted a considerable growth of power systems through the EES systems due to the aforementioned advantages. The studies around energy storage technologies in power systems have focused on different subjects, for which the important ones are sizing and placement of EES in power systems [8], energy management of EES-based power systems [9], integration of EES in power markets [10], and various types of EES technologies [11]. This main objective of this chapter is to provide an overview of the application of ESSs for managing energy in power systems. Accordingly, different types of energy storage technologies are investigated and their influence in energy saving and optimization of power systems are discussed. Additionally, the operation model of energy storage systems (ESS) in power markets is studied. Finally, the current condition of the application of ESSs in the world is investigated.

An EES generally consists of several components for storing and releasing energy within an electrical energy system. The main components of an EES include batteries that consist of the racking and battery management system, conversion facilities consisting of inverters and transformers, the contractor/integrator supplying software, and the building/containers to house the system. The battery management system (BMS) provides a means for checking battery banks and improving life span, reliability, safety, and performance level of the EES systems. Such a system utilizes software and sensors to attain real-time performance information of the battery as well as collecting data. Controlling the charge level and monitoring the battery health are responsibilities of the BMS. The main functions of such a system can be classified as protection of battery cells against damage, meeting real-time energy load, extending the battery life, and attaining a suitable state of the battery. The monitoring of fundamental system parameters, including voltage, current, temperature, and charge/discharge time duration, is a basic process of a BMS. In electric vehicles (EVs), the BMS plays a joint role with on-board units of the car, such as engine management, climate control, communication, and safety systems. Inverter configurations are necessary for the integration of DC EES to single-phase AC grids. Inverters are basic elements of the operation of EES that play the main role in a storage project. The inverter is responsible for managing, optimizing, and finally driving project performance and returns. Two basic power converter topologies exist for EESs connected to a grid [12]. In one of the topologies, a boost DC/DC converter and a DC/AC converter are utilized for the integration of an EES to an AC system [13,14]. In the other one, direct DC/AC conversion is performed utilizing converters with both boosting and inversion abilities or applying cascaded converter configurations [15,16].

2. Types of EES technologies

The EES will occur through the conversion of electrical energy into other forms, such as mechanical, thermal, electrical, chemical, electrotechnical, and electrochemical. The important point is that the kind of energy supply, duration of storage, and the final application have an influence on the features and properties of the ESS. Investigating the function and form of ESSs can be two key criteria in classifying these systems [17–19]. ESSs have a variety of technologies based on their application and utilizes, and various types of energy storage technologies will be investigated in this section [20–24]. Fig. 2.1 is provided to demonstrate a comprehensive viewpoint for various types of EES technologies used for energy management.

2.1 Electrotechnical such as batteries and fossil fuel storage

Electrotechnical EESs are among the most widely utilized types of EES technologies and are also classified as electrochemical and chemical energy storage technologies such as batteries and fuel cell storages. In electrochemical storage systems (ECSS), the chemical energy contained in the active substance inside the system is converted into electricity. This type of conversion is performed through chemical interaction and the energy is stored as an

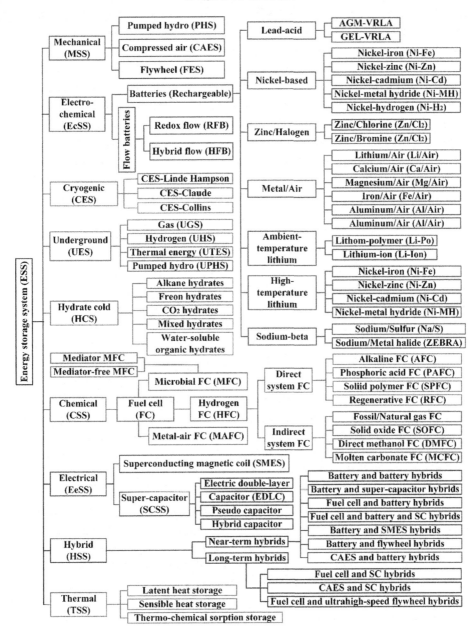

FIGURE 2.1 Various types of EES technologies.

electric current for a certain voltage and time [25,26]. The internal connections of the cells, such as the parallel connection or series, determine the voltage and current level. For example, for systems that store energy as electrochemicals, we can name ordinary rechargeable batteries and flow batteries (FBs). Diverse sizing of ECSS is one of the most important advantages of this technology. These technologies come in a variety of forms, including lead—acid (LA) [27], sodium—sulfur (NaS) [28,29], lithium-ion (Li-ion) [30], FBs [30,31], nickel—cadmium (NiCd) [32], and nickel—metal hydride (NiMH) [33].

On the other hand, chemical storage systems (CSS), as secondary energy carriers, store or release energy in the system through a series of chemical reactions and form other compounds in this way. These systems are concentrates on hydrogen technology due to their remarkable properties as fuel and the capability to store large amounts of electrical energy [34]. CSS is a hydrogen electrolysis process that can also be made with carbon dioxide in natural gas, such as methane, during reactions in which hydrogen or artificial gas is considered. With this clean, green technology that stores hydrogen energy, energy can be stored on a large scale. Coal, gasoline, propane, ethanol, diesel, and hydrogen are among the most widely utilized chemical fuels in power production and energy transportation system [35,36]. Fuel cell (FC) is one of the most well-known types of CSS technology, which continuously converts the chemical energy of the fuel to power. The method of making a power supply can be the most important difference between a battery and an FC [34]. In the following, the types of each of the electrochemical and chemical storage technologies will be introduced.

2.1.1 Lead—acid (LA) batteries

Among all electrolyte batteries, LA is the most widely utilized rechargeable storage device with different designs and sizes, which has the highest efficiency and cell voltage. Due to their advanced technologies of LA batteries, they provide energy storage with high performance and fast response time. These types of batteries have some other advantages such as safe operation, excellent cycle efficiency (~63%—90%), high-temperature tolerance, and low maintenance and operation costs, which leads to their selection and widespread use compared to other batteries [27]. Ambient temperature, the number of charges and discharges, and the current rate for charge and discharge are the factors on which the LA batteries' life depends. Using a series or parallel connection in LA batteries, the required power and energy ratings are obtained. Today, the development of LA batteries is focused on innovative materials to improve the performance of applications in renewable energy such as wind, PV power, and EVs [24]. The charging phenomenon of LA batteries is done by converting $PbSO_4$ to Pb or PbO_2. During the discharge phenomenon, $PbSO_4$ and HC ions are produced by passing HSO_4^- ions within the negative electrode and forming a chemical reaction with Pb. Hydrogen is produced in the negative electrode and oxygen in the positive electrode. In general, the electrochemical reactions that occur when charging/discharging in acid batteries can be inferred as [37]:

$$2PbSO_4 + 2H_2O \xrightarrow{\text{Charging}} PbO_2 + Pb + 2H^+ + 2HSO_4^- \qquad (2.1)$$

$$PbO_2 + Pb + 2H^+ + 2HSO_4^- \xrightarrow{\text{Dicharging}} 2PbSO_4 + 2H_2O \qquad (2.2)$$

2.1.2 Sodium—sulfur (NaS)

In the composition of NaS batteries, sodium and sulfur are used as molten electrodes in a liquid state and nonaqueous beta-alumina electrolyte is utilized as electrolyte and isolator. Sodium and sulfur are used as negative and positive electrodes, respectively [35]. Long lifetime nearly 15 years, almost zero self-discharge rate i.e., 0.05%–1%, good energy density such as 150–300 Wh/L, recyclability due to cheap battery materials, and high energy efficiency are the main advantages of NaS batteries. Despite all these advantages, these types of batteries suffer from some problems such as high operating temperature, very corrosive nature of molten cathodes, and very high internal resistance. So that any fracture in the electrolyte ceramic will eventually lead to flame and battery explosion [24,38]. During the discharge phenomena, Na^+ ions are produced by the interface of sodium (Na) and beta alumina, then Na^+ ions pass through the solid electrolyte and are recombined with sulfur at the anode. The reverse of this process occurs during the charging process of these batteries. In general, the electrochemical reaction in a NaS can be described as follows [24]:

$$At\ cathode: 2Na \leftrightarrow 2Na^+ + 2e^- \tag{2.3}$$

$$At\ anode: xS + 2Na^+ + 2e^- \leftrightarrow Na_2S_x \tag{2.4}$$

$$Overall: 2Na + xS \leftrightarrow Na_2S_x \tag{2.5}$$

where the value of x should be within three to five.

2.1.3 Lithium ion (Li ion)

Li-ion battery belongs to the family of electrochemical batteries and works on porous electrodes containing foil to make electrical isolation on the basis of reversible extraction of ions. So that such electrodes and foils are immersed in the electrolytic solution of Li^+ ions before extraction [39]. The small size, lightweight, and potential for energy storage make Li-ion batteries a portable product. In addition, high energy density, stable cycle, low power consumption, rapid charge/discharge, long lifetime, and high power density (500–2000 W/kg) are other advantages of Li-ion batteries. However, the life cycle of these batteries is affected by temperature and deep discharging leads to a reduction in their life span [30,34]. Increasing the energy density and specific power of lithium-ion batteries significantly reduces production costs, so that less material is required to produce the same amount of energy cells, and this is a good economic cycle for these batteries. Lithium ion and lithium polymer are the raw materials for Li-ion batteries so that each cell of the lithium-ion battery has a voltage of 4 V. The process of discharging lithium-ion batteries is such that lithium ion is transferred from the anode to the cathode. But in the charging process, this phenomenon is reversed, so that the Li^+ ions move through the electrolyte from the cathode to the anode to form lithium atoms that are located between the carbon layers of the electrode. The electrochemical reactions that occur in electrolytic batteries are as follows [40]:

$$\text{At cathode: } \text{Li}_x\text{C}_6 \leftrightarrow x\text{Li}^+ + xe^- + \text{C}_6 \tag{2.6}$$

$$\text{At anode: } \text{Li}_{1-x}\text{CoO}_2 + x\text{Li}^+ + xe^- \leftrightarrow \text{LiCoO}_2 \tag{2.7}$$

$$\text{Overall: } \text{LiC}_6 + \text{CoO}_2 \leftrightarrow \text{C}_6 + \text{LiCoO}_2 \tag{2.8}$$

Based on the positive electrode, lithium-ion batteries are economically dependent on lithium manganese oxide (LiMn_2O_4), lithium nickel manganese cobalt oxide (LiNiMnCoO_2), lithium cobalt oxide (LiCoO_2), lithium iron oxide phosphate (LiFePO_4) phosphate. Iron phosphate batteries are economical options between all the lithium-ion batteries considering high power density and economical discharge capacity. Their reliable chemical and thermal properties have led to many applications in EVs [24].

2.1.4 Nickel-based batteries

Nickel-based batteries are widely developed since the 1990s and are used in applications of uninterruptible power supply (UPS), electronic equipment, and telecommunication devices. In nickel batteries, active materials comprising nickel oxyhydroxide as anode and cathode are one of the following: Cd, MH, Fe, Zn, or H_2 [41]. Nickel-based batteries include nickel−iron (Ni-Fe), nickel−cadmium (Ni-Cd), nickel−hydrogen (Ni-H_2), nickel−metal hydride (Ni-MH), and nickel−zinc (Ni-Zn). In the meantime, Ni-Cd and Ni-MH batteries are more commonly used in commercial applications. Ni-Fe and Ni-Zn are also used to a limited extent considering their low specific energy, very high cost, and short life cycle. Among the aforementioned batteries, the Ni-Cd is by far the most successful product due to its long life cycle, ideal efficiency i.e., 70%−90%, nominal cell voltage of 1.2 V, and the ability to work at very extreme temperatures of −20 to −40°C. The electrochemical discharge reactions of nickel-based batteries are as follows [24,42]:

$$\text{At cathode: } \text{NiOOH} + \text{H}_2\text{O} + e^- \rightarrow \text{OH}^- + \text{Ni(OH)}_2 \tag{2.9}$$

$$\text{At anode: } \text{Cd} + 2\text{OH}^- \rightarrow \text{Cd(OH)}_2 + 2e^- \tag{2.10}$$

$$\text{Overall: } \text{Cd} + 2\text{NiOOH} + 2\text{H}_2\text{O} \rightarrow \text{Cd(OH)}_2 + 2\text{Ni(OH)}_2 \tag{2.11}$$

2.1.5 Fuel cell (FC)

FC is a common CSS and the most suitable solution for using hydrogen. FC converts the chemical energy of fuel into power continuously. As long as active fuel and external oxidants are available, FC will supply the electricity continuously, decreasing the consumption of fossil fuel and CO_2 emissions by reducing the emission of harmful gases [24,43]. In FC, the anode is liquid fuel or gas and the cathode can be air and chlorine. FC technologies are significantly used in renewable energy due to the use of hydrogen. Hydrogen-based FCs (HFCs), which are an integration of hydrogen and oxygen to generate electricity, are popular and more efficient than other types of FCs [44]. HFCs are classified into two types, direct and indirect, depending on the type of fuel. Hydrogen and methanol can be considered as fuel for direct systems, and fossil fuels and natural gas as fuels for indirect systems. The reaction of fuel in direct systems is direct, but in indirect systems, the fuel is first converted to hydrogen-rich gas and then supplied to the cell to react. The combination of fuel and oxidant, operating temperature, and electrolyte type are the criteria that categorize FCs into different categories. Different types of FCs include alkaline FC (AFC), molten carbonate FC (MCFC), phosphoric acid FC (PAFC), direct methanol FC (DMFC), solid polymer fuel cell—proton exchange membrane FC (SPFC-PEMFC), solid oxide FC (SOFC), regenerative FC (RFC). The principle chemical reaction of FC is defined as follows [44,45]:

$$2H_2 + O_2 \leftrightarrow 2H_2O + \text{electricity} \tag{2.12}$$

2.2 Mechanical storage systems

Today, mechanical energy storage systems (MSS) are commonly used around the world to generate electricity. Converting and storing energy using flexible resources is one of the major benefits of MSS. Electricity is consumed in the course of off-peak hours from the system and stored mechanically according to the principle of forced spring, kinetic energy, pressurized gas, and potential energy until it is needed and transferred back to the grid [46,47]. MSS classification can be based on two principles. If the MSS categorization is based on the working basics, they can be classified as compressed gas, forced spring, kinetic energy, and potential energy. Also, from a technological point of view, MSS can be divided into four categories: CAES, flywheel energy storage (FES), pumped hydro storage (PHS), and gravity energy storage systems (GES) [46]. Among these four systems, PHS, with its long life cycle, accounts for approximately 96% of the electrical storage capacity in the world or about 3% of global electricity production capacity [48]. In the following section, each of the MSS technologies is introduced.

2.2.1 Pumped hydraulic storage (PHS)

PHS systems with 125 GW power storage capacity are known as the largest ESS in the world. PHS stores electrical energy through the pumping of water from the bottom of the tank to the height of the tank as potential energy. PHS has flexible operation characteristics and can be useful in smoothing the intermittent output power of renewable energy resources such as wind and PV [49]. An efficiency of about 76%—85%, a very long lifetime of nearly 50 years or more are the most important benefits of PHS [50]. Despite this, PHS suffers from some problems such as high cost of capital, a negative impact on the environment, and reduced geological performance, which has limited the future development of PHS. Among all these limitations, the PHS size is its most important weakness, which is very large and not comparable to the new ESS technologies, which have a smaller scale [24].

2.2.2 Flywheel energy storage (FES)

FES systems have made considerable progress in recent decades in terms of material technology and power electronic structure. As a result of these advances, FES systems have become a widely used technology in the EVs and power systems industry today. The flywheel is a massive rotating cylinder that is supported by magnetic bearings on the stator [19,51]. FES systems are divided into two categories based on high and low speeds. A flywheel is used to the smooth running of machines and can mechanically store kinetic energy from the rotation of the high-speed rotor mass. Based on the principle of speed, it can be said that low-speed FESS contains a steel disk with high inertia and low speed. But, the high-speed FES systems have a high speed and composite disk with relatively lower inertia. The stored energy is based on the rotor speed, and increasing the rotor speed increases the stored energy [46,51]. High-speed and low-speed FES systems can be compared from density perspective, so that the energy density in low-speed FES systems is about 2000 W/kg, but the energy density in these systems is about 5 Wh/kg. Also, high-speed FES systems have a very high energy density and a higher energy density in the range of 200 Wh/kg. FES systems have an efficiency of 90%−95% and rated power ranges of 0−50 MW. The energy from FES technologies is obtained as [24]:

$$E = \frac{1}{2} m r^2 \left(w_{max}^2 - w_{min}^2 \right) \tag{2.13}$$

where E depicted usable energy within the maximum angular speed (w_{max}) and minimum angular speed (w_{min}), m is concentrated mass of the flywheel at the rim, and r shows the radius.

2.2.3 Compressed air energy storage (CAES)

The CAES stores power by compressing the air in the reservoir and then converting it into modified gas. In these systems, modified (compressed) gas is used to rotate the turbine coupled with a generation unit so that the turbine can generate electricity by expanding the compressed gas [52]. CAES systems have advantages such as high capacity and longevity, moderate geographical dependence, and low cost per kilowatt. Having these benefits has become a significant reason for the realistic replacement of these systems with PHS systems. Complex designs of CAES systems based on processes such as exothermic and endothermic operations involved, extension and compression of the air, and heat exchange are divided into three categories: Isothermal, adiabatic, and diabatic systems. Each of these systems has a unique way of working so that isothermal and adiabatic systems have performed well in terms of small power density requirements. Diabatic storage systems have also used a larger share of CAES systems due to their excellent system flexibility and high power density [53,54]. CAES systems have an installed commercial capacity in the range of 35−300 MW and have significant usage for grid such as voltage and frequency control. Developments that have recently occurred in the hybrid CAES with off-shore and on-shore wind turbines have significantly decreased fluctuations in the power output. Despite these advances, however, these technologies suffer from some limitations, such as the appropriate air storage tank geographical location or underground cavern. To deal with this limit, solutions such as increasing the pressure on the carbon fiber reservoir on the ground for advanced CAES systems on a small scale have been proposed [55,56].

2.3 Thermal energy storage (TES)

According to the report of the U.S. Department of Energy, the capacity of TESs accounts for about 1.9% of the world's energy. TES systems store energy for use in power plants or other purposes in an isolated depository from the solar or electric heater. Thermal storage tanks, heat transfer mechanisms, and containment control systems are the three main components of TES systems. The heat transfer mechanism system directly or indirectly extracts the stored heat to generate electricity or heat energy consumption by the engine cycle. The restraint control system is responsible for controlling the insulation, operation of heat transfer medium, and storage reservoir [19,24,57]. TES systems have some advantages such as good energy density (80–250 Wh/kg specific energy), low investment cost, and environmental friendliness. While these systems only suffer in terms of the low-efficiency cycle of the whole system, which is in the range of 30%–50%. TES technologies can be classified into two types based on the operating temperature of energy storage materials: high-temperature TES and low-temperature TES [19,57]. Latent heat storage, sensible heat storage, and thermochemical sorption storage systems are methods through which thermal energy storage can be achieved. Organic materials, inorganic materials, and phase change materials (PCM) are used in hidden heat storage systems as storage intermediaries to change the heat exchange in providing storage environment during phase change. Efficient heat transfer and high energy density are the most important advantages of latent heat storage systems. In sensible heat storage systems, thermal storage depends on the change in temperature in the storage environment, and its capacity depends on the specific heat and mass of the environment [58,59].

2.4 Other types of EES such as superconductive magnetic and supercapacitors

The EES systems modify the electrical or magnetic field through capacitors or superconducting magnets and thus store electrical energy directly as electricity. EESS technologies store energy by changing magnetic or electrical fields through superconducting magnets or capacitors. Given that today's power system is facing the combination of renewable resources with the transmission and distribution network, EESS can be used as an ideal technology to reduce this problem. This performance can be of great help in operating the power system, improving power quality, reducing energy import requirements during peak hours, and load balance [57,60]. Superconductive magnetic energy storage (SMES) and supercapacitor energy storage are the two types of EESS, which are introduced in the continuation of this section.

2.4.1 Superconductive magnetic energy storage (SMES)

The main operation of SMES systems is based on electrodynamics. In the SMES, energy is stored in a magnetic field using an AC-to-DC converter, which can also transfer to the network via a DC-to-AC converter [61]. In SMES technologies, ohmic losses can generate heat and thus lead to SMES thermal instability. To prevent these losses and reduce them, the coil temperature must be maintained below its superconducting temperature [62]. SMES technologies are available for commercial use in the range of 0.1—10 MW, so that with the advancement of technology in the coming decades, this capacity is expected to reach about 100 MW. In addition to all the benefits, SMESs suffer from factors such as high installation costs ($10,000/kWh), the complexity of the cooling system, and coil materials. Today, most studies on these storage systems are based on reducing the cost of coils and cooling systems, so that SMESs can be of great interest to consumers by solving these problems [35,63].

Considering the superconductor material and cryogenic conditioning system type, SMES are divided into two categories: low-temperature superconductor (LTS) and high-temperature superconductor (HTS). LTS SMES due to its proper performance and advantages such as high energy density (4 kW/L), fast response to charging and discharging phenomena, high efficiency (95%—98%), and a long lifetime in the range of 30 years can be widely used than HTS SMES [64]. The energy stored in SMES is as follows:

$$W_{LS} = \frac{1}{2}L \times I^2 \tag{2.14}$$

where L is the self-inductance of the coil, I denotes the value of flowing current via the coil, and WLS shows the stored energy in the coil.

2.4.2 Supercapacitor energy storage (SCES)

The supercapacitor energy storage (SCES) is one of the applications of EESS that stores energy in the form of electrostatic fields. SCESs have been replaced with traditional capacitors used in electronics and batteries based on some features such as high specific energy, high charging capability with high currents, and low internal resistance and wide temperature range [65,66]. In addition to these features, SCESs have a key structural difference from classical capacitors, with the very short gap between the electrode and the electrolyte, enabling SCES to have very high capacitance amounts in the range of many thousand farads.

The SCES system is divided into two types based on the electrode materials used: symmetrical and asymmetric. In the symmetric SCES system, both electrodes are made of activated carbon, appropriate for application in small and medium-sized projects. In asymmetric type, one of the electrodes is made of nickel hydroxide metal, which dramatically increases energy density and reduces leakage current. However, asymmetric SECSs are recommended for use in large-scale applications [24,67]. Different types of SCES systems are available in 10—75 kW power and 4—70 W energy rating based on power rating, and they can be connected in series or parallel without any problems and can be used for collective capacity. Robustness, usability in a variety of environments (hot, cold and humid), reliability, very long lifetime are the important advantages of SCES technology [24,64,68].

3. The operation of EES in power markets

In recent decades, the investment and use of renewable energies and EVs have made significant progress in the power system and electricity markets. Although the penetration of these clean energies has had more benefits, in most cases, it has reduced the stability and reliability of the power system. Today, energy storage is an important alternative to increasing the level of renewable energy for improving the reliability and flexibility of the power system, reducing operating costs, and balancing the electricity market in increasing the level of renewable energy and their investments [69]. So that an ESS can be used to store and maintain energy in periods of low load and then transfer the stored energy to grid during periods of high load. With high levels of renewable energies penetration in power systems, storage systems will be able to store energy when energy production is high and release it when required. This performance is considered a cost-effective mechanism in the economic issues of the power system, but it should be noted that the economic viability of energy storage depends on the details of how the storage mechanisms work and how the related services are displayed. The benefits that energy storage can have depend on the characteristics or structure of the power system in which the storage systems operate [70,71]. For example, a study has shown that in addition to all the benefits of storing energy in a power system, it can be very economical. Such benefits are obtained by integrating renewable energy, delaying production and transmission costs, ancillary services, and sustainability of voltage and network frequency [72].

Recently, EES for renewable energy applications and demand response has received significant attention in academic and industry research. The main factor in these interests and orientations can be considered the rapid development of renewable energy generation, which has significantly met the needs of the power system [73]. Today, beyond all these benefits, ESSs have had a significant impact on a variety of energy markets. Thus, for both producers and consumers of energy it has been profitable and has had a significant impact on energy prices in different types of electricity markets. For energy storage owners, the main goal is maximizing the revenue of the EES by participating in various markets such as day-ahead, real-time, and ancillary service markets. Thus, increase in the energy storage participation in electricity markets has made their scheduling model an important challenge for EES owners in order to minimize investment costs and maximize revenue [74,75]. Table 2.1 is provided to highlight the role of energy storage technologies in electrical energy markets.

4. The current stage of energy storage systems in the world

In the previous sections, the types and technologies of ESSs were reviewed. But given that the ESS is related to various aspects of the power system, the field of research on ESS can also be very broad. Therefore, in this section, we will have an overview of the aspects of the development and establishment of the ESS in today's world.

Today, most research and policies related to ESS focus on issues such as investment and applications of these technologies. Achieving a standard relationship can be very important in this regard and is an important factor in providing other ESS benefits.

The EES is most widely used for grid-connected renewable energy projects. In recent years, the frequency regulation service has been the cumulative installed capacity of energy storage with the fastest growth. In this regard, the United States has the largest installed capacity, commercial projects, and frequency regulation market. Generally, ESSs today have the largest share of applications in projects such as solar power-storage power charging station for EV, renewable energies, transactive energy, peer-to-peer energy trading. Compared to the aforementioned applications, ESSs have less applications in the fields of transmission and distribution of electricity.

TABLE 2.1 Participation of energy storage technologies in the power market.

References	Market type	Storage type	Objective
[76]	Day-ahead electricity market	Wind-PHS	Maximizing the profit to the wind-PSH owner
[77]	Pool based wholesale electricity market	CAES	Increasing the overall pool revenues for most power producers and decreasing CO_2 emissions by 3%
[78]	Deregulated markets	FES	Frequency regulation
[79]	Day-ahead electricity market	Hydrogen	Optimal sizing of a storage plant
[80]	Real-time electricity market	TES	Minimizing cost and improving the efficiency of a polygeneration district energy system
[81]	Day-ahead market	Hydrogen	Optimal operation of a hybrid plant with wind power and hydrogen storage
[82]	Day-ahead energy market, spinning and regulation reserve markets	PHS	Increasing the profits of wind resources
[83]	Balancing power market	Battery	Estimating the profitability of energy storage operation in the balancing power market

5. Conclusions

Expanding the power system, increasing the influence of renewable energies, and other developments in the performance and structure of energy systems necessitate the need for an ESS. Due to the variety of ESS, a large number of choices can be made in using these systems. Choosing a suitable case for a particular function requires knowing the basic technologies, the functional role of the ESS, and some other issues. In this chapter, various technologies of ESSs along with their structure and functional capabilities with future trends are examined. So that the practical fields of these technologies in the electrical system were expressed based on their functional characteristics and capacity. In addition, the most challenging issues related to the installation and operation of ESSs were considered, taking into account the advantages and disadvantages of each of these technologies. After getting acquainted with ESS technologies, literature was expressed in order to influence energy storage in electricity markets. The study found that energy storage contributes significantly to the electricity market, with energy storage owners seeking to minimize investment costs and increase their revenue. Finally, the current stage of ESS in the world and their progress were examined. It is safe to conclude that ESS is expanding and innovative approaches to many of these technologies will emerge.

References

[1] A. Moradzadeh, O. Sadeghian, K. Pourhossein, B. Mohammadi-Ivatloo, A. Anvari-Moghaddam, Improving residential load disaggregation for sustainable development of energy via principal component analysis, Sustainability 12 (8) (April 2020) 3158, https://doi.org/10.3390/su12083158.

[2] M. Nazari-Heris, M.A. Mirzaei, B. Mohammadi-Ivatloo, M. Marzband, S. Asadi, Economic-environmental effect of power to gas technology in coupled electricity and gas systems with price-responsive shiftable loads, J. Clean. Prod. 244 (January 2020) 118769, https://doi.org/10.1016/j.jclepro.2019.118769.

[3] M.A. Mirzaei, M. Nazari-Heris, B. Mohammadi-Ivatloo, K. Zare, M. Marzband, A. Anvari-Moghaddam, Hourly price-based demand response for optimal scheduling of integrated gas and power networks considering compressed air energy storage, in: Demand Response Application in Smart Grids, Springer International Publishing, Cham, 2020, pp. 55–74.

[4] B.C. Erdener, K.A. Pambour, R.B. Lavin, B. Dengiz, An integrated simulation model for analysing electricity and gas systems, Int. J. Electr. Power Energy Syst. 61 (October 2014) 410–420, https://doi.org/10.1016/j.ijepes.2014.03.052.

[5] Q. Zeng, J. Fang, J. Li, Z. Chen, Steady-state analysis of the integrated natural gas and electric power system with bi-directional energy conversion, Appl. Energy 184 (December 2016) 1483–1492, https://doi.org/10.1016/j.apenergy.2016.05.060.

[6] L. Bai, F. Li, H. Cui, T. Jiang, H. Sun, J. Zhu, Interval optimization based operating strategy for gas-electricity integrated energy systems considering demand response and wind uncertainty, Appl. Energy 167 (April 2016) 270–279, https://doi.org/10.1016/j.apenergy.2015.10.119.

[7] C. He, T. Liu, L. Wu, M. Shahidehpour, Robust coordination of interdependent electricity and natural gas systems in day-ahead scheduling for facilitating volatile renewable generations via power-to-gas technology, J. Mod. Power Syst. Clean Energy 5 (3) (May 2017) 375–388, https://doi.org/10.1007/s40565-017-0278-z.

[8] B. Khaki, P. Das, S. Member, Sizing and Placement of Battery Energy Storage Systems and Wind Turbines by Minimizing Costs and System Losses, 2019 arXiv preprint arXiv:1903.12029.

[9] J.D. Vergara-Dietrich, M.M. Morato, P.R.C. Mendes, A.A. Cani, J.E. Normey-Rico, C. Bordons, Advanced chance-constrained predictive control for the efficient energy management of renewable power systems, J. Process Contr. 74 (February 2019) 120–132, https://doi.org/10.1016/j.jprocont.2017.11.003.

[10] C. Olk, D.U. Sauer, M. Merten, Bidding strategy for a battery storage in the German secondary balancing power market, J. Energy Storage 21 (February 2019) 787–800, https://doi.org/10.1016/j.est.2019.01.019.

[11] S. Hajiaghasi, A. Salemnia, M. Hamzeh, Hybrid energy storage system for microgrids applications: a review, J. Energy Storage 21 (February 2019) 543–570, https://doi.org/10.1016/j.est.2018.12.017.

[12] D.B.W. Abeywardana, B. Hredzak, J.E. Fletcher, G. Konstantinou, A cascaded boost inverter based battery energy storage system with reduced battery ripple current, in: IECON 2017 - 43rd Annual Conference of the IEEE Industrial Electronics Society, vol. 2017, 2017, pp. 2733–2738, https://doi.org/10.1109/IECON.2017.8216460. Janua.

[13] N. Mukherjee, D. Strickland, Control of second-life hybrid battery energy storage system based on modular boost-multilevel buck converter, IEEE Trans. Ind. Electron. 62 (2) (February 2015) 1034–1046, https://doi.org/10.1109/TIE.2014.2341598.

[14] A. Lahyani, P. Venet, A. Guermazi, A. Troudi, Battery/supercapacitors combination in uninterruptible power supply (UPS), IEEE Trans. Power Electron. 28 (4) (April 2013) 1509–1522, https://doi.org/10.1109/TPEL.2012.2210736.

[15] S. Danyali, S.H. Hosseini, G.B. Gharehpetian, New extendable single-stage multi-input DC-DC/AC boost converter, IEEE Trans. Power Electron. 29 (2) (February 2014) 775–788, https://doi.org/10.1109/TPEL.2013.2256468.

[16] W. Jiang, L. Huang, L. Zhang, H. Zhao, L. Wang, W. Chen, Control of active power exchange with auxiliary power loop in a single-phase cascaded multilevel converter-based energy storage system, IEEE Trans. Power Electron. 32 (2) (February 2017) 1518–1532, https://doi.org/10.1109/TPEL.2016.2543751.

[17] J. Rugolo, M.J. Aziz, Electricity storage for intermittent renewable sources, Energy Environ. Sci. 5 (5) (2012) 7151–7160, https://doi.org/10.1039/c2ee02542f.

[18] H. Ibrahim, A. Ilinca, J. Perron, Energy storage systems-characteristics and comparisons, Renew. Sustain. Energy Rev. 12 (5) (2008) 1221–1250, https://doi.org/10.1016/j.rser.2007.01.023.

[19] H. Chen, T.N. Cong, W. Yang, C. Tan, Y. Li, Y. Ding, Progress in electrical energy storage system: a critical review, Prog. Nat. Sci. 19 (3) (2009) 291–312, https://doi.org/10.1016/j.pnsc.2008.07.014.

[20] R.F. Abdo, H.T.C. Pedro, R.N.N. Koury, L. Machado, C.F.M. Coimbra, M.P. Porto, Performance evaluation of various cryogenic energy storage systems, Energy 90 (October 2015) 1024–1032, https://doi.org/10.1016/j.energy.2015.08.008.

[21] C. Cheng, et al., Review and prospects of hydrate cold storage technology, Renew. Sustain. Energy Rev. 117 (January 2020) 109492, https://doi.org/10.1016/j.rser.2019.109492.

[22] C.R. Matos, J.F. Carneiro, P.P. Silva, Overview of large-scale underground energy storage technologies for integration of renewable energies and criteria for reservoir identification, J. Energy Storage 21 (February 2019) 241–258, https://doi.org/10.1016/j.est.2018.11.023.

[23] F. Kalavani, B. Mohammadi-Ivatloo, K. Zare, Optimal stochastic scheduling of cryogenic energy storage with wind power in the presence of a demand response program, Renew. Energy 130 (January 2019) 268–280, https://doi.org/10.1016/j.renene.2018.06.070.

[24] F. Nadeem, S.M.S. Hussain, P.K. Tiwari, A.K. Goswami, T.S. Ustun, Comparative review of energy storage systems, their roles, and impacts on future power systems, IEEE Access 7 (2019) 4555–4585, https://doi.org/10.1109/ACCESS.2018.2888497.

[25] K.C. Divya, J. Østergaard, Battery energy storage technology for power systems-An overview, Elec. Power Syst. Res. 79 (4) (2009) 511–520, https://doi.org/10.1016/j.epsr.2008.09.017.

[26] J.O. Besenhard, Handbook of Battery Materials, 2007.

[27] S.M. Lukic, J. Cao, R.C. Bansal, F. Rodriguez, A. Emadi, Energy storage systems for automotive applications, IEEE Trans. Ind. Electron. 55 (6) (2008) 2258–2267, https://doi.org/10.1109/TIE.2008.918390.

[28] F. Díaz-González, A. Sumper, O. Gomis-Bellmunt, R. Villafáfila-Robles, A review of energy storage technologies for wind power applications, Renew. Sustain. Energy Rev. 16 (4) (2012) 2154–2171, https://doi.org/10.1016/j.rser.2012.01.029.

[29] O. Palizban, K. Kauhaniemi, Energy storage systems in modern grids—matrix of technologies and applications, J. Energy Storage 6 (2016) 248–259, https://doi.org/10.1016/j.est.2016.02.001.

[30] R. Amirante, E. Cassone, E. Distaso, P. Tamburrano, Overview on recent developments in energy storage: mechanical, electrochemical and hydrogen technologies, Energy Convers. Manag. 132 (2017) 372–387, https://doi.org/10.1016/j.enconman.2016.11.046.

[31] D. Parra, et al., An interdisciplinary review of energy storage for communities: challenges and perspectives, Renew. Sustain. Energy Rev. 79 (2017) 730–749, https://doi.org/10.1016/j.rser.2017.05.003.

[32] J. Baker, New technology and possible advances in energy storage, Energy Pol. 36 (12) (2008) 4368–4373, https://doi.org/10.1016/j.enpol.2008.09.040.

[33] M. Verbrugge, E. Tate, Adaptive state of charge algorithm for nickel metal hydride batteries including hysteresis phenomena, J. Power Sources 126 (1–2) (2004) 236–249, https://doi.org/10.1016/j.jpowsour.2003.08.042.

[34] M.A. Hannan, M.M. Hoque, A. Mohamed, A. Ayob, Review of energy storage systems for electric vehicle applications: issues and challenges, Renew. Sustain. Energy Rev. 69 (2017) 771–789, https://doi.org/10.1016/j.rser.2016.11.171.

[35] M. Faisal, M.A. Hannan, P.J. Ker, A. Hussain, M. Bin Mansor, F. Blaabjerg, Review of energy storage system technologies in microgrid applications: issues and challenges, IEEE Access 6 (2018) 35143–35164, https://doi.org/10.1109/ACCESS.2018.2841407.

[36] L. Yao, B. Yang, H. Cui, J. Zhuang, J. Ye, J. Xue, Challenges and progresses of energy storage technology and its application in power systems, J. Mod. Power Syst. Clean Energy 4 (4) (2016) 519–528, https://doi.org/10.1007/s40565-016-0248-x.

[37] D.A.J. Rand, P.T. Moseley, Lead-acid battery fundamentals, in: Lead-Acid Batteries for Future Automobiles, 2017, pp. 97–132.

[38] B. Dunn, H. Kamath, J.M. Tarascon, "Electrical energy storage for the grid: a battery of choices, Science 334 (6058) (2011) 928–935, https://doi.org/10.1126/science.1212741.

[39] G. Zubi, R. Dufo-López, M. Carvalho, G. Pasaoglu, The lithium-ion battery: state of the art and future perspectives, Renew. Sustain. Energy Rev. 89 (2018) 292–308, https://doi.org/10.1016/j.rser.2018.03.002.

[40] J. Cho, S. Jeong, Y. Kim, Commercial and research battery technologies for electrical energy storage applications, Prog. Energy Combust. Sci. 48 (2015) 84–101, https://doi.org/10.1016/j.pecs.2015.01.002.

[41] C.S. Lai, Y. Jia, L.L. Lai, Z. Xu, M.D. McCulloch, K.P. Wong, A comprehensive review on large-scale photovoltaic system with applications of electrical energy storage, Renew. Sustain. Energy Rev. 78 (2017) 439–451, https://doi.org/10.1016/j.rser.2017.04.078.

[42] G.M. Ehrlich, Linden's Handbook of Batteries, fourth ed., 2002.

[43] D. Collins, Handbook of batteries and fuel cells by David Linden, published by McGraw-Hill Book Company GmbH., Hamburg, F.R.G., 1984; 1024 pp.; price DM 258.80, J. Power Sources 17 (4) (1986) 379–384, https://doi.org/10.1016/0378-7753(86)80059-3.

[44] K.T. Chau, Y.S. Wong, C.C. Chan, Overview of energy sources for electric vehicles, Energy Convers. Manag. 40 (10) (1999) 1021–1039, https://doi.org/10.1016/S0196-8904(99)00021-7.

[45] S. Mekhilef, R. Saidur, A. Safari, Comparative study of different fuel cell technologies, Renew. Sustain. Energy Rev. 16 (1) (2012) 981–989, https://doi.org/10.1016/j.rser.2011.09.020.

[46] M.S. Guney, Y. Tepe, Classification and assessment of energy storage systems, Renew. Sustain. Energy Rev. 75 (2017) 1187–1197, https://doi.org/10.1016/j.rser.2016.11.102.

[47] V.C. Prantil, T. Decker, The captains of energy: systems dynamics from an energy perspective, Synth. Lect. Eng. 9 (1) (2015) 1–220, https://doi.org/10.2200/S00610ED1V01Y201410ENG024.

[48] International Energy Agency, "Tracking Clean Energy Progress 2013," Technology, 2012, pp. 1–82, https://doi.org/10.1787/energy_tech-2014-en.

[49] M. Daneshvar, B. Mohammadi-Ivatloo, K. Zare, S. Asadi, Two-stage stochastic programming model for optimal scheduling of the wind-thermal-hydropower-pumped storage system considering the flexibility assessment, Energy (2020), https://doi.org/10.1016/j.energy.2019.116657.

[50] Energy storage: program planning docoument, in: Lightning in a Bottle: Electrical Energy Storage, 2011, pp. 261–289.

[51] M.G. Molina, Distributed energy storage systems for applications in future smart grids, in: 2012 Sixth IEEE/PES Transmission and Distribution: Latin America Conference and Exposition (T&D-LA), 2012, pp. 1–7, https://doi.org/10.1109/TDC-LA.2012.6319051.

[52] M. Abbaspour, M. Satkin, B. Mohammadi-Ivatloo, F. Hoseinzadeh Lotfi, Y. Noorollahi, Optimal operation scheduling of wind power integrated with compressed air energy storage (CAES), Renew. Energy 51 (2013), https://doi.org/10.1016/j.renene.2012.09.007.

[53] J. Cheng, F.F. Choobineh, A comparative study of the storage assisted wind power conversion systems, in: 2017 6th International Conference on Clean Electrical Power: Renewable Energy Resources Impact, ICCEP 2017, 2017, pp. 608–613, https://doi.org/10.1109/ICCEP.2017.8004751.

[54] M. Jadidbonab, A. Dolatabadi, B. Mohammadi-Ivatloo, M. Abapour, S. Asadi, Risk-constrained energy management of PV integrated smart energy hub in the presence of demand response program and compressed air energy storage, IET Renew. Power Gener. (2019), https://doi.org/10.1049/iet-rpg.2018.6018.

[55] F. Crotogino, K.-U. Mohmeyer, R. Scharf, Huntorf CAES: more than 20 years of successful operation, in: Solution Mining Research Institute (SMRI) Spring Meeting, no. April, 2001, pp. 351–357.

[56] E. Jannelli, M. Minutillo, A. Lubrano Lavadera, G. Falcucci, A small-scale CAES (compressed air energy storage) system for stand-alone renewable energy power plant for a radio base station: a sizing-design methodology, Energy 78 (2014) 313–322, https://doi.org/10.1016/j.energy.2014.10.016.

[57] X. Luo, J. Wang, M. Dooner, J. Clarke, Overview of current development in electrical energy storage technologies and the application potential in power system operation, Appl. Energy 137 (2015) 511–536, https://doi.org/10.1016/j.apenergy.2014.09.081.

[58] R. Guerrero-Lemus, L.E. Shephard, Executive Summary, 2017.

[59] M.R. Anisur, M.A. Kibria, M.H. Mahfuz, I.H.S.C. Metselaar, R. Saidur, Latent heat thermal storage (LHTS) for energy sustainability, Green Energy Technol. 201 (2015) 245–263, https://doi.org/10.1007/978-81-322-2337-5_10.

[60] T. Kousksou, P. Bruel, A. Jamil, T. El Rhafiki, Y. Zeraouli, Energy storage: applications and challenges, Sol. Energy Mater. Sol. Cell. 120 (Part A) (2014) 59–80, https://doi.org/10.1016/j.solmat.2013.08.015.

[61] H. Kiehne, Battery Technology Handbook, CRC Press, 2003.

[62] K. Gong, J. Shi, Y. Liu, Z. Wang, L. Ren, Y. Zhang, Application of SMES in the microgrid based on fuzzy control, IEEE Trans. Appl. Supercond. 26 (3) (April 2016) 1–5, https://doi.org/10.1109/TASC.2016.2524446.

[63] A.H. Moghadasi, H. Heydari, M. Farhadi, Pareto optimality for the design of smes solenoid coils verified by magnetic field analysis, IEEE Trans. Appl. Supercond. 21 (1) (2011) 13–20, https://doi.org/10.1109/TASC.2010.2089791.

[64] M.V. Aware, D. Sutanto, Improved controller for power conditioner using high-temperature superconducting magnetic energy storage (HTS-SMES), IEEE Trans. Appl. Supercond. 13 (1) (2003) 38–47, https://doi.org/10.1109/TASC.2003.811352.

[65] R. Mishra, R. Saxena, Comprehensive review of control schemes for battery and super-capacitor energy storage system, in: 2017 7th International Conference on Power Systems, ICPS 2017, 2018, pp. 702–707, https://doi.org/10.1109/ICPES.2017.8387381.

[66] C. Zhu, R. Lu, L. Tian, Q. Wang, The development of an electric bus with super-capacitors as unique energy storage, in: 2006 IEEE Vehicle Power and Propulsion Conference, 2006, pp. 1–5, https://doi.org/10.1109/VPPC.2006.364372.

[67] V. Ganesh, S. Pitchumani, V. Lakshminarayanan, New symmetric and asymmetric supercapacitors based on high surface area porous nickel and activated carbon, J. Power Sources 158 (2 Spec. iss) (2006) 1523–1532, https://doi.org/10.1016/j.jpowsour.2005.10.090.

[68] A. Burke, Ultracapacitors: why, how, and where is the technology, J. Power Sources 91 (1) (2000) 37–50, https://doi.org/10.1016/S0378-7753(00)00485-7.

[69] R. Fioravanti, K. Vu, W. Stadlin, Large-scale solutions, IEEE Power Energy Mag. 7 (4) (2009) 48–57, https://doi.org/10.1109/MPE.2009.932869.

[70] C.A. Silva-Monroy, J.P. Watson, Integrating energy storage devices into market management systems, Proc. IEEE 102 (7) (2014) 1084–1093, https://doi.org/10.1109/JPROC.2014.2327378.

[71] C. Opathella, A. Elkasrawy, A.A. Mohamed, B. Venkatesh, A novel capacity market model with energy storage, IEEE Trans. Smart Grid 10 (5) (September 2019) 5283–5293, https://doi.org/10.1109/TSG.2018.2879876.

[72] EPRI, Electric energy storage technology options: a white paper primer on applications, costs and benefits, Epri (2010) 1–170. EPRI 1020676.

[73] M. Cui, J. Zhang, H. Wu, B.M. Hodge, Wind-friendly flexible ramping product design in multi-timescale power system operations, IEEE Trans. Sustain. Energy 8 (3) (2017) 1064–1075, https://doi.org/10.1109/TSTE.2017.2647781.

[74] X. Fang, F. Li, Y. Wei, H. Cui, Strategic scheduling of energy storage for load serving entities in locational marginal pricing market, IET Gener. Transm. Distrib. 10 (5) (2016) 1258–1267, https://doi.org/10.1049/iet-gtd.2015.0144.

[75] X. Fang, B.M. Hodge, L. Bai, H. Cui, F. Li, Mean-variance optimization-based energy storage scheduling considering day-ahead and real-time lmp uncertainties, IEEE Trans. Power Syst. 33 (6) (2018) 7292–7295, https://doi.org/10.1109/TPWRS.2018.2852951.

[76] D.K. Khatod, V. Pant, J. Sharma, Optimized daily scheduling of wind-pumped hydro plants for a day-ahead electricity market system, in: 2009 International Conference on Power Systems, ICPS '09, 2009, pp. 1–6, https://doi.org/10.1109/ICPWS.2009.5442767.

[77] A. Foley, I. Díaz Lobera, Impacts of compressed air energy storage plant on an electricity market with a large renewable energy portfolio, Energy 57 (August 2013) 85–94, https://doi.org/10.1016/j.energy.2013.04.031.

[78] M.L. Lazarewicz, T.M. Ryan, Integration of flywheel-based energy storage for frequency regulation in deregulated markets, in: IEEE PES General Meeting, PES, vol. 2010, 2010, pp. 1–6, https://doi.org/10.1109/PES.2010.5589748.

[79] C. Brunetto, G. Tina, Optimal hydrogen storage sizing for wind power plants in day ahead electricity market, IET Renew. Power Gener. 1 (4) (2007) 220–226, https://doi.org/10.1049/iet-rpg:20070040.

[80] K.M. Powell, et al., Thermal energy storage to minimize cost and improve efficiency of a polygeneration district energy system in a real-time electricity market, Energy 113 (October 2016) 52–63, https://doi.org/10.1016/j.energy.2016.07.009.

[81] M. Korpås, A.T. Holen, Operation planning of hydrogen storage connected to wind power operating in a power market, IEEE Trans. Energy Convers. 21 (3) (September 2006) 742–749, https://doi.org/10.1109/TEC.2006.878245.

[82] A.K. Varkani, A. Daraeepour, H. Monsef, A new self-scheduling strategy for integrated operation of wind and pumped-storage power plants in power markets, Appl. Energy 88 (12) (December 2011) 5002–5012, https://doi.org/10.1016/j.apenergy.2011.06.043.

[83] N. Belonogova, J. Haakana, V. Tikka, J. Lassila, J. Partanen, Feasibility studies of end-customer's local energy storage on balancing power market, in: IET Conference Publications, vol. 2016, 2016, p. CP686.

3

Overview of energy storage systems for wind power integration

Roghayyeh Pourebrahim[1], Sajjad Tohidi[1],
Hossein Khounjahan[2]

[1]Faculty of Electrical and Computer Engineering, University of Tabriz, Tabriz, East Azarbaijan, Iran;
[2]Azarbijan Regional Electric Company, Tabriz, East Azarbaijan, Iran

1. Introduction

Renewable energy resources such as solar systems, wind turbines, tidal force, biomass, geothermal, etc., play an important role in providing energy for modern human societies. Due to renewability, widespread availability, and pollution-free features, wind energy is one of the most regarded energy resources. One solution to exploit wind energy is to convert it to electrical energy through wind turbines. Wind turbines have been altered during the last decades and global wind energy generation capacity increases daily. Fig. 3.1 shows the global wind energy power generation capacity from 2013 up to 2019.

Among various power plants, the wind power generation systems stand out for the input power control scheme (turbine drive actuator). In conventional fossil-fuel-based power plants, the active and reactive powers are, respectively, controlled by the input fuel injection system (governor) and the automatic voltage regulation. However, because wind velocity is intermittent under different climate conditions, the conventional active power control method cannot be applied in wind turbines. Besides, due to abrupt changes in wind speed, the output power of the wind turbines is not certain; this can give rise to many problems and bring about system instability.

One of the most important issues in power systems is nonflat power demand. This necessitates using reserve power generation systems to assist the power system during peak periods. Therefore, the continuous power supply is provided at the cost of installation and exploitation of many reserve systems. The problem can be tackled by using storage systems, which can also lead to increasing the renewable energy penetration level, facilitating active

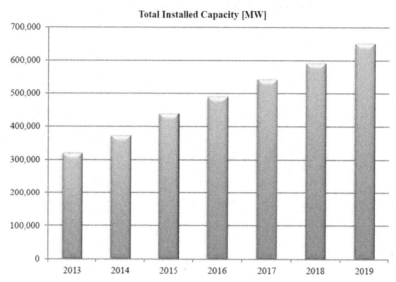

FIGURE 3.1   Global wind power installation capacity. *Source: World Wind Energy Association.*

and reactive power control, reduction in voltage fluctuations, and improving power quality and reliability of the systems. In the forthcoming sections, various energy storage systems with an emphasis on storage for wind power applications will be discussed.

2. Electrical energy storage systems

An electrical energy storage system is a system in which electrical energy is converted into a type of energy (chemical, thermal, electromagnetic energy, etc.) that is capable of storing energy and, if needed, is converted back into electrical energy.

The energy storage system value is for the services it can provide for power system networks. This technology can be used all over the power networks. Energy storage systems particularly on large scale have various applications. These applications include power quality improvement for reliability to long-term power management in power systems.

For high-power applications such as power quality and emergency power applications, the energy should be discharged in a fraction of a second. On the other hand, for high-energy applications such as energy management, including load curve leveling and peak shaving, the energy should be discharged within several hours.

There are several different methods for energy storage systems in large scales. All energy storage system methods are expensive, so economic calculations are necessarily required. Fig. 3.2 depicts the yearly cost of energy storage systems.

In Fig. 3.2 we acquire that by 2035, the total energy storage market will grow to $546 billion in yearly income and 3046 GWh in annual deployments.

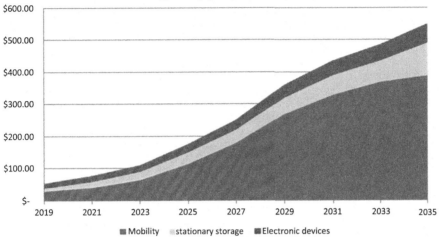

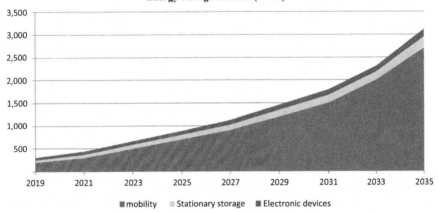

FIGURE 3.2   Total market projection. *Source: www.luxresearchinc.com.*

3. Energy storage system application

3.1 Frequency regulation

An unbalance in generation and consumption of electric power can destabilize the frequency. In the case that the generated power is higher (lower) than the demanded power, the frequency would be higher (lower) than the nominal value. Therefore, the frequency regulation is executed automatically by controlling and coordinating the generated and consumed powers.

3.2 Voltage level improvement

Voltage drop emerges due to load unbalance or faults in the electric grid. The voltage level can be kept in nominal value by injecting or absorbing the reactive power from the grid.

3.3 Load following

Load flow is used by the power system operator to regulate the electrical power. In general, the load flow is employed to manage the power fluctuation of a system during a period from 15 min to 24 h. In such a system, the extra energy is stored during low power demand periods and injected to the grid during peak times.

3.4 Demand shifting and peak shaving

Demand shifting and peak shaving in order to regulate the demand and supply of power and to help integrate the various sources of supply, the time of energy demand can be transferred and relocated.

3.5 Seasonal power storage

The seasonal power storage is the ability to store energy for a daily, weekly, or monthly duration, which is used to compensate for the energy loss of long-term supply or seasonal variation in the supply and demand sides of a grid. Since the seasonal power storage is used only once a year, it can be considered as a high-priced system. Seasonal storage can be competitive only for low-energy systems with very high penetration of certain types of renewable energy.

3.6 Black start

Under rare conditions, when a power system collapses and a blackout occurs, the black start capability of energy storage systems makes it possible to reboot the system without using any external energy.

3.7 Off-grid

Off-grid energy consumers generally use fossil fuels or renewable energy to generate heat and electricity. In order to improve the reliability of off-grid energy supplies and support local energy sources, energy storage systems can be used to compensate for the energy shortage.

3.8 Arbitrage/storage trades

In this case, cheap energy is stored during a period when demand is low and then it is sold back when the energy demand price is high. This is executed through the energy market.

3.9 Nonspinning reserve

The energy storage capacity is used to stabilize the system whenever all or some of the energy generation systems are lost. This capacity concerning its response time can be categorized into rotary (response time less than 15 min) and nonrotary types (response time more than 15 min). The applications of energy storage systems are illustrated and classified in Table 3.1.

4. Different types of energy storage systems

As tabulated in Table 3.2, the electricity storage mechanism is divided into five types as (i) chemical, (ii) thermal, (iii) mechanical, (iv) electrical, and (v) electrochemical. In the following, different energy storage systems are briefly explained.

4.1 Superconducting Magnetic Energy Storage (SMES)

The SMES systems consist of three parts as (i) superconductor coil unit, (ii) power improving system, and (iii) cooling system. The superconductor winding functions as an inductor and the electrical energy is stored as a magnetic power via a direct current in the

TABLE 3.1   The electrical energy storage application.

Energy to power ratio	Short-time (s to min) storage systems	Daily storage systems	Long-time (weekly to monthly) daily storage systems
Applications	1. Voltage control 2. Spinning reserve 3. Peak shaving 4. Black start capability 5. Electromobility (hybrid electric vehicle) 6. Island grids	1. Load leveling 2. Standing reserve 3. Tertiary frequency Control 4. Electromobility (full electric vehicle)	1. Storage for "Dark calm" periods 2. Island grids

TABLE 3.2   Classification of EES technologies.

Electricity storage system	Thermal	Electrical	Mechanical	Chemical	Electro-chemical
Types	1. Thermochemical 2. Planet-thermal 3. Sensible-thermal	1. Supercapacitor 2. Superconducting magnetic energy	1. Pumped storage 2. Liquid air storage 3. Flywheel 4. Compressed air	1. Hydrogen storage 2. Synthetic storage	1. NaS battery 2. Lead–acid battery 3. lithium-ion battery 4. Redox flow battery

magnetic field of the mentioned inductor. This technique is employed during off-peak hours of a power system. A copper winding brings about significant power loss due to the high resistance. However, superconductor windings have tiny power losses, leading to more stable energy storage.

The cooling system is used to decrease the temperature of the superconductor to maintain its superconductivity. The optimum temperature for the mentioned system is 50–77 K. A superconductor magnetic-based energy storage system and its components are shown in Fig. 3.3.

The stored energy in the system can be injected into the power system by discharging the electricity. Power managing systems or the power improving systems, in other words, convert the ac voltage to the dc voltage and vice versa by using inverters and rectifiers in charging and SMES discharging modes, respectively.

The stored energy depends on the current flowing through the winding and inductance. To store a large amount of energy, a winding with high inductance is required. Since the inductance of an inductor is proportional to the turning of winding and core volume, the size and cost noticeably increase for a powerful energy storage system. Table 3.3 shows the advantages and disadvantage of SMES.

4.2 Supercapacitor Energy Storage System

Supercapacitors are specific types of electrochemical capacitors, but no chemical reactions are used to store electrical energy. Supercapacitors have a higher storage capacity than typical capacitors. Supercapacitors are made of two metal plates that are covered by different materials. These porous materials create a larger surface for energy storage. These two metal plates are then put into an electrolyte liquid, which includes positive and negative ions. If a voltage is applied to the metal plates, these ions are separated from each other. Supercapacitors are

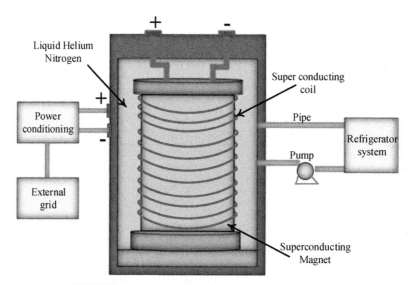

FIGURE 3.3   System description of an SMES facility [2].

TABLE 3.3 Advantages and disadvantages of SMES.

SMES advantages	SMES disadvantages
- Long life span	- Extra energy consumed for cooling system
- High efficiency, more than 95%	- The high cost of the superconductor
- High power density	- Low energy density
- Fast discharging ability	- Design complexity of converters

also referred to as two-layer electrical capacitors. Fig. 3.4 shows the schematic structure of supercapacitors, as shown in this figure, the energy is stored in the electrical field between two dielectrics. These capacitors are considered as low-voltage equipment where their nominal voltage is lower than 3V. In order to make these devices suitable for high-voltage applications, the capacitors are connected in series.

Supercapacitors are used in medical and military systems, laser and microwave applications, power suppliers, as a backup for security and intelligence systems, high-power LEED drivers, wind turbines, electrical automatic doors under power-off conditions, as power regeneration unit in the brake system of electric vehicles, voltage stabilizers, and the applications in which fast charging/discharging capabilities are required.

4.2.1 Supercapacitor advantages
➤ High efficiency
➤ Extremely low internal resistance
➤ Long life span
➤ No overcharging stress

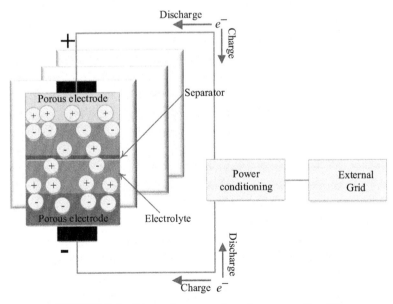

FIGURE 3.4 Schematic construction of supercapacitors [2].

4.2.2 *Supercapacitor disadvantages*
➤ Low energy density
➤ Expensive per unit energy capacity
➤ High-rate self-discharge characteristic limits application in long-term energy storage

4.3 Compressed Air Energy Storage (CAES)

CAES, which is depicted in Fig. 3.5, is counted as a mechanical energy storage system that is structured with the following components:

1. Motor–generator system, which changes the turbine and compressor operations using clutches.
2. Air compressors that use a set of cooling systems to make the system cost-efficient and reduce the humidity of the compressed air.
3. An underground capsule to store the compressed air.
4. System control equipment and combustion chamber.

In this approach the extra electrical energy during off-peak hours is used to run the compressors to compress air in underground capsules, during the demand peak, the compressed air is released to provide needed energy by running the turbines. The compressed air is preheated either by the recycled heat of the compression reactions or burning fossil fuels. This system is considered as a long-term energy storage system, which can supply the demand for several days.

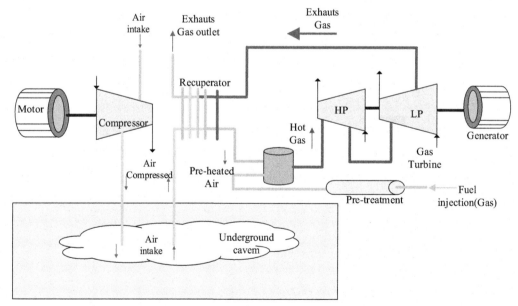

FIGURE 3.5   Basic components for a CAES system [1].

The usual capacity of CAES systems is about 50–300 MW, which is the highest among all storage methods except the pumped storage method. Due to low losses, the storage period in this system reaches to 1 year.

The main advantage of using the CAES is to provide ancillary services such as reducing the peak demand, supplying the reactive power, and frequency regulation of the power system. On the contrary, the main disadvantage of this system is requiring extra energy to provide heat during the expansion cycle. Therefore, any variation in fuel prices can adversely affect the system from an economic point of view. Moreover, extra fuel increases energy generation costs. Nevertheless, the power loss of the expansion cycle can be reduced by recycling the remaining heat to preheat the air. Some advantages and disadvantages of the CAES are shown in Table 3.4.

4.4 Flywheel Energy Storage (FES)

A flywheel stores the electrical energy as kinetic energy in a rotating object. The main components of a flywheel system are motor/generator, flywheel, bearings, power electronic devices, and vacuum chamber to minimize the friction and power losses. Fig. 3.6 illustrates the overall structure of a flywheel system. This system absorbs electrical energy from the grid in off-peak hours and rotates the flywheel using an electrical motor. Flywheels are categorized into two low-speed and high-speed groups. Low-speed flywheels are made of steel that rotate with a velocity lower than 6000 rpm. On the other hand, high-speed flywheels, which are made of very light carbon fiber composites, can rotate with a velocity higher than 100,000 rpm. Due to the low weight of carbon fiber flywheels, the tension is low in these types of storage systems. Usually, flywheels are used to create virtual inertia for low-inertia power networks to improve the stability and power quality of the system.

4.4.1 FES advantage
➢ Low maintenance requirements
➢ Fast charge capabilities
➢ High power density, largely independent of stored energy level
➢ Fast response times

4.4.2 FES disadvantage
➢ Low energy density compared with battery systems
➢ Requiring precisely designed components
➢ High cost

TABLE 3.4 Different type of advantages and disadvantages of CAES.

CAES advantages	CAES disadvantages
- Very high energy and power capacity	- Adverse environmental impact
- Long life span	- Low efficiency

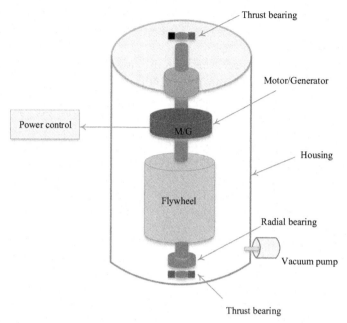

FIGURE 3.6   Overall structure of a flywheel storage system. *Source: energy storage.org.*

4.5 Pumped Storage

Pumped storage power plants store the potential water in dams to produce the electrical energy in peak demand periods. In this method, pump stations make use of the extra energy during off-peak periods to store water in upper-hand storage tanks. When electrical power is required, the water flow path reverses, and the potential energy is converted to electrical energy. The efficiency of the system, which depends on the power plant capacity, diameter of water pipe, water turbine type, and the height difference between the highest and lowest point, could reach above 80%.

4.6 Batteries

In batteries, the electrical energy is stored through chemical reactions. Batteries are considered as a long-term energy storage system. However, they are expensive and have a short life span.

Lead—acid batteries are very well-known and mature rechargeable battery types. The main components of these types of batteries are negative and positive lead electrodes, which are separated from each other through an isolator. To improve the performance of these types of batteries, a lead—antimony alloy is used instead of pure lead electrodes. The feature of the low discharge rate makes the mentioned batteries suitable for long-time charging. Nowadays, researches, which are focused on the material improvement of these batteries, have led to advancements in deep discharging ability making them suitable for wind and solar energy generation systems.

Lithium-ion batteries were first used in small-scale applications such as laptops and cell-phones. In late 2017, the price of a battery pack for electric vehicles reduced to 209$/kWh. Nickel cadmium batteries are used in small-scale applications. Their main advantage over other battery technologies is the ability to operate in low temperatures. Sodium nickel chloride batteries, which are known as Zebra, are suitable for high-temperature environments. Recent researches are conducted on a high-energy Zebra battery for renewable energy systems.

Rechargeable current batteries are state-of-the-art technology in the battery field. In these types of batteries, the energy is stored in one or many active electrodes, which float in electrolyte liquid. An electrolyte is stored in an outer tank, which is then pumped into the stack cell by an electrochemical reactor. The chemical energy is converted to electrical energy and vice versa. The main advantage of these batteries is that their power capacity is independent of their energy storage capacity. The power capacity is proportional to the number of cells and the electrode sizes; however, the energy storage capacity is related to the tank capacity and the amount of stored electrolyte. Other advantage is their total discharging capability that has no adverse effect on the system. What is more, they have a low self-discharge rate owing to the fact that the electrolyte is kept in some separate tanks. On the contrary, they have demerits such as low energy density, high construction, and installation costs, as well as high complexity compared to conventional batteries.

5. Energy storage systems in wind turbines

With the rapid growth in wind energy deployment, power system operations have confronted various challenges with high penetration levels of wind energy such as voltage and frequency control, power quality, low-voltage ride-through, reliability, stability, wind power prediction, security, and power management.

5.1 Power characteristic in terms of wind turbine speed

The wind turbine's power characteristic is actually the same as the wind turbine's mechanical power diagram in terms of wind speed. This feature actually guarantees the efficiency of the wind turbine by the manufacturer. To obtain this characteristic, the mechanical power of the wind turbine must be calculated first.

The kinetic energy (E_k) in the wind is obtained by the following:

$$E_k = \frac{1}{2}mV_w^3 \tag{3.1}$$

According to Eq. (3.1), wind power is equal to:

$$P_w = \frac{dE_k}{dt} = \frac{1}{2}\rho A V_w^3 \tag{3.2}$$

where P_w is wind power, ρ is the air density (kg/m^3), A is the sweep area (m^2), m is the mass of the object (kg) and V_w is linear wind speed (m/s). According to Eq. (3.2), it can be concluded that wind power is proportional to:

✔ Air density: The air density is lower in the highlands and mountainous areas. But on the other hand, the average density of areas in cold weather is up to 10% higher than it is in warm areas.
✔ The cross-sectional area through which the wind passes.
✔ Wind speed cube: wind speed has a significant effect on power output.
✔ Although Eq. (3.2) expresses the power in the wind, the power transmitted to the wind turbine is reduced by the power coefficient (cp). Therefore, the relationship between wind power and mechanical power transferred to the turbine shaft is:

$$P_m = C_p P_w \tag{3.3}$$

C_p depends on the wind speed, the rotation speed of the turbine rotor, and the angular position of the turbine rotor blades (β). This coefficient has a theoretical maximum value of 0.59, according to the Betz limits. In other words, in theory, about 59% of the wind's kinetic energy can be converted into mechanical energy. Modern three-bladed wind turbines have an optimal efficiency factor in the range of 0.45–0.55.

In various designs, Cp is expressed as a function of tip speed ratio (λ) and pitch angle (β). The tip speed ratio is expressed as the following equation.

$$\lambda = \frac{r_T \omega_m}{v_w} \tag{3.4}$$

where ω_m and r_T are the rotating speed of the blade and the radius of the turbine rotor, respectively. By combining Eqs. (3.2)–(3.4), the mechanical power transferred to the turbine shaft is written as Eq. (3.5).

$$P_m = \frac{1}{2}\rho A C_p(\lambda, \beta) V_w^3 \tag{3.5}$$

The power coefficient is expressed as Eqs. (3.6) and (3.7) for the most wind turbines:

$$C_p(\lambda, \beta) = c_1 \left\{ \frac{c_2}{\lambda_i} - c_3\beta - c_4 \right\} e^{-\frac{c_5}{\lambda_i}} + c_6\lambda \tag{3.6}$$

$$\frac{1}{\lambda_i} = \frac{1}{\lambda + 0.08\beta} - \frac{0.035}{\beta^3 + 1} \tag{3.7}$$

5.2 Taxonomy of most general electric generators in wind energy conversion systems

In general, wind turbines can be categorized into fixed and variable speed types (Fig. 3.7).

The wind turbine whose generator is directly connected to the grid is called fixed-speed wind turbines. In other words, there is not any electrical control system for this wind turbine. In addition, any variations in the wind speed will affect the output generated power quickly. Due to the power variations, these changes are not favorable for wind turbines, which are directly connected to the power system, and cause mechanical tension on the turbine. As a result, lifetime and quality will be reduced. Fixed-speed wind turbines are often operated at nonoptimum points and cannot extract the maximum power from the wind. On the other hand, variable-speed wind turbines are connected to the grid via power electronic converters, and it is possible to independently control the injected reactive and active power. Such wind turbines can operate in a wide range of wind speeds. The most popular generators that are widely used in variable-speed wind turbines are doubly-fed induction generator (DFIG) and permanent magnet synchronous generator (PMSG).

Wind turbine type has effect on the power system steady-state voltage level. Since fixed-speed wind turbines cannot generate reactive power, higher demand values will decrease the system voltage level. However, variable-speed wind turbines, due to their ability to control reactive power, can improve the voltage profile of a distribution network.

Due to the intermittent nature of wind speed, the generated power of the wind energy generation systems is variable. These changes in the output power of the system can influence the stability of the system. Various strategies can be used to mitigate the negative effect of the wind speed changes and to improve the reliability of the system such as spreading wind turbines in a wide area and using energy storage systems along with the wind turbines.

Widespread siting of wind turbines can reduce the effect of the wind speed variations on the output power generation. Considering the climate conditions in a wider area, it can be observed that when small wind turbines are distributed, the output power fluctuation is less than the case where larger and fewer wind turbines with the same output power are

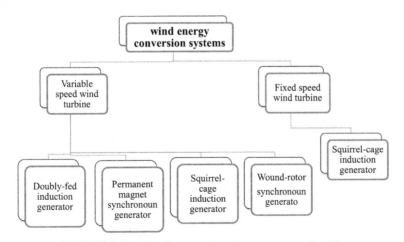

FIGURE 3.7 Classification of wind energy conversion [3].

employed. In brief, by increasing the number of wind turbines and distributing them in a larger area, the output power fluctuations will be mitigated. Although this method is useful and effective, the output power of the wind turbines is not yet consistent in comparison with the fossil fuel power plants, so the reliability is still low.

Another method is installing an energy storage system in a wind farm. When the generated power is more than the demand, the energy can be stored in the storage packs, and when the generated power is low, the energy storage system can inject the stored energy into the system. In other words, energy storage systems can absorb or inject active power to fixed- or variable-speed wind turbines to reduce the output power fluctuations. In addition, output voltage fluctuations in the fixed-speed wind turbines can be mitigated by controlling the reactive power when the energy storage system is connected. Two parameters are important in the energy storage systems; the first one is the amount of energy it can store, and the second one is the power transfer rate of the energy storage system. Another crucial points in the energy storage systems are their technological characteristics, cost, and maintenance considerations.

Energy storage systems are expensive and have a fixed efficiency during their operation interval. Hence, it is important to consider their nominal capacity for the optimal design of the system. Furthermore, closed-loop feedback control systems are required to control the output power and state of the charge of the system.

There are two common methods to connect energy storage systems in wind farms. The first technique is that energy storage systems can be connected to the common bus of the wind power plant and the network (PCC). Another method is that each wind turbine unit can have a small energy storage system proportional to the wind turbine's size, which is called the distributed method Fig. 3.8. Research has shown that the first undistributed method is much better than the distributed scheme due to its lower cost and effectiveness in damping the output power fluctuations, improving the system stability.

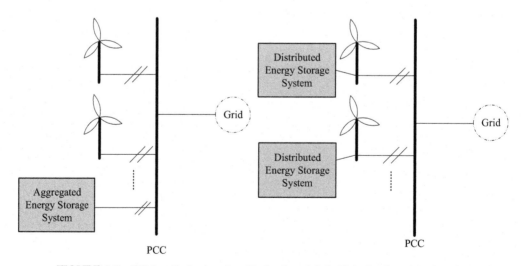

FIGURE 3.8   ESS installation location. Undistributed (left side), distributed (right side).

Several energy storage systems are available for wind energy applications such as batteries, magnetic energy storage systems, superconductors, supercapacitors, flywheel, and combinations of various aforementioned methods.

5.3 SMES connected to wind farms

SMES is used to improve the overall operation of the DFIG to flatten the output power level in various references. These generators have different advantages such as fractional nominal power of power electronic devices, lower costs, variable speed operation, and independent active and reactive power control. However, the existence of brushes, fluctuations in the output power, and drawbacks with low-voltage ride-through are the main disadvantages of such generators.

The energy stored in the superconductor winding and nominal real power can be calculated as follows:

$$E_s = \frac{1}{2} I_s^2 Ls \tag{3.8}$$

$$P_s = \frac{dE_s}{dt} = L_s I_s \frac{dI_s}{dt} = V_s I_s \tag{3.9}$$

where E is the stored energy, P_s is nominal real power, L_s is the inductance of the superconductor, and I_s and V_s are the DC current and the voltage across the SMES coil, respectively.

SMES consists of a superconductor winding and a DC/DC chopper to charge/discharge the winding. The DC/DC chopper comprises two diodes and two IGBTs (Insulated-Gate Bipolar Transistor), as shown in Fig. 3.9. In the charge state, the switches S_1 and S_2 turn ON by the controller and in the discharge state, the switches turn OFF and SMES is discharged through the diodes.

Various reference signal-based control systems are available to charge and discharge the SMES. The most common method is to use a PI controller to control the duty cycle of the DC/DC chopper. In normal operation mode, the duty cycle is 0.5, and energy is not transferred between the SMES and the grid. When a fault such as low-voltage fault occurs in the power system, the PI controller adjusts the duty cycle between 0 and 0.5, and the energy stored in the winding is discharged and transferred to the system. On the other hand, when the duty cycle is between 0.5 and 1, the charging state occurs.

The relation between V_{smes} and $V_{DC,smes}$ is written as:

$$V_{smes} = (1 - 2D)V_{DC,smes} \tag{3.10}$$

where V_{smes} is the average voltage across the SMES coil, $V_{DC,smes}$ is the average voltage of the dc-link capacitor of the *SMES* configuration, and D is duty cycle.

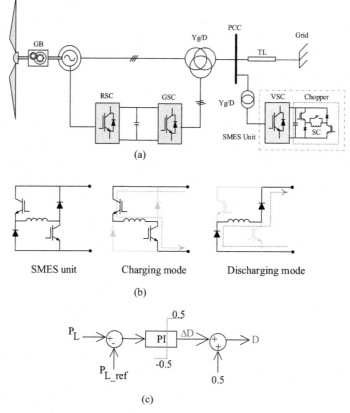

FIGURE 3.9 (A) DFIG equipped with SMES unit. (B) SMES unit. (C) DC chopper control.

5.4 FES connected to wind farms

Flywheels due to their long lifetime, high energy and power density, high charge and discharge capability, and high efficiency can be used in wind energy applications to flatten the power and frequency fluctuations. As shown in Fig. 3.10, the flywheel is connected to the DC link via a power electronic converter and an induction machine. Three operation modes are possible for flywheels, i.e., charging more, standby mode, and discharging mode.

FES systems consist of a large cylinder that is held by magnetic bearings in the stator. Magnetic bearings can improve the system's lifetime and reduce the tensions in the bearings.

In the charging mode, the flywheel rotates at high speeds to absorb the energy. This increases the speed and as a result, the stored energy of the flywheel. The stored kinetic energy in the flywheel can be expressed as Eq. (3.11).

$$E = \frac{I\omega^2}{2} = \frac{mr^2\omega^2}{2} = \frac{mv^2}{2} \tag{3.11}$$

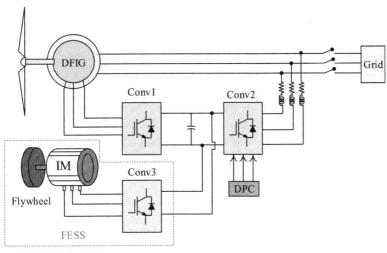

FIGURE 3.10 Typical schematic diagram of FES connected to DFIG.

where I is the inertia of rotating flywheel, m is the cylinder mass, v is the linear rim velocity, and ω is the rotational speed. Considering Eq. (3.11), the stored energy is proportional to the mass and square of the velocity of the flywheel. Therefore, designers are willing to increase the rotational speed of the flywheel so that they could store more energy. In the charging state, converter 2 operates as a rectifier and converter 3 operates as an inverter.

The rotor is used to increase the speed of the shaft to its nominal value. Once the shaft is in its nominal speed, the shaft is disconnected and the flywheel continues rotating due to its inertia. This shows that the electrical energy has been converted into the kinetic energy and stored in the rotating mass. In this way, the system is ready to discharge to the system (standby mode). In order to improve the efficiency of the system, the friction in the bearing, aerodynamic friction, and other mechanical power losses should be mitigated. For instance, the flywheel can be kept within a vacuum chamber to eliminate the effect of the aerodynamic friction. The main drawback of the flywheel-based systems is their high standby loss, which is about 5% of the nominal power.

In order to discharge the stored energy, the rotational energy is transferred to the mechanical load in the flywheel. This will reduce the rotational speed of the flywheel and the store energy will be reduced. In this case, the electric motor acts as a generator and converts the kinetic energy of the rotating cylinder to the electrical energy. In this case, converter 3 works as a rectifier and converter 2 is an inverter.

5.5 Supercapacitor connected to wind farms

Supercapacitors are used along with wind turbines to ensure the DC link voltage and to mitigate the output power fluctuations. In other words, it is possible to compensate for the fluctuations by employing a supercapacitor with a control loop.

The electrical structure of supercapacitors in wind turbines is illustrated in Fig. 3.11, which consists of a supercapacitor bank and a two-switch DC/DC converter connected to a DFIG via DC link. The main duty of the buck-boost converter is to absorb or inject active power. At first, the reference power is recognized. If the network power is more than the reference power, the supercapacitor is charged and the converter operates in buck mode. In this mode, the switch G_2 is turned OFF. On the other hand, if the network power is less than the reference power, the supercapacitor will be discharged and the DC/DC converter will operate in boost mode where the switch G_1 is turned OFF. In buck operation mode, where the switch G_1 is turned ON, the duty cycle of the switch G_1 can be expressed as D_1, which is the ratio of the capacitor bank voltage to the DC link voltage.

$$D_1 = \frac{V_{sc}}{V_{dc}} \tag{3.12}$$

In addition, in the boost operation mode where the switch G_2 is turned ON, duty cycle of the switch G_2 is expressed as D_2.

$$D_2 = 1 - D_1 \tag{3.13}$$

Capacity of the supercapacitor bank is calculated as [4].

$$C_{ess} = \frac{2P_n T}{v_{sc}^2} \tag{3.14}$$

where C_{ess} is in Farad, P_n is the rated power of the DFIG in watts, V_{sc} is the voltage rating of the supercapacitor bank in volts, and T is the desired time period in second that the ESS can supply/store energy at the rated power of DFIG.

5.6 Battery connected to wind farm

Methods such as step angle control, inertial use, and energy storage systems are used to reduce wind power output fluctuations. Batteries are also used as storage in combination with wind farms to control the frequency and reduce the power fluctuations. Like an energy buffer, they absorb wind power when it is more than the amount of threshold, Moreover, when there is a shortage of network power, they return the stored energy to the network, thereby stabilizing the network power.

The battery energy storage system (BESS) includes a battery bank and a bidirectional DC-DC converter, as shown in Fig. 3.12A.

The DC converter controls the power flow from the battery to the network and vice versa. There are various control methods for this converter, in which we use the conventional cascaded power and current control method. According to Fig. 3.12B, the reference and real powers are compared and the error is given to a PI controller, whose output is the reference value of the battery current then, in the current control loop, the value of reference current is compared to the actual value of battery current, and their difference enters the PI controller. Finally, the output of this controller is the amount of duty cycle to drive the DC-DC converter.

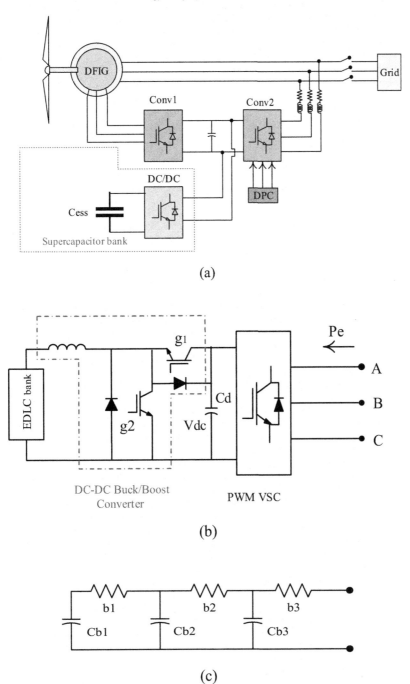

FIGURE 3.11 (A) SCES connected to DFIG, (B) SCES unit, (C) detailed model of ELDC bank.

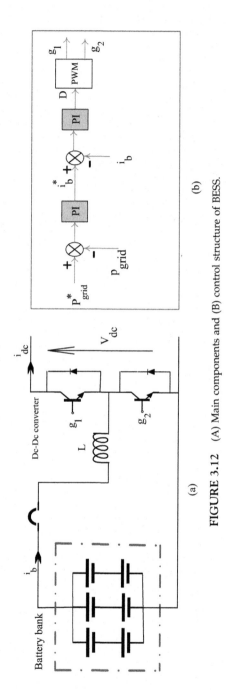

FIGURE 3.12   (A) Main components and (B) control structure of BESS.

The value of smooth power transfer reference to the grid is obtained by filtering the output power of the wind turbine by a low-pass filter for the first time at the time constant of τ.

There are two common structures for combining batteries and wind turbines. In Fig. 3.13A, the PMSG is connected to the grid via a back-to-back converter. The battery is also connected to the link via a bidirectional DC-DC converter. This bidirectional converter is capable of delivering negative and positive power to discharge and charge the battery, respectively. In the second type of structure Fig. 3.13B, the generator and battery are independently connected to the grid via the converters 1 and 2 and through the point of common coupling (PCC).

The battery capacity is usually specified in terms of power and energy capacity. The energy capacity describes the amount of energy taken or stored in the battery. Nominal power is the power that can be stored or delivered at a specified time interval (T) in charge or discharge mode. Battery investment costs are determined based on energy capacity and power. In the buffer system, the battery power (p_b), which is the difference between the expression of the dispatched power (p_g) and the output power of wind turbine (p_w), is obtained from Ref. [5].

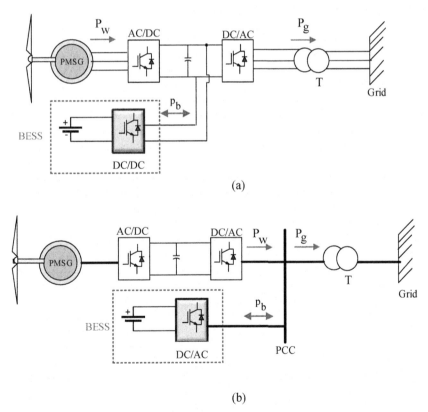

FIGURE 3.13   Combined model of BESS and wind turbine. (A) BESS connected to dc link, (B) BESS connected to PCC.

$$p_b = p_w - p_g \tag{3.15}$$

From Eq. (3.15), for a constant value of the p_g, the battery power (p_b) will change similar to the output power of the turbine. Suppose that the system operates over a period of time T and the nominal power is defined as Eq. (3.16). By integrating the battery power over time t, the amount of energy stored or extracted from the battery is obtained as [6].

$$p_b^{rate} = \underset{0 \leq t \leq T}{\text{Max}} |p_b(t)| = \underset{0 \leq t \leq T}{\text{Max}} |p_w(t) - p_d(t)| \tag{3.16}$$

$$E_b(t) = \int_0^t p_b(\tau)d\tau = \int_0^t [p_w(\tau) - p_d(\tau)]d\tau \tag{3.17}$$

Similarly, nominal energy is the maximum energy that can be released or stored by the BESS while system operation in T, which is defined as Eq. (3.18).

$$E_b^{rate} = \underset{0 \leq t \leq T}{\text{Max}} |E_b(t)| \tag{3.18}$$

5.7 Design criteria and comparison of energy storage system technologies

Each energy storage system technology has its unique characteristics depending on its applications and energy storage scale. The main parameters to select a proper energy storage system are the charge and discharge rate, nominal power, storage duration, power density, energy density, initial investment costs, technical maturity, lifetime, efficiency, energy storage capacity, and the environmental effects. In Table 3.5, parameters of selecting a suitable energy storage system are listed.

According to Table 3.5, the flywheel, SMES, and supercapacitors have fast response. Therefore, it can be used to improve power quality enhancement, such as instantaneous voltage and frequency response. The usual nominal power for this type of application is less than 1 MW.

Efficiency is higher than 90% in SMES, flywheels, supercapacitors, and lithium-ion batteries, which are known as high-efficiency systems. The energy density refers to the amount of energy stored per unit volume. In flywheels, it is more than supercapacitors and superconductors. On the other hand, the power density, which is the nominal output power divided by the volume, is lower in flywheels than supercapacitors and conductors.

According to Fig. 3.14, energy storage systems are divided into three categories based on their discharge time: short-term, medium-term, and long-term.

CAES, PHS, and rechargeable batteries have very little spontaneous discharge; hence they are suitable for long-term storage. Supercapacitors and superconductors have discharge rate of 10%–40%, which is in the medium and short-term storage facilities, and flywheels will discharge 100% of their stored energy if the storage period is more than 1 day.

TABLE 3.5 Comparison of energy storage characteristics [1].

ESS technologies	SMES	FES	SC	CAES	PHC	Lead–acid	Nas	Li-ion
Efficiency (%)	80–95	70–95	80–98	40–75	70–85	80–90	70–90	85–98
Charge time	s–h	s–min	s–h	h–months	h–months	min–days	s–h	min–days
Discharge time	ms–8 s	ms–15 min	ms–60 min	1–24 h	1–24	s–h	s–h	min–h
Cycling	100,000	20,000–100,000	10,000–100,000	20–40 years	30–60 years	6–40 years	2,500–4,400	1,000–10,000
Power rating (MW)	0.1–1	0.001–1	0.01–1	10–1000	100–100	0.001–100	10–100	0.1–100
Response time	<100 ms	10–20 ms	10–20 ms	s–min	s–min	<s	10–20 ms	10–20 ms
Storage duration (h)	ms–min	sec–h	ms–min	2–30 h	4–12 h	1 min–8 h	1 min–8 h	1 min–8 h
Self-discharge	10–15	1.3–100	20–40	Approx. 0	0.2–2	0.1–0.3	0.05–20	0.1–0.3
Energy density (Wh/L)	Approx. 6	20–80	10–20	2–6	0.2–2	50–80	150–300	200–400
Power density(W/L)	1000–4000	5000	40,000–120,000	0.2–0.6	0.1–0.2	90–700	120–160	1,300–10,000

h, hours; *min*, minutes; *ms*, milliseconds; *MW*, MegaWatt; *S*, Seconds.

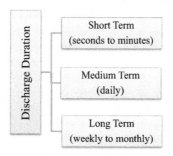

FIGURE 3.14 EES technologies according to the discharge duration.

It is worth noting that the cost of investment is one of the most important factors for the industrial usage of energy storage systems.

These factors include the position, system size, market variability, environmental considerations, efficiency, and useful life of the system.

6. Chapter summary

In this chapter, first, the basic applications of energy storage systems are introduced and then the structure, advantages, and disadvantages of some of the most widely used energy storage systems, such as SMES, supercapacitor energy storage, CAES, FES, pumped storage, and batteries are discussed. As mentioned, due to the intermittent nature of wind speed, the generated power of the wind energy generation systems is variable. Therefore, energy storage systems are used to smooth the fluctuations of wind farm output power. In this chapter, several common energy storage systems used in wind farms such as SMES, FES, supercapacitor, and battery are presented in detail. Among these energy storage systems, the FES, SMES, and supercapacitors have fast response. Therefore, it can be used to improve power quality enhancement, such as instantaneous voltage and frequency response.

References

[1] Energy Storage Monitor Latest Trends in Energy Storage, 2019. Available from: http://www.worldenergy.org.
[2] X. Luo, J.W.M. Dooner, J. Clarke, Overview of current development in electrical energy storage technologies and the application potential in power system operation, Appl. Energy 137 (October 2015), https://doi.org/10.1016/j.apenergy.2014.09.081, 551–536.
[3] B. Wu, Y. Lang, N. Zargari, S. Kouro, Power Conversion and Control of Wind Energy Systems, Wiley, 2011.
[4] J. Zhan, O.A. Ansari, W. Liu, G.Y. Chung, An accurate bilinear cavern model compressed air energy storage, Appl. Energy (March 2019) 752–768.
[5] M. Cao, Q. Xu, X. Qin, J. Cai, Battery energy storage sizing based on a model predictive control strategy with operational constraints to smooth power the wind power, Electr. Power & Energy Syst. (August 2019) 1–10.
[6] W. Liu, Y. Liu, Hierarchical model predictive control of wind farm with energy storage system for frequency regulation during black-start, Electr. Power & Energy Syst. (January 2020) 1–11.

Optimization of energy storage systems in energy markets

Mohammad Taghi Ameli, Sasan Azad, Mohammad Hosein Asgharinejad Keisami

Department of Electrical Engineering, Shahid Beheshti University, Tehran, Iran

Sets and indices

i,j Index for buses
N total number of buses
N_b set of buses
S_{DG} set of distributed generation (DG) units
S_{ESS} set of energy storage system (ESS)
t Index for hours

Variables and parameters

MUT,MDT minimum up time and minimum down time of dispatchable DG
$\rho_{E,i,t}$ energy market price
$\rho_{R,i,t}$ spinning reserve market price
η_{str} efficiency of ESS
r_t probability of reserve delivery
$P_{DG_i^t}, Q_{DG_i^t}$ Active/reactive power injection of node i at time period t with DG
$R_{i,t}$ amount of active power generated by ith DG unit at hour t for reserve market
$P_{PV_i^t}, Q_{PV_i^t}$ Active/reactive power injection of node i at time period t with PV
$P_{WT_i^t}, Q_{WT_i^t}$ Active/reactive power injection of node i at time period t with WT
$P_{G_i^t}, Q_{G_i^t}$ Active/reactive power injection of upstream network
P_{load_t}, Q_{load_t} Active/reactive demand at time period t
$P_{ch_i^t}, P_{Dch_i^t}$ Charge/discharge power of ESS at node i at time period t
$P_{str_i^t}$ amount of charged/discharged capacity of ith ESS at hour t in MW
$P_{str_i}^{min}, P_{str_i}^{max}$ minimum and maximum capacity of ith ESS in MWh

R_{ch}, R_{Dch} maximum charge/discharge rate of ith ESS in MW

Cap_i^t state of charge of the ith ESS at hour t

$C\left(P_{DG_i^t}\right)$ generation cost function of ith DG unit

$C\left(P_{WT_i^t}\right)$ generation cost function of ith WT unit

$C\left(P_{str_i^t}\right)$ operation cost function of ith ESS

$V_{i,t}$ Voltage magnitude of bus i at time t

$\delta_{i,t}$ Voltage angle of bus i at hour t

$S_{i,j}^t$ apparent power between buses i and j at hour t

$\mathbf{SUC}, \mathbf{SDC}$ start-up and shutdown cost of dispatchable DG

1. Introduction

Renewable energy resources are rising phenomena in the energy industry. No part of the energy and electricity sector is not affected partially or entirely by the emergence of new clean and reliable resources of energy. From the more known resources such as wind and solar to the more experimental forms of renewables such as tidal power, researchers are studying the challenges and benefits of using renewables as an alternative for fossil fuels.

All aspects of modern daily life are affected by renewables from the global perspective of an increase in the carbon and other forms of greenhouse gas emissions to the local effects of competing for the cheapest most reliable resource; renewables are an inseparable part of the modern world.

The main objective of researching the renewables is to get as close as possible to a clean, reliable free resource of energy to replace the conventional resources.

Ongoing evermore increasing demand for renewables can be understood by careful examination of the advantages involved in using them. The first and most important motive for using renewables is that they are free. Costly processes of extracting fossil fuels and the geopolitical challenges regarding obtaining them have led many countries, especially those who lack the considerable amount of fossil fuel resources, members of the EU, for example, to find solutions in renewables.

The geopolitical aspects of using fossil fuels are important, and many of the political problems in the fossil fuel—rich areas of the world could have been avoided if there was a balanced approach to the use of conventional resources.

However, political issues come second to the real threat of fossil fuels, climate change. The importance of climate change cannot be overstated. The summer of 2019 was the hottest summer ever recorded in the northern hemisphere [1], and it seems the temperature is only going to be increased.

Electricity generation made for about 26.9% of 2018 greenhouse gas emissions, making it the second most source of greenhouse gases. Approximately 63% of all of the electricity is generated by burning fossil fuels, mostly coal and natural gas [2,3].

Climate change is an existential crisis and should be taken seriously, thus using energy resources that help reduce carbon emissions is preferred. The science and engineering community must look for a better cleaner and more reliable solutions for electricity generation. According to the International Renewable Energy Agency, by 2050, two-thirds of the energy generated sources should be supplied by renewable energy resources [4].

With the rapid growth and development of renewable energy in the world, renewables are being used in a variety of fields. For example, in the power generation sector, refrigeration, the automotive industry, the telecommunications industry, and the construction sector. Power generation is the most significant use of renewable energy. Thus, the main focus of the discussion on renewable energy generators is their participation in the electricity industry.

The relationship between the use of renewable energy and the economies of countries is inseparable, as there is a direct relationship between GDP growth and the development of renewable energy. Renewables are growing more than ever, and likewise, the challenges and opportunities are rising accordingly, among them is their participation in frequency and voltage control, which is not discussed in this chapter, but can be seen in Ref. [5] for further study.

Also, the development of renewable energies requires financial investments. Therefore, the relationship between renewable energies and the economy is a two-way street, participation in the energy market is used to maximize profits and to make the most of renewable energy.

These markets are highly competitive, and participating in them requires a review of various dimensions for renewable energy generators.

The use of renewable energy resources has been an essential subject in studying power markets and energy generation. In recent years, the presence of energy, reserve, and auxiliary markets make the process of selling energy a much more dynamic, unpredictable, and, of course, profitable endeavor.

However, as much as the new technologies for producing renewable energy have evolved, still, there are fundamental properties of renewables that have to be addressed. One, for example, in the intermittent nature of such resources, as the solar and wind generation and most other forms of renewables depend heavily on weather conditions. This characteristic of RES, of course, limits the capabilities of renewables, making them economically challenging. If the RES providers are not participating in the market, the intermittent nature of the resources would be much more tolerated by the retailers, and the magnitude of the problems is smaller than that of participating in the energy and reserve market. However, as mentioned before, the current structure of the power sector is majorly shaped by the markets, and the traditional forms of energy distribution are fading. Therefore, the participation of renewables in the energy markets is a significant problem.

The renewable energy companies that desire to participate in the energy and reserve market need to know the exact amount of generation both for the day ahead and in real time.

Hence, renewable energy providers are at a disadvantage compared to their conventional competitors. For example, suppliers usually consider penalty coefficients for generating less than the promised amount. Therefore there is always a risk involved in participation.

In short, the main goal in this chapter is to answer the question of "How to implement the great potential of renewables in the energy and reserve market atmosphere without being overwhelmed by the uncertainties created by the intermittent nature of renewable resources?". Finding a solution to this problem is what we are aiming at and together review the proposed answers.

The solution for overcoming this challenge is a popular subject in the engineering community, making it a crucial question to be answered. There are many methods proposed to tackle this problem.

One of these methods for overcoming this challenge is using energy storage system (ESS). These systems provide a stable source of energy, making participation in the electricity markets more feasible. Investing in storage units could potentially benefit renewable energy companies.

However, investing in storage costs a considerable amount. Therefore, these investments depend on the amount of revenue generated from the electricity market to justify the use of ESS. To increase the profit from the participation of ESS in the energy market, ESS must be optimally planned to increase the profit of the company.

Various studies have been conducted on the types of ESS used in renewable energy plants. Renewable energy resources with the help of ESS can store the energy produced in the hours when the energy price is low and sell it in the hours when the energy price is higher. Some papers show that the performance of ESS affects market prices, especially during times when energy prices are high. Of course, using ESS is not limited to market control, and utilization of ESS is an excellent subject in the literature, for example, one application of ESS in frequency control in microgrids can be seen in Ref. [6].

The following sources can give a good overview on the subject of participation of renewables in the energy market and are briefly listed.

In [23], a two-step and stochastic approach to optimizing a wind farm combined with a hydro pumped storage unit in the energy market is presented. This model is an effective way to participate in wind farms in the real-time market under uncertainty.

There have been different methods and algorithms studied in the literature for using ESS to optimize the presence of renewable energy resources with wind farms. For example, a multiobjective optimization algorithm is proposed in Ref. [7] to reduce the wind fluctuation and ESS size. This study can help with the understanding of the nature of wind fluctuation and power production.

There are several other papers regarding the use of ESS in the wind farms, including [8], where there is a market-based ESS dispatch plan, which provided a financial improvement plan for wind farms. Also, in Ref. [9], a risk-based approach is presented for evaluating the strategy for the participation of battery storage units in the electricity markets. This market includes the day-ahead energy market, spinning reserve market, and the regulation market.

On the economic aspects of implementing ESS, there is a sufficient amount of research available. For example, in Ref. [10], a whole-systems approach for evaluating the value of the electricity storage is presented.

Optimizing the investments while minimizing the system operation cost considering reserve and security markets is the objective of [10]. Also, in Ref. [11], an optimal bidding strategy for battery storage in the electricity market is introduced. Incorporating the battery life cycle model for determining the optimal bids is one of the subjects of this paper.

The subject of uncertainties is also discussed in the literature, as it can be seen in Ref. [12], where a stochastic mixed-integer linear programming method for maximizing total profit in a pool-based electricity market that has introduced the uncertainty about the competitor's offer is represented by a conditional value at risk. A method based on Markov decision processes is also discussed in Ref. [13] for optimally scheduling the energy storage devices in the distribution network with the presence of renewable resources. For further study of the electricity market and uncertainty, there is an interesting paper available in Ref. [14], which can give a sense of market behavior in the presence of distributed generation.

To study the battery-based ESS, we can find excellent research in Ref. [15] where an optimization model for market participation of battery storage units that take the uncertainties of energy prices into account has been proposed.

For the planning and operation of electricity retailers and their associated storage resources in the market, there is a variety of research done, and the problems have been modeled, taking into account the uncertainties of the load side, one good paper on this subject it can be found in Refs. [16,17], different types of ESS have been studied from an economic point of view to be used in the power grid, and a model based on the presence of storage devices for participation in the electricity market has been considered that its objective is to increase the profit of ESS owners for participating in the energy market. In Ref. [18], an analytical solution for reserve planners to participate in the electricity market is presented. This is done by introducing the allowable price range for the participation of ESS in the electricity market.

Various sources have examined the planning of storage facilities for the use of arbitrage. In most areas, wind production at night, when demand is lower, is higher [19]. In this case, the primary purpose of ESS from the private owner's point of view is to make more profit using arbitrage [20]. This is done by optimally planning ESS to store energy at low prices and supply it to the market at higher prices. In Refs. [21–27], the optimal utilization of the storage unit for increasing income and profit in combination with solar or wind farms has been discussed. In these studies, the ESS is considered as a part of the wind farm, and the owner of the wind farm must invest in the storage. In Refs. [28,29], optimization methods and programs for the participation of storage in the energy market as a unit are examined. In these methods and plans, the goal is to generate financial benefits for the reserves resulting from participating in the energy market.

When using battery storage, several conditions and constraints should be accounted for, and a good study can be found in Ref. [30] where the author provides an optimal battery discharge program to increase profits that do not violate battery performance limitations. For other forms of storage, there is an abundance of papers available, including [31], where researchers have investigated the optimal use of compressed air ESS in energy markets and reservation markets. This study shows that the outputs of arbitrage energy alone do not meet the need for investment in CAS systems in most US markets. However, the conventional CAS approach would be justified if we consider the earnings from participation in the reservation market. But if we use adiabatic energy storage methods, even if we participate in the reservation and arbitrage markets, their use would not be economically justified [31].

The battery technology has its limitations, including its lifetime, and an excellent paper on this subject can be seen in Ref. [32], in which a Pareto optimal arbitrage policy is used to balance the economic value and the lifetime trade-off of the battery storage units. A stochastic model for characterizing the lifetime value of the unit is introduced.

Above all, the methods mentioned earlier of using ESS has the fundamental problem of economic feasibility. Examining the possibilities and challenges of using ESS in the electric grid and electricity market requires a deep understanding of the economy and mathematical theories such as game theory. It is argued in the literature that using ESS might not be economically justifiable by default.

The mentioned issues are also studied in the literature, as in Ref. [33], where the authors argue that although using ESS increases the reliability and reduces the uncertainties of the participation of renewables in the energy market, investing in ESS reduced uncertainty might in contrast to common knowledge be unfavorable once the consumer demand passes a certain threshold, the revenues of both suppliers that invested in ESS approach zero.

Also, a higher penalty and a higher storage cost can be favorable to the suppliers in the market and competition framework. In that sense, the first supplier who invests in energy storage can be disadvantaged [33]. Therefore, the implementation of ESS is a challenge for both governments and private investors.

These economic problems arise when there are two or more providers of renewables that are competing against each other in the same market. If governments require the energy providers to reduce their carbon emissions, the penalties involved with the usage of conventional energy resources may be compelling enough for the investors to approach renewables and therefore with an increased share of renewables in the energy market the overall reliability standards of the network require the implementation of ESS and other forms of energy storage such as hydro pump plants.

These challenges are essential and make for a great discussion in the literature. We examine the economic aspects of investing in ESS technology in the following sections.

By reviewing various papers in the previous section revealed that with the influence of renewable energy sources in the electricity industry, we found that one of the major problems is the intermittent generation of renewable resources, so system operators are forced to use cost-effective methods to balance supply and demand. One way to solve problems is to use ESS. However, it is not yet clear whether the presence of ESS in the network is sufficient for its owners or not, given the various uncertainties, such as the price of different energy markets. This study examines the optimal planning of ESS in the presence of distributed and renewable generation resources to participate in the day-ahead energy market, as well as the impact of uncertainty on energy prices on the optimal behavior of ESS.

The discussed topics in this chapter are as follows:

- Providing a model for optimal storage programming to participate in the energy market
- Investigating the effect of price uncertainty on the energy market on the optimal behavior of ESS
- Investigating the impact of price uncertainty on the energy market and reservations on dispatchable DG performance
- Investigating the effect of ESS on the behavior of DGs to participate in the energy market

2. Problem formulation

2.1 Objective functions

In this section, the objective is to plan ESS to participate in the day-ahead energy market optimally. Therefore, the objective function is presented for the owners of distributed generation sources and renewable energy to be examined in different conditions of change in the optimal behavior of ESS, for participating in the energy market.

- **Maximizing the profitability of dispatchable DG and ESS owners**

    In this section, it is assumed that DG and ESS are owned by a private company, and the objective is to maximize the company's profits.

$$F_1 = Benefit_1 = \sum_{t=1}^{24} \rho_{E,i,t} \times E_{i,t} + \sum_{t=1}^{24} \left[r_t \cdot (\rho_{E,i,t} + \rho_{R,i,t}) + (1 - r_t) \cdot \rho_{R,i,t} \right] R_{i,t}$$
$$- \sum_{i=1}^{N_{DG}} \sum_{t=1}^{24} \left[r_t \cdot C \left(P_{DG_i^t} + R_{DG_i^t} \right) + (1 - r_t) \cdot C \left(P_{DG_i^t} \right) \right] \tag{4.1}$$
$$- \sum_{i=1}^{N_{DG}} \sum_{t=1}^{24} \left[SUC_{DG_i^t} + SDC_{DG_i^t} \right] - \sum_{t=1}^{24} C \left(P_{str_i^t} \right)$$

Where $\rho_{E,i,t}$ and $\rho_{R,i,t}$ are the price in the energy and the reservation market in the bus i and at hour t, respectively. $E_{i,t}$ is the power supplied to the energy market from bus i at hour t, $R_{i,t}$ is equal to the sum of power provided by DG for the reservation market at bus i and at hour t and r_t is the probability of delivery reservation at time t. $C \left(P_{DG_i^t} \right)$ and $C \left(P_{str_i^t} \right)$ are the operating costs of ESS are DG, which are formulated as follows:

$$C \left(P_{DG_i^t} \right) = A_i \times P_{DG_i^t}^2 + B_i \times P_{DG_i^t} + C_i \tag{4.2}$$

$$C \left(P_{str_i^t} \right) = 0.1 \times \left| P_{str_i^t} \right| + 3.5 \tag{4.3}$$

Where A_i, B_i, and C_i are the operating coefficients of DG and $P_{str_i^t}$ the amount of charge or discharge of ESS per hour t.

- **Maximizing the profit of WT and ESS owners**

    In this section, the objective is to maximize the profits of the owners of these resources to participate in the energy market, and it is assumed that these resources are only involved in the energy market and are owned by a private company.

$$F_2 = Benefit_2 = \sum_{t=1}^{24} \rho_{E,i,t} \times E_{i,t} - \sum_{t=1}^{24} C \left(P_{WT_i^t} \right) - \sum_{t=1}^{24} C \left(P_{str_i^t} \right) \tag{4.4}$$

Where $\rho_{E,i,t}$ is the price in the energy market in bass i and hour t, $E_{i,t}$ is the power offered to the energy market in bass i per hour. $C \left(P_{str_i^t} \right)$ and $C \left(P_{WTs_i^t} \right)$ are the operating costs of ESS and WT. The costs are formulated as follows:

$$C \left(P_{str_i^t} \right) = 0.1 \times \left| P_{str_i^t} \right| + 3.5 \tag{4.5}$$

$$C \left(P_{WF_i^t} \right) = k_{WT} \times P_{WT_i^t} \tag{4.6}$$

The $P_{WT_i^t}$ is the output power of WT at hour t.

- **Maximizing the profitability of PV and ESS owners**

  In this section, the objective is to maximize the profits of PV and ESS owners, which are formulated as follows:

$$F_3 = Benefit_3 = \sum_{t=1}^{24} \rho_{E,i,t} \times E_{i,t} - \sum_{t=1}^{24} C\left(P_{str_i^t}\right) \tag{4.7}$$

$C\left(P_{str_i^t}\right)$ is the operating cost of ESS and is formulated as follows:

$$C\left(P_{str_i^t}\right) = 0.1 \times \left|P_{str_i^t}\right| + 1 \tag{4.8}$$

The cost of PV operation is not considered in this study.

- **The final objective function**

  The final objective function is a combination of all three objective functions with weight coefficients w_1, w_2, w_3 in which the final objective is to maximize the profits of the owners of all resources.

$$F_4 = max(w_1 \times F_1 + w_2 \times F_2 + w_3 \times F_3) \tag{4.9}$$

Where the weight coefficients of considered equal to 1.

2.2 Constraints

- Voltage constraint

$$V_{i,t}^{min} < V_{i,t} < V_{i,t}^{max}, \quad i \in N_b, \quad t = 1, 2, ..., 24. \tag{4.10}$$

- Detachable DG capacity

$$P_{DG_i}^{min} \leq (P_{DG_i} + R_{DG_i}) \leq P_{DG_i}^{max}, \quad i \in N_b \tag{4.11}$$

- Maximum output limit for dispatchable DG

$$\left(P_{DG_i^{t+1}} - P_{DG_i^t}\right) \leq R_i^{Up}, \quad i \in N_b, \quad t = 1, 2, ..., 24. \tag{4.12}$$

- Minimum output limit for dispatchable DG

$$\left(P_{DG_i^t} - P_{DG_i^{t+1}}\right) \leq R_i^{Down}, \quad i \in N_b, \quad t = 1, 2, ..., 24. \tag{4.13}$$

- Line power flow limit

    Limiting the power to flow through the lines is another constraint of the problem, which is applied as follows. According to this constraint, the apparent power of the lines must be within a specified range.

$$S_{i,j}^t \leq S_{i,j,t}^{\max}, \quad i,j \in N_b, \quad t = 1,2,...,24. \tag{4.14}$$

- Power balance constraints

$$P_{G_i^t} + R_{DG_i^t} + P_{DG_i^t} + P_{WT_i^t} + P_{PV_i^t} + \left(P_{ch_i^t} - \eta_{str} P_{Dch_i^t} \right)$$

$$-P_{load_t} = \sum_{j=1}^{N} |V_{i,t}||V_{j,t}||Y_{ij}|\cos(\delta_{j,t} - \delta_{i,t} + \theta_{ij}) \quad i,j \in N_b, \quad t = 1,2,...,24. \tag{4.15}$$

$$Q_{G_i^t} + Q_{DG_i^t} + Q_{WT_i^t} + Q_{PV_i^t}$$

$$-Q_{load_t} = -\sum_{j=1}^{N} |V_{i,t}||V_{j,t}||Y_{ij}|\sin(\delta_{j,t} - \delta_{i,t} + \theta_{ij}) \quad i,j \in N_b, \quad t = 1,2,...,24. \tag{4.16}$$

- Chemical storage constraints [34].

$$-\left(P_{str_i^t}^{\min} - Cap_i^{t-1} \right) \leq P_{str_i^t} \leq \left(P_{str_i^t}^{\max} - Cap_i^{t-1} \right) \tag{4.17}$$

$$Cap_i^{t-1} - Cap_i^t \leq R_{Dch} \tag{4.18}$$

$$Cap_i^t - Cap_i^{t-1} \leq R_{ch} \tag{4.19}$$

$$-\left(P_{str_i^t}^{\min} - Cap_i^{t-1} \right) \leq \sum_{t=1}^{24} P_{str_i^t} \leq \left(P_{str_i^t}^{\max} - Cap_i^{t-1} \right) \tag{4.20}$$

$$P_{str_i^t} \leq R_{Dch} \tag{4.21}$$

$$P_{str_i^t} \leq R_{ch} \tag{4.22}$$

3. Modeling uncertainties

In this study, due to the effect of different network uncertainties on optimal DG and ESS planning, the price uncertainty in the energy market and the reserve market was considered. Also, due to the profitability of participation in the reserve market for DG owners, uncertainty in the delivery of the reservation has been considered to investigate this issue. The log-normal probability distribution function (PDF) has been used to generate random price data in the energy and reserve market [35].

3.1 Uncertainty in energy prices

In this study, to produce random price data, the average price for each scenario is equal to the average of predicted prices and the standard deviation used is related to the price prediction process.

First, 2500 scenarios are generated randomly, and then by using the forward scenario reduction method, the total number of scenarios is reduced to 25. Each scenario includes two price ranges in the energy market and reservations for a 24-h time horizon.

3.2 Uncertainty in reservation delivery

Uncertainty in reservation delivery is one of the factors influencing the participation of energy generators in the reservation market. In this study, the objective function is modeled r_t as the uncertainty parameter in reservation delivery. r_t is acquired using the following equation for a time horizon of 24 h.

$$r(1,24) = 0.9 + 0.1 \times rand(1,24) \tag{4.23}$$

3.3 Uncertainty in the final objective function

In this study, to determine the effect of random parameters on the final profit of resource owners, the maximum expected profit is obtained as follows.

$$F_{4_{expected}} = \sum_{S=1}^{s} P_S \times F_4 \tag{4.24}$$

In this case, the price is chosen as a random parameter, and the probability of each scenario depends on the probabilities of the price scenarios in the energy markets.

4. Systems understudy

To evaluate the optimal planning of ESS in the presence of DGs to participate in the day-ahead energy market and their impact on the profits of the owners of these resources, IEEE 33 standard bus network is used. The network consists of 33 buses and 32 lines. The nominal voltage of the network is 12.66 kV, and its active and reactive power consumption is 3715 KW and 2300 KVAr, respectively. The one-line diagram of this network with the tools added to it can be seen in Fig. 4.1. It is assumed that the network load is predicted according to Fig. 4.2 over the next 24 h.

In this study, as shown in Fig. 4.1, a dispatchable DG and ESS unit are installed on bus 32, a WT and ESS unit are installed on bus 14, and a PV and ESS unit are installed on bus 24. Dispatchable DG information and ESS added to the network can be seen in Tables 4.1 and 4.2, respectively.

It should also be noted that the PV installed on the 24 bus has a capacity of 0.5 MW, which has been neglected. WT has a capacity of 1 MW. The WT and PV output patterns are respectively shown in Fig. 4.3.

5. Simulation and numerical results

In this section, the numerical results of simulating the proposed model for a 24-h time horizon on the 33-bus IEEE network are discussed. It is assumed that DisCo sells energy to the consumer in the distribution network directly. The projected energy prices in the energy and reservation market are shown in the figure. Also, the dispatchable DG unit can participate in the reservation market, and PV, WT, and ESS cannot participate in the reservation market. The simulation was performed for different scenarios in the **MATLAB** software environment using the **TLBO** algorithm.

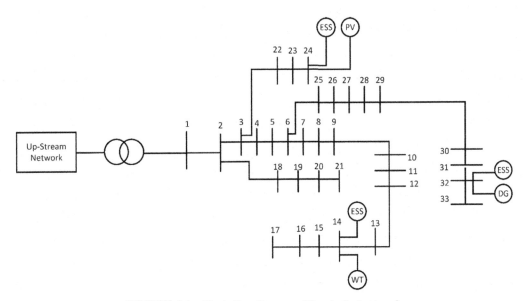

FIGURE 4.1   Single line diagram of the studied network.

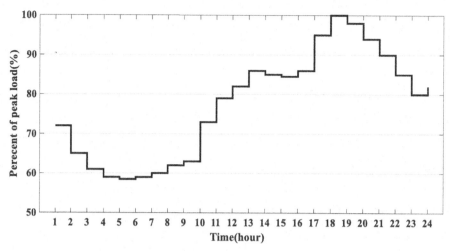

FIGURE 4.2   Forecasted day-ahead load curve.

TABLE 4.1   Technical information and DG cost function.

SDC(UC)	SUC(CU)	MDT(h)	MUT(h)	P_{max} (MW)	P_{min} (MW)	C	B	A	DG unit
13	24	2	3	3	0.5	4	38	0.0058	Bus 32

TABLE 4.2   Storage system information.

η_{srt}	R_{Dch}	R_{ch}	P_{str}^{max}(MW)	P_{str}^{min}(MW)	ESS unit
96	1.5	1.5	8	0.5	Bus 14,32
96	0.12	0.12	0.65	0.05	Bus 24

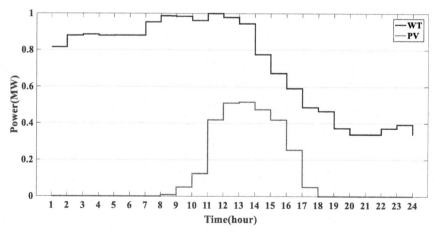

FIGURE 4.3   Forecasted hourly power output of WT and PV.

The scenarios examined in this chapter are as follows:
Scenario 1:
In this scenario, to examine the effect of the presence of ESS on profits generated by DGs, it is assumed that ESS are not available in the network, and there is no uncertainty in the price of energy in the energy and reservation market. Also r_t is equal to 1 in all 24 h.
Scenario 2:
In this scenario, the effect of the presence of ESS in the network is investigated and compared with the first scenario. There is no uncertainty in this case.
Scenario 3:
In this scenario, the effect of price uncertainty on the energy and reserve market, uncertainty on the delivery of reservations on the optimal planning of ESS for participation in the energy market, and their impact on dispatchable DG planning are examined.

5.1 Scenario 1: No ESS presence and no uncertainty

In this scenario, there is no ESS in the studied network, and the price in the energy and reservation market is predetermined, and there is no uncertainty in the price. Reserve is also 100% delivered. In this case, the results of the optimal planning of dispatchable DGs for participation in the energy and reservation market can be seen without the presence of ESS in Fig. 4.4.

Examination of the results shows that the dispatchable DG participates in the energy market with its maximum power at all times except hours 21–11. In the hours 16–21, the cost of power generation is lower than the energy price in the market, and it is economically viable for DG not to participate in the market. It is also assumed that the reservation must be delivered within 10 min, and since the reservation delivery speed is 50 kW/min, DG can supply a maximum of 0.5 MW of power to the energy market. A review of the results shows that in all the hours that DG is generating power, it participates with the maximum capacity in the

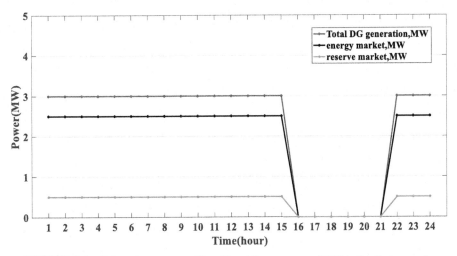

FIGURE 4.4 DG performance results without the presence of ESS in the first scenario.

reservation market, which is due to the profits earned from participating in the reservation market with 100% reservation delivery. In the following sections, the effect of uncertainty on reservation delivery on the amount of power allocated to the reserve market is examined. The hourly profit of DG owners can be seen in Fig. 4.5 within 24 h. The total profit of DG owners in 24 h is 838.33 monetary units.

In this scenario, the output pattern of PV and WT is already known, and profit is obtained only according to the price in the energy market, and there is no flexibility in the behavior of PV and WT to increase the profit of resource owners.

An examination of the results of Fig. 4.6 shows that WT is forced to sell the energy produced at any cost, as it does not benefit from the presence of ESS to store energy at times when energy prices are low.

The profit for WT is seen hourly in Fig. 4.7. The total profit of WT owners in 24 h is 746.53 monetary units.

Fig. 4.8 shows the results of PV participation in the energy market. These results are obtained in the absence of ESS, which can increase the profits of PV owners. Fig. 4.9 shows the hourly profits of PV for 24 h. The total profit of PV owners is 189.84 monetary units.

5.2 Scenario 2: with the presence of ESS and without uncertainty

In this scenario, all the resources added to the network under study are included in the simulation, and the optimal planning of ESS and their role in increasing the profit generated by the company's participation in the energy market are discussed. Dispatchable DG performance and accompanying ESS can be seen in Fig. 4.10.

ESS stores 1.5 MW of electrical energy per hour when energy prices are low and reaches its maximum capacity at hour 6. At hour 11–15, when the price of energy in the market rises, it sells energy and makes ESS and DG owners profitable.

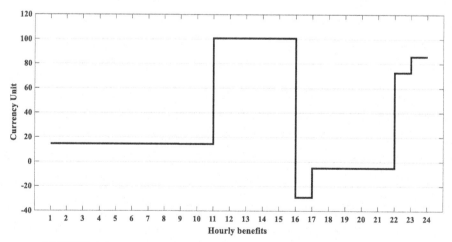

FIGURE 4.5   Hourly profits of DG owners without ESS in the first scenario.

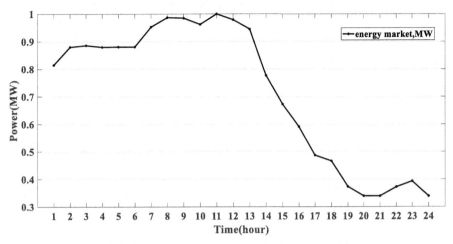

FIGURE 4.6   WT performance results without ESS in the first scenario.

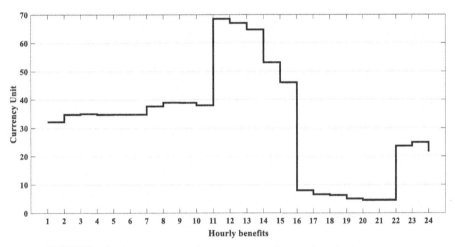

FIGURE 4.7   Hourly profits of WT owners without ESS in the first scenario.

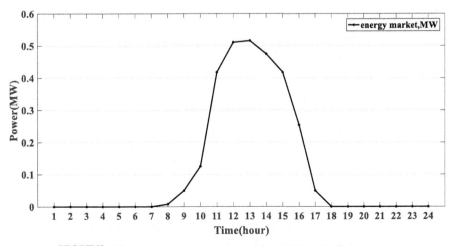

FIGURE 4.8   PV performance results without ESS in the first scenario.

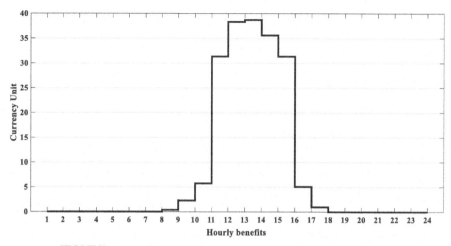

FIGURE 4.9 Hourly profits of PV owners without ESS in Scenario 1.

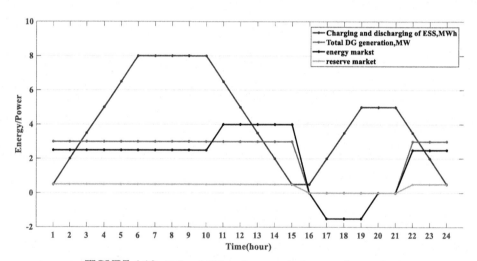

FIGURE 4.10 DG and ESS performance in the second scenario.

The results also show that DG has no production during the hours 16–21 when energy prices are low, and only ESS saves energy by buying from the energy market to sell at hours 22–24 to increase the profit. Examining the behavior of ESS in the energy market, it is clear that ESS performance was optimal. Note that DG participates in the reserve market at its maximum capacity.

The hourly profit of the owners of the resources is visible in Fig. 4.11. The total profit from this case for DG and ESS is 1251.7 monetary units, which is 413.37 monetary units more than the previous case where there was no ESS.

In this scenario, the effect of ESS presence on increasing the profit of WT and PV owners will be examined. With optimal ESS planning, the profits of their affiliates increase. Fig. 4.12 shows the WT performance and the optimal ESS behavior for participating in the energy market.

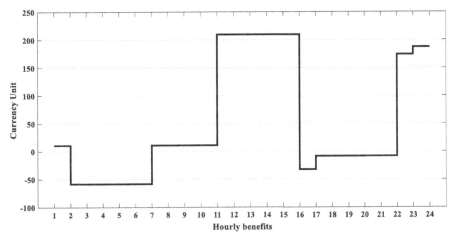

FIGURE 4.11   DG and ESS owners' hourly profits in the second scenario.

Examination of the results of Fig. 4.12 shows that ESS tries to increase the profit of WT owners and itself with optimal performance and tries to increase the profit by storing energy in the hours when the energy price is low. The hourly profit of WT and ESS owners can be seen in Fig. 4.13. With the presence of ESS and its optimal planning to participate in the market, the profit of WT owners in this scenario is 1080 monetary units.

In this scenario, PV-related ESS charges when the market price of energy is low and sells for hours 10–15 and hours 22–24 when the price of energy in the market is high. An examination of energy prices at different hours shows the optimal performance of ESS over 24 h. Fig. 4.14 shows the performance of PV and ESS. The profit from the participation of PV and ESS can be seen in Fig. 4.15 within 24 h. In this case, the storage word has increased the profits of PV and ESS owners, and the total profit in 24 h has reached 337,048 monetary units.

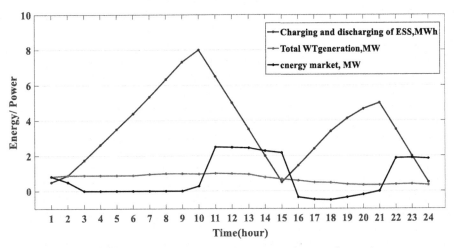

FIGURE 4.12   WT and ESS performance in the second scenario.

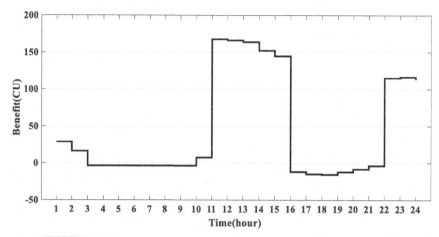

FIGURE 4.13   WT owners' hourly profits with ESS in the second scenario.

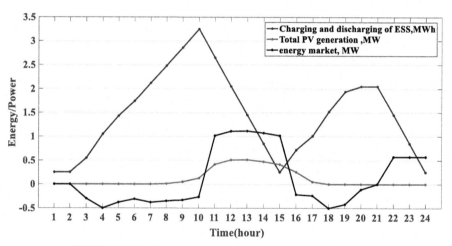

FIGURE 4.14   PV and ESS performance in the second scenario.

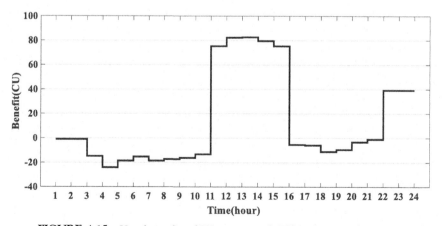

FIGURE 4.15   Hourly profits of PV owners with ESS in the second scenario.

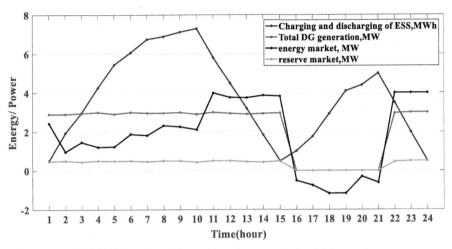

FIGURE 4.16 DG and ESS performance in the third scenario.

5.3 Scenario 3: uncertainty in energy prices and reservation delivery

In the second scenario, optimized planning for ESS and dispatchable DGs took place in a situation where there was no uncertainty in the projected price in the energy and reserve markets, as well as the delivery of reservations. In the energy market, the impact of these resources has been examined. Fig. 4.16 shows the DG function of the disk and its associated ESS.

The results in the table show that the existing uncertainties have affected the optimal ESS planning more. The uncertainty in the reservation delivery has caused changes in the amount of DG's participation in the reservation market. In this study, the reservation is 10% likely not to be delivered in the worst-case scenario. The hourly profits graph shows in Fig. 4.17 total profit of the DG and ESS owners, which in this case is 1154.4 monetary units.

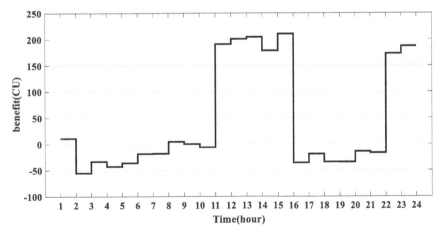

FIGURE 4.17 DG owners' hourly profits with ESS in the third scenario.

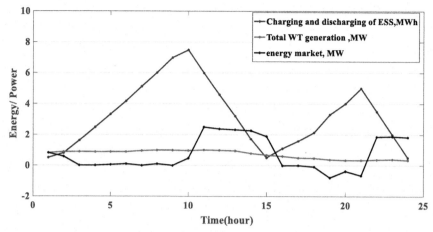

FIGURE 4.18   WT and ESS performance in the third scenario.

The following scenario examines the impact of uncertainties on the optimal planning of ESS associated with WT and PV. Given that in this scenario, the output pattern of WT and PV is predetermined, the uncertainty of energy prices in the energy and reserve markets only affects the optimal behavior of ESS and the profit obtained from participation in the energy and reserve market. The performances of WT and ESS are shown in Fig. 4.18.

In this scenario, uncertainty in energy prices prevented ESS from reaching its maximum charge level due to uncertain prices in the energy market. The hourly profits of resource owners are shown in Fig. 4.19. Uncertainty in energy prices has led to a 55.4-unit drop in profits compared to the second scenario. It is important to note that uncertainty in energy prices does not always reduce profits, and resource owners may gain more profits as prices rise in the energy market.

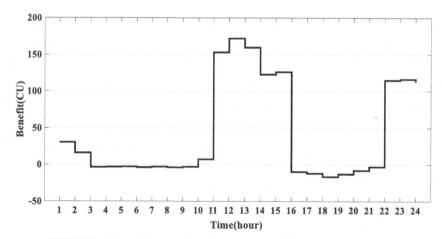

FIGURE 4.19   WT owners' hourly profits with ESS in the third scenario.

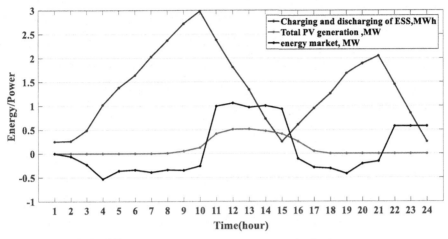

FIGURE 4.20 PV and ESS performance in the third scenario.

Fig. 4.20 shows that uncertainty in energy prices has led to changes in ESS optimization planning, leaving 0.27 MW of ESS capacity depleted. The results obtained in this scenario clearly show the effect of uncertainty on energy prices on storage behavior. To participate in the energy market, ESS needs to predict a reasonable price to increase the profit of the collection according to the obtained results, and they may cause a loss of the collection by predicting an inappropriate price. Examining Fig. 4.21, which shows the hourly profits of PV and ESS owners, the total profit of the collection has reached 289,14 monetary units, and the uncertainty has reduced the profit of PV and ESS.

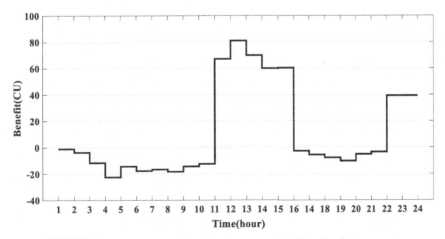

FIGURE 4.21 Hourly profits of PV owners with ESS in the third scenario.

6. Conclusion

In this chapter, the optimal planning of ESS to help distributed generation and renewable resources to participate in the energy market was examined. The impact of price uncertainty on the energy and reserve market, as well as uncertainty on the delivery of reservations on the optimal behavior of ESS and DG, was discussed. In this study, the IEEE 33 bus network was used.

The results show the effect of ESS on increasing the profits of different resource owners as well as them being influenced by the uncertainties.

It is shown that despite the empty capacity, ESS has not been charged due to uncertainty in energy prices, reducing the profitability of resource owners. ESS s have dual mode when participating in the market, meaning they can buy or sell energy to the network.

In the scenario with no certainty, the DG participated in delivering reservations with its maximum capacity, and uncertainty in delivering the reserve had little to no effect on the power delivered.

References

[1] U.S Department of Commerce, National Oceanic and Atmospheric Administration. https://www.noaa.gov/news/summer-2019-was-hottest-on-record-for-northern-hemisphere.

[2] United States Environmental Protection Agency. https://www.epa.gov/ghgemissions/sources-greenhouse-gas-emissions.

[3] United States Department of Energy, The U.S. Energy Information Administration's (EIA). https://www.eia.gov/energyexplained/electricity/electricity-in-the-us.php.

[4] X. Xiaofeng, et al., Global renewable energy development: influencing factors, trend predictions and counter-measures, Resour. Pol. 63 (2019) 101470.

[5] A. Hossein, et al., A fuzzy-logic—based control methodology for secure operation of a microgrid in intercon-nected and isolated modes, Int. Trans. Elec. Energy Syst. 27 (11) (2017) e2389.

[6] A. Hossein, M. Taghi Ameli, S. Hossein Hosseinian, Multi-stage frequency control of a microgrid in the presence of renewable energy units, Elec. Power Compon. Syst. 45 (2) (2017) 159—170.

[7] C. Luan, H. Qi, C. Chi, Strategy on wind power smoothing subject to target parameters and energy-storage ca-pacity, Electr. power 6 (2013).

[8] L. Meng, F.L. Quilumba, W.-J. Lee, Dispatch scheduling for a wind farm with hybrid energy storage based on wind and LMP forecasting, IEEE Trans. Ind. Appl. 51 (3) (2014) 1970—1977.

[9] K. Mostafa, et al., Operation scheduling of battery storage systems in joint energy and ancillary services markets, IEEE Trans. Sustain. Energy 8 (4) (2017) 1726—1735.

[10] P. Danny, et al., Whole-systems assessment of the value of energy storage in low-carbon electricity systems, IEEE Trans. Smart Grid 5 (2) (2013) 1098—1109.

[11] H. Guannan, et al., Optimal bidding strategy of battery storage in power markets considering performance-based regulation and battery cycle life, IEEE Trans. Smart Grid 7 (5) (2015) 2359—2367.

[12] M.I.P. Hugo, et al., Risk-constrained scheduling and offering strategies of a price-maker hydro producer under uncertainty, IEEE Trans. Power Syst. 28 (2) (2012) 1879—1887.

[13] G. Samuele, A. Pievatolo, E. Tironi, Optimal storage scheduling using Markov decision processes, IEEE Trans. Sustain. Energy 7 (2) (2015) 755—764.

[14] S. Azad, M. Mehdi Amiri, M.T. Ameli, Contribution of virtual power plants in electric market considering un-certainty in virtual power plant connection with upstream network, J. Energy Manag. Technol. 2 (3) (2018) 70—80.

[15] A. Juan, H. Zareipour, A price-maker/price-taker model for the operation of battery storage systems in elec-tricity markets, IEEE Trans. Smart Grid 10 (6) (2019) 6912—6920.

[16] X. Yixing, L. Xie, C. Singh, Optimal scheduling and operation of load aggregator with electric energy storage in power markets, 2011 North American Power Symposium (2011) 1–7, https://doi.org/10.1109/NAPS.2011.6024888.

[17] M. Kloess, Electric storage technologies for the future power system—an economic assessment, in: 2012 9th International Conference on the European Energy Market. IEEE, 2012.

[18] J. Qin, et al., Optimal electric energy storage operation, in: 2012 IEEE Power and Energy Society General Meeting. IEEE, 2012.

[19] S. Zhen, P. Jirutitijaroen, Optimal operation strategy of energy storage system for grid-connected wind power plants, IEEE Trans. Sustain. Energy 5 (1) (2013) 190–199.

[20] A.-H. Hossein, H. Mohsenian-Rad, Optimal operation of independent storage systems in energy and reserve markets with high wind penetration, IEEE Trans. Smart Grid 5 (2) (2013) 1088–1097.

[21] C. Brunetto, G. Tina, Optimal hydrogen storage sizing for wind power plants in day ahead electricity market, IET Renew. Power Gener. 1 (4) (2007) 220–226.

[22] G.-G. Javier, et al., Stochastic joint optimization of wind generation and pumped-storage units in an electricity market, IEEE Trans. Power Syst. 23 (2) (2008) 460–468.

[23] P. Emilio, et al., Predictive power control for PV plants with energy storage, IEEE Trans. Sustain. Energy 4 (2) (2012) 482–490.

[24] B. Hector, et al., Daily solar energy estimation for minimizing energy storage requirements in PV power plants, IEEE Trans. Sustain. Energy 4 (2) (2012) 474–481.

[25] K. Arash, et al., Improving wind farm dispatch in the Australian electricity market with battery energy storage using model predictive control, IEEE Trans. Sustain. Energy 4 (3) (2013) 745–755.

[26] T. Jen-Hao, et al., Optimal charging/discharging scheduling of battery storage systems for distribution systems interconnected with sizeable PV generation systems, IEEE Trans. Power Syst. 28 (2) (2012) 1425–1433.

[27] C. Brendan, et al., Assessing the economic benefits of compressed air energy storage for mitigating wind curtailment, IEEE Trans. Sustain. Energy 6 (3) (2015) 1021–1028.

[28] L. Ning, J.H. Chow, A.A. Desrochers, Pumped-storage hydro-turbine bidding strategies in a competitive electricity market, IEEE Trans. Power Syst. 19 (2) (2004) 834–841.

[29] F.C. Figueiredo, P.C. Flynn, Using diurnal power price to configure pumped storage, IEEE Trans. Energy Convers. 21 (3) (2006) 804–809.

[30] R.L. Fares, M.E. Webber, A flexible model for economic operational management of grid battery energy storage, Energy 78 (2014) 768–776.

[31] D. Easan, P. Denholm, R. Sioshansi, The value of compressed air energy storage in energy and reserve markets, Energy 36 (8) (2011) 4959–4973.

[32] T. Xiaoqi, Y. Wu, D.H.K. Tsang, Pareto optimal operation of distributed battery energy storage systems for energy arbitrage under dynamic pricing, IEEE Trans. Parallel Distr. Syst. 27 (7) (2015) 2103–2115.

[33] Z. Dongwei, et al., Storage or no storage: duopoly competition between renewable energy suppliers in a local energy market, IEEE J. Sel. Areas Commun. 38 (1) (2019) 31–47.

[34] M. Elaheh, S.M. Moghaddas-Tafreshi, Mathematical modeling of electrochemical storage for incorporation in methods to optimize the operational planning of an interconnected micro grid, J. Zhejiang Univ. Sci. C 11 (9) (2010) 737–750.

[35] A.J. Conejo, F. Javier Nogales, J. Manuel Arroyo, Price-taker bidding strategy under price uncertainty, IEEE Trans. Power Syst. 17 (4) (2002) 1081–1088.

IGDT-based optimal low-carbon generation dispatch of power system integrated with compressed air energy storage systems

Alireza Akbari-Dibavar[1], Behnam Mohammadi-Ivatloo[1,2], Kazem Zare[1], Vahid Hosseinnezhad[3]

[1]Faculty of Electrical and Computer Engineering, University of Tabriz, Tabriz, East Azerbaijan, Iran; [2]Department of Energy Technology, Aalborg University, Aalborg, Denmark; [3]School of Engineering, University College Cork (UCC), Cork, Ireland

1. Introduction

It is proved that one of the reasons behind global warming is carbon dioxide (CO_2) pollution and the main resource of CO_2 production is the large-scale conventional generators, i.e., gas or coal power plants [1]. Considering the consensus on the reduction of CO_2 emission, the power system operators have to provide practical solutions for this challenge. Power systems have a large share regarding CO_2 production and the emission resource is concentrated, so the extenuation is much easier than other industries [2].

The basic idea behind power plants with the ability of CO_2 reduction is to use carbon capturing and storing assets besides the conventional power plants, which are named carbon capture power plants (CCPPs) equipped with carbon capture system (CCS) [3]. This technology can capture and store a significant portion of CO_2 produced by the power plant when generating electricity. With this intuition, the produced emission will be isolated from the atmosphere. According to various politics made by the European Commission, the CCPPs will be important parts of future power systems [4]. This is coming along with high carbon pricing, which forces generation companies to produce less or even zero emission. So, the global trend is to use more green energy as well as conventional generators equipped with CCSs. According to a report by the Global CCS Institute, in the year 2017, 37 large-scale CCPPs were under operation or construction over the world and this number is planned to be increased every year.

In addition to operational complexity and high investment costs due to carbon capture system mechanisms, an amount of produced electric power should be consumed for CO_2 reduction, and this can increase the operational cost of the system in the presence of CCPPs [5]. In this manner, a thread of researches has been conducted on modeling and operating of CCPP. In Ref. [6], the mechanism of CCPP is studied and a feasible operational region for the power output of generation units equipped with carbon capture systems is deliberated. The dynamic behavior of a pulverized coal power plant integrated into a post combination carbon capture system is investigated in Refs. [7]. The operational flexibility and efficiency assessment of future CCPP was the subject of [8]. Various CCS technologies, as well as the technoeconomical evolution of these systems, can be found in Ref. [9]. The sensitivity analysis of resources on low carbon unit commitment is done in Ref. [10]. The generation expansion planning problem considering low carbon and clean generation is addressed by Ref. [2]. The state-of-the-art review on flexibility enhancement of CCS systems has been done by Refs. [11], and various techniques are introduced.

On the other hand, there is a trend to utilizing renewable energies as main energy production sources since they are low emission resources and their investment cost is going to reduce year by year. However, the energy production of RES is uncertain and the prediction of production is a very challenging task. In this regard, the authors of [12] discussed the flexibility requirement of the power systems in the presence of renewable generation and CCPPs using technoeconomic evolutions. Some options for flexibility enhancement including utilization of demand response resources, fossil-fueled power plants equipped with CCS, energy storage systems, and increment of interconnector capacity were tested. They have concluded the power storage is a costly option and increases the system costs. Similarly, the authors of [13] have evaluated the effects of electricity storage and hydrogen technologies on renewable generation deployment under carbon-constrained situations using integrated assessment modeling. They found out that without considering policies regarding carbon production, the pessimistic costs related to storage and hydrogen will reduce the large deployment of RESs. The power to gas (PtG) storage is another promoting option for RES integration in a carbon constraint environment. Using PtG technology, the required hydrogen for industrial applications can be produced without releasing a huge amount of carbon pollution [14]. In this regard, a thorough problem of dispatching of PtG facility and fossil-fueled gas power plants equipped with CCS has been studied by Ref. [15]. The authors of [16] presented a framework for generation dispatch of CCPPs in the presence of responsive demands and wind power uncertainty. The uncertainty of wind generation is modeled by stochastic programming. In Ref. [17], the effects of ESS on optimal unit commitment problems in a low-carbon constrained case study, including wind generation, are studied. However, the CCPP modeling, as a vital part of low-carbon systems, is not considered. In Ref. [18], the authors presented a framework for optimal power dispatching problems considering the effects of CCPPs, battery storage systems, and wind generation uncertainty. The adjustable robust optimization is proposed to handle the wind uncertainty.

In this chapter, a cost minimization framework is proposed for the generation scheduling problem in which a DC power flow calculation is proposed to investigate the effects of compressed air energy storage (CAES) units combined with wind turbines and postcombustion CCPPs. The CAES systems have the ability to be exploited in large-scale power systems. The cycle of charging and discharging of these storage systems includes a compression stage and an expansion stage; i.e., during low-demand periods, the air is compressed using a rotary compressor and injected into the vessel of the CAES, which can be an underground cavern. Then during high-demand periods, the pressurized air is released to regenerate electrical power in the senses of an air turbine. The simple mechanism as well as its high efficiency and low reservoir cost of the CAES systems make them a promoting storage technology for power system applications such as integrating renewable generation or participation in power markets [19]. For example, in Ref. [20], the CAES systems are employed to compensate for the deviations of the wind generation in an electrical network coupled with gas networks. In Ref. [21], the self-scheduling problem of the merchant CAES system is conducted to find the optimal bidding strategy of the CAES system in a power market environment.

Furthermore, in this chapter, wind generation uncertainty is modeled with the Information Gap Decision Theory (IGDT), which results in robust decisions regarding system scheduling. The IGDT is not a probabilistic method and does not require the probability distribution function of the uncertain parameters. However, the idea is to model the gap between the actual and forecasted amounts of the uncertain parameter by finding an interval, which is adequate with respect to a predefined critical objective value [22].

The proposed optimization problem is formulated as mixed-integer linear programming (MILP) and modeled using GAMS and solved with CPLEX solver. A modified version of IEEE 24-bus RTS is employed to verify the validation of the proposed optimization problem in order to reveal the effects of CCPPs and CAESs on the operational cost.

2. Problem formulation

This section presents the formulation of generation dispatch problem considering the effects of CCPPs and ESS operation on total emission production, system cost in both deterministic and robust cases. The schematic of the system is shown in Fig. 5.1. As it can be seen, the fuel and air are the input of the power plants and the power and gas pollution are the outputs. However, for the CCPPs the exhausted gas is purified using a CO_2 separation and capturing stage. If the power plant is equipped with the CCS, a portion of the generated power will be consumed for CO_2 purification.

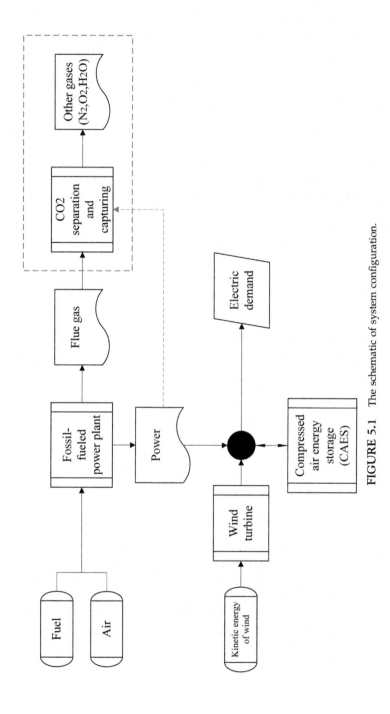

FIGURE 5.1   The schematic of system configuration.

2.1 Deterministic approach

The optimization is formulated as a DC power flow calculation, where the objective function (1) tends to minimize the costs related to daily power generation by conventional generators and emission production, the daily generation cost of CCPPs, emission production, and abatement costs, respectively. The results of the deterministic case are used as a criterion for the IGDT method would be mentioned next.

$$
\begin{aligned}
OF = & \sum_t \sum_{i \in \Omega_G} \left(P_{i,t}^N \times \lambda_i^{en} \right) + \left(P_{i,t}^N \times e_i^G \times \lambda^{CO_2} \right) + \\
& \sum_t \sum_{i \in \Omega_{ccpp}} \left(P_{i,t}^N \times \lambda_i^{en} \right) + \left(E_{i,t}^N \times \lambda^{CO_2} \right) + \left(E_{i,t}^S \times \lambda^{ccpp} \right)
\end{aligned}
\tag{5.1}
$$

In (1), i is used for indicating bus number and t is the time index. For power plants without CCS, the first item indicates the cost of power generation according to the net output power $P_{i,t}^N$ and unit cost λ_i^{en}, the second item calculates the cost of produced emission according to the amount of net output power, which is equal to actual generated power, $P_{i,t}^G$ and emission intensity factor of the output power of each power plant, e_i^G. λ^{CO_2} is the price of CO_2 emission production and dispersion. The third, fourth, and fifth terms, respectively, calculate the power generation cost according to power output $P_{i,t}^N$ and generation cost λ_i^{en}, the penalty cost of the CO_2 dispersal according to net emission production $E_{i,t}^N$ and CO_2 price λ^{CO_2}, and the cost of CO_2 abatement for the CCPPs based on the value of rescued emission $E_{i,t}^S$ and the abatement price of the CCPP, λ^{ccpp}.

$$
P_{i,t}^N = P_{i,t}^G - P_{i,t}^{EP}, \forall t, i
\tag{5.2}
$$

$$
E_{i,t}^N = E_{i,t}^G - E_{i,t}^S, \forall t, i
\tag{5.3}
$$

$$
\underline{P^G} \leq P_{i,t}^G \leq \overline{P^G}, \forall t, i
\tag{5.4}
$$

For the power plants equipped with CCS, the net power output, $P_{i,t}^N$, is not equal to the generated power $P_{i,t}^G$ and a portion of generated power is consumed during the CO_2 abatement process indicated by $P_{i,t}^{EP}$, as in Eq. (5.2). Apparently, $P_{i,t}^{EP}$ is zero for the power plants without CCS. Based on Eq. (5.3), instead of initially produced emission $E_{i,t}^G$, the net carbon emission diffusion (after CO_2 reduction) is taken into account indicated by $E_{i,t}^N$ and $E_{i,t}^S$ shows the amount of purified release for the units equipped with CCS. Constraint (4) poses a limitation on the power output of thermal plants. Constraints (2)–(4) are consistent for all thermal plants.

Regarding the CCPP's model, the consumed power of the CCS ($P_{i,t}^{EP}$) includes two terms, basic power consumption of the CCS and operational power consumption of the CCS, $P_{i,t}^{EPo}$. Usually, the basic power consumption of the CCS is assumed constant, which is ignored in this chapter. Similarly, the initial emission produced by CCPPs (i.e., $E_{i,t}^G$) is related to their total power generation ($P_{i,t}^G$) and emission intensity factor e_i^G. On the other hand, the operational power consumption of the CCS is proportional to the amount of emission being treated ($E_{i,t}^P$) and the energy efficiency of the CCPP, i.e., η^{ccpp}. However, the captured emission $E_{i,t}^S$ is proportional to the treated CO_2 and carbon capture rate of CCS, i.e., α^C [7]. Thus, Eqs. (5.5)–(5.8) model the operation region of CCPPs.

$$P_{i,t}^{EP} = P_{i,t}^{EPo}, \quad \forall t, i \in \Omega_{ccpp} \tag{5.5}$$

$$P_{i,t}^{EPo} = \eta^{ccpp} E_{i,t}^P, \quad \forall t, i \in \Omega_{ccpp} \tag{5.6}$$

$$E_{i,t}^G = e_i^G P_{i,t}^G, \quad \forall t, i \tag{5.7}$$

$$E_{i,t}^S = \alpha^C E_{i,t}^P, \quad \forall t, i \in \Omega_{ccpp} \tag{5.8}$$

In the following, the power flow problem is modeled as a DC power flow calculation considering active power exchange and balance. Where, constraint (9) shows the active power balance of the network considering the effects of wind turbines, conventional generators, CCPPs, ESS, and deterministic loads. The right-hand side shows the exchanged power between connected buses. Eq. (5.10) calculates the power flow ($P_{i,j,t}^l$) through lines between connected buses based on the voltage angles ($\delta_{i,t}$) of the corresponding buses and the line reactance ($x_{i,j}$). For reference bus, the voltage angle is set to zero by (11). Finally, the exchanged power between connected buses is limited by (12).

$$P_{i \in \Omega_{wind},t}^W + P_{i \in \Omega_G,t}^N + P_{i \in \Omega_{ccpp},t}^N + P_{i \in \Omega_{ess},t}^{dis} - P_{i \in \Omega_{ess},t}^{ch} - P_{t,i}^{load} = \sum_{j \in I_{i,j}}^{N_j} P_{i,j,t}^l, \forall t, i \tag{5.9}$$

$$P_{i,j,t}^l = \frac{\delta_{i,t} - \delta_{j,t}}{x_{i,j}}, \quad \forall t, i, j \tag{5.10}$$

$$\delta_{i,t} = 0, \quad \forall t, \text{slack bus} \tag{5.11}$$

$$-\overline{P_{i,j}^l} \leq P_{i,j,t}^l \leq +\overline{P_{i,j}^l} \tag{5.12}$$

Besides, the mathematically CAES system model given by Refs. [23] is adopted here. The CAES system works based on the frequent air injection/pumping into/from a vessel (for example, salt caves). Constraints (13) and (14) limit the minimum and maximum volumes of the injected air ($V_{i,t}^{inj}$) and pumped air ($V_{i,t}^{pump}$) into/from the vessel, each of which respectively models the charged and discharged powers of the ESS. Simultaneously charging and discharging are prohibited for the ESS and this is done using binary variables ($u_{i,t}^{ch}$ and $u_{i,t}^{dis}$) in (15). Eqs. (5.16) and (5.17) determine the equivalent electric energy of injected and pumped air according to the efficiency (η^{inj} and η^{pump}) of the ESS that is named charged power ($P_{i,t}^{ch}$) and discharged power ($P_{i,t}^{dis}$), respectively. The stored air ($E_{i,t}^{ess}$) in the vessel of the CAES is considered based on (18) and (19) for $t = 1$ and $t > 1$, respectively. Finally, the restriction on the lower and upper amounts ($\underline{E^{ess}}$ and $\overline{E^{ess}}$) of the air stored in the ESS is stated by (20).

$$0 \leq V_{i,t}^{inj} \leq \overline{V_i^{inj}} \times u_{i,t}^{ch}, \forall t, i \in \Omega_{ess} \tag{5.13}$$

$$0 \leq V_{i,t}^{pump} \leq \overline{V_i^{pump}} \times u_{i,t}^{dis}, \forall t, i \in \Omega_{ess} \tag{5.14}$$

$$u_{i,t}^{ch} + u_{i,t}^{dis} \leq 1, \quad \forall t, i \in \Omega_{ess} \tag{5.15}$$

$$V_{i,t}^{inj} = \eta^{inj} P_{i,t}^{ch}, \quad \forall t, i \in \Omega_{ess} \tag{5.16}$$

$$V_{i,t}^{pump} = \eta^{pump} P_{i,t}^{dis}, \forall t, i \in \Omega_{ess} \tag{5.17}$$

$$E_{i,t}^{ess} = E_{i,t=0}^{ess} + V_{i,t}^{inj} - V_{i,t}^{pump}, \forall t = 1, i \in \Omega_{ess} \tag{5.18}$$

$$E_{i,t}^{ess} = E_{i,t-1}^{ess} + V_{i,t}^{inj} - V_{i,t}^{pump}, \forall t > 1, i \in \Omega_{ess} \tag{5.19}$$

$$\underline{E^{ess}} \leq E_{i,t}^{ess} \leq \overline{E^{ess}}, \forall t > 1, i \in \Omega_{ess} \tag{5.20}$$

2.2 IGDT-based robust generation dispatch

Till here, the deterministic power dispatching optimization problem is modeled by (1)−(20), which results in a criterion for the IGDT method. Despite scenario-based stochastic programming, which is used extensively in power system problems, the IGDT is based on the definition of the uncertainty interval, which is targeted to be maximized when the system operators tend to find the maximum operation cost. In other words, the system operator tries to find a generation scheduling that is robust against undesired effects of uncertainties realization. As the wind generation is associated with the natural condition and the wind velocity is considered as an uncertain parameter, the IGDT is a reliable method to provide risk-averse decisions in the absence of information regarding the uncertain parameters. The mathematical model of the IGDT can be mentioned as (21)−(23) according to Ref. [24].

$$\min_{x} z(x, \phi) \tag{5.21}$$

Subject to

$$H_n(x, \phi) = 0 \tag{5.22}$$

$$G_m(x, \phi) \leq 0 \tag{5.23}$$

Where $z(x, \phi)$ models the system's objective function, x contains decision variables, ϕ is the set of uncertain parameters, H_n and G_m are equality and inequality constraints of the mentioned problem. For the presented work, the IGDT is turned into (24)–(28).

$$\max_{x} \widetilde{\alpha} \tag{5.24}$$

Subject to

$$H_n(x, \phi) = 0 \tag{5.25}$$

$$G_m(x, \phi) \leq 0 \tag{5.26}$$

$$\widetilde{\alpha} = \left\{ \max_{\alpha} |z(x, \phi) \leq Z_C \right\} \tag{5.27}$$

$$\phi \in U(\overline{\phi}, \alpha) = \left\{ \phi : \left| \frac{\phi - \overline{\phi}}{\overline{\phi}} \right| \right\} \leq \alpha \tag{5.28}$$

Where, α is the decision variable indicating the deviation of the uncertain parameter (i.e., uncertainty radius), which is targeted to be maximized for a given value of x, in order to give the robust solutions and $\widetilde{\alpha}$ is the maximum amount of α. Z_C is the critical operation cost, which is obtained from solving the deterministic problem and $\overline{\phi}$ indicates the forecasted amount of the uncertain parameter.

3. Input data and results

The presented optimization is modeled as MILP in GAMS optimization software and solved using CPLEX solver [25]. The well-known IEEE 24 bus test system has been used to verify the presented optimization problem considering minor modifications. The schematic of the employed test system is depicted in Fig. 5.2. Lines' specifications are available in Ref. [26]. The required data regarding generation units are presented in Table 5.1. The location of the thermal units is revealed in Fig. 5.2. Without loss of generality, it is assumed that

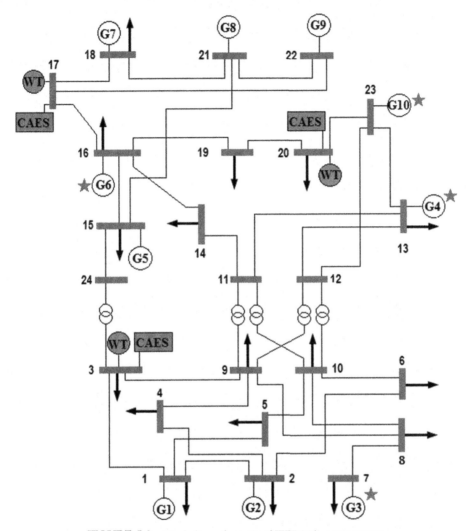

FIGURE 5.2 Single line schematic of IEEE 24-bus test system.

the thermal plants located at buses 7, 13, 16, and 23 are equipped with CCS, which are emphasized with a red (gray in printed version) star in Fig. 5.2. The penalty cost of CO_2 production is set at 100 \$/ton and CO_2 capturing and storing cost is 25 \$/ton. The efficiency of the CCPP is 0.27 and the carbon capture rate is 0.9, according to Ref. [16].

Moreover, three wind turbines are installed on buses 3, 17, and 20. The location of these turbines and the output power are randomized. Fig. 5.3 illustrates the forecasted power generation from wind turbines. In order to deliver a constant power, three CAES systems with a

TABLE 5.1 Information on generation units.

Generator	$\underline{P}^G$ (MW)	$\overline{P}^G$ (MW)	λ_i^{en} ($/MW)	e_i^G
1	0	152	20.32	0.65
2	0	152	20.32	0.7
3	0	350	20.7	0.75
4	0	590	30.93	0.6
5	0	155	16	0.75
6	0	155	30.52	0.75
7	0	200	15.47	0.65
8	0	200	25.47	0.65
9	0	100	10	0.65
10	0	150	30.52	0.8

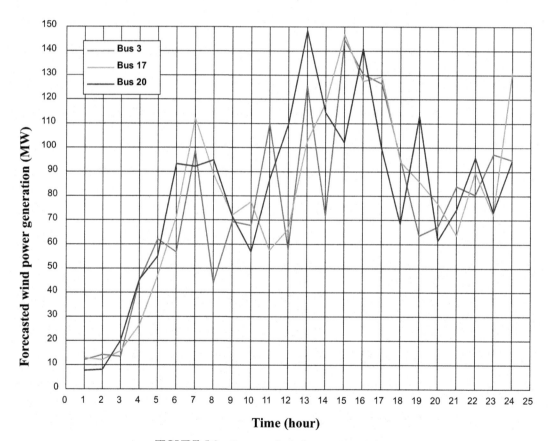

FIGURE 5.3 Forecasted wind power generation.

TABLE 5.2 Cost comparison.

Case number	Deterministic cost ($)	Robust cost ($)
1	3,223,045.603	3,233,045.603
2	2,441,024.068	2,451,024.068
3	2,423,887.292	2,433,887.292

total capacity of 200 MWh and with the rated power of 50 MW are installed besides wind turbines. The charging and discharging efficiency of the CAES systems is assumed to be 0.95. The lower bound of stored energy and charging/discharging powers are assumed zero. However, to reflect the effects of the CCPPs and ESS, three cases are studied. Case 1 is a base case without neither the ESS nor CCS systems. Case 2 includes CCPPs without ESSs. Case 3 considers the effects of CCPPs and ESS units cooperatively.

3.1 Results

As mentioned before, three case studies are investigated in this chapter to check how the CCPPs and ESS affect the total costs of the system. The first case is a basic load flow in a network with conventional plants and wind turbines without CCS and ESS. Table 5.2 reports the operational costs for all cases in both deterministic and robust cases. The cost step of the IGDT method is $ 1000.

From Table 5.2, the integration of CCS in case 2 has led to a decrease of operation cost about $ 782,021 (i.e., 24%) in comparison with case 1, in a sample day operation due to a significant decrease in the amount of CO_2 emission. It should be noted that the obtained results are confirmed for the presented case study and the costs will be changed according to the CO_2 emission price and generation cost. Comparing the cases 2 and 3 reveals the effectiveness of the CAES deployment in the system operation. From the comparison, the daily system cost is reduced by $ 17,136 (i.e., 0.7%). The percentage of cost reduction seems not significant. However, by means of cost–benefit evaluation and considering the investment cost of CAES, which is about 50,000 $/MWh and more than 30 years lifetime, the utilization of CAES systems is economically viable and shows an approximated rate of return of 20%. In the following, the cost comparison between cases 1 and 3 considering the impact of uncertainty realization is approved. Fig. 5.4 illustrates the system cost changes versus the robustness function for case 3. From Fig. 5.4, being about 2% robust against undesired effects of wind generation uncertainty will charge $ 10,000. According to strategies, the system operator picks the required scheduling, considering cost evaluation based on Fig. 5.4. Any point on the line represents different scheduling regarding units' commitment. To clarify the effectiveness of the proposed optimization framework, Fig. 5.5 shows the generation dispatch of case 3 in the deterministic (the system operator takes the first strategy) and robust case (the last strategy taken by the system operator). The differences in the power generation between

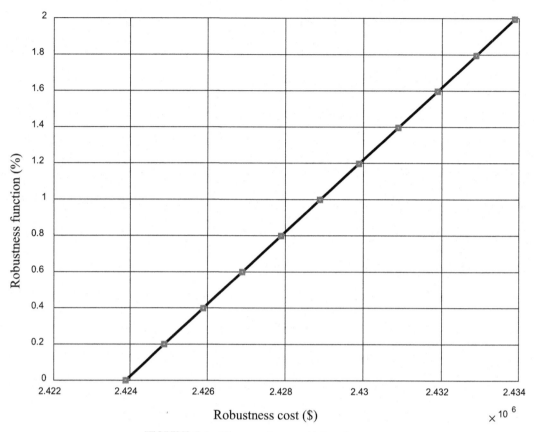

FIGURE 5.4 The exceed cost paid for robustness.

deterministic and robust cases are evident in hours 18 and 19. However, for the other hours, the differences are in the range of 1–10 MW, which cannot be seen accurately in Fig. 5.5, due to large scale requested power from generators. Another important point with Fig. 5.5 is the utilization of the high-cost generation units such as units #4, #6, and #10. In fact, the penalty cost of CO_2 emission is the main reason behind this kind of unit commitment. As the CCS of these units has eliminated a high percentage of the CO_2 dispersal (even full emission reduction occurs in some cases). Fig. 5.6 is intended to show the generation dispatch of case 1, without considering CCSs. For instance, in the base case, generation units #1 and #2 are fully deployed during the day; on the other hand, the generation unit #10 is not committed at all despite case 3. The differences between cases 1 and 3 are clear from Figs. 5.5 and 5.6.

As mentioned before, the IGDT method is employed to consider the effects of the uncertainty of the wind generation and the CAES units are in the vicinity of the wind turbines in the network to reduce the generation deviations. Hence, the charging and discharging pattern of the CAES systems would remarkably be dissimilar, which is shown in Fig. 5.7 for the deterministic and robust cases. The differences between deterministic and robust cases are crystal clear in Fig. 5.7.

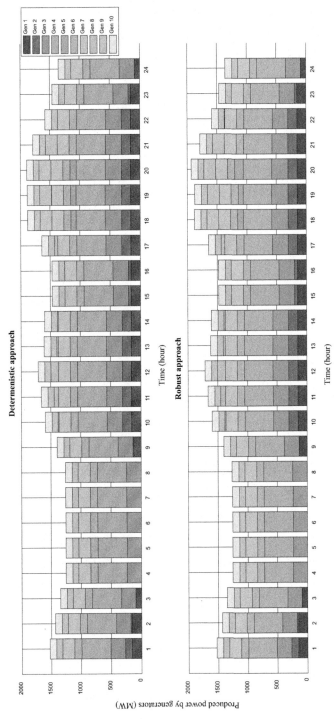

FIGURE 5.5 Generation dispatch in both deterministic and robust approached in case 3.

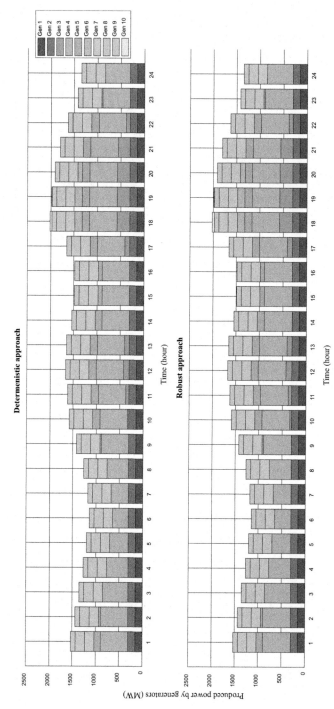

FIGURE 5.6 Generation dispatch in both deterministic and robust approached in the base case.

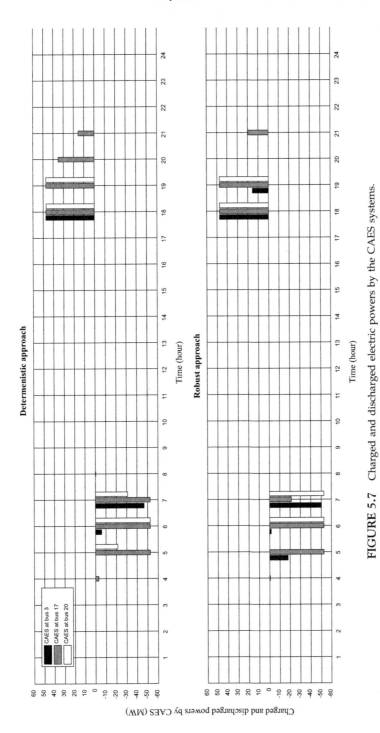

FIGURE 5.7 Charged and discharged electric powers by the CAES systems.

4. Conclusion

Power plants are the main source of CO_2 emission in the electrical industry. Considering the global tendency for carbon reduction, the power system should take decarbonization policies. One usual solution is utilizing the CCSs beside the fossil-fueled power plants that capture the carbon dioxide emission and lead to carbon reduction. However, the efficiency of CCPPs and the power consumption of the CCS are the main problems. On the other hand, the renewable generation sector is a promoting and advancing alternative for green power generation. The main challenge with these resources is their uncertain power output, which can be compensated with storage systems and strong uncertainty management decision-making tools. In this chapter, the effects of CCPPs and CAES systems operation on the system costs have been investigated. The uncertainty of the wind power generation was tackled with the IGDT technique and various decision-making strategies are developed according to the different risk-aversion levels. Deterministic and robust strategies are shown in this chapter to illustrate the effectiveness of the proposed method. Moreover, the effects of employing CCS were magnificent and resulted in high-cost savings due to emission reduction (approximately the system cost was reduced by 25%). Considering the high capital cost of CAES systems, the cost—benefit evaluation showed a 20% rate of return for the considered CAES systems for the studied cases. It is worth noting that the costs are directly related to the generation cost offered by power plants and penalty cost of emission, which can be various in different case studies.

References

[1] A. Akbari-Dibavar, B. Mohammadi-Ivatloo, Security interactions of food, water, and energy systems: a stochastic modeling BT - food-energy-water nexus resilience and sustainable development: decision-making methods, planning, and trade-off analysis, in: S. Asadi, B. Mohammadi-Ivatloo (Eds.), Cham, Springer International Publishing, 2020, pp. 305—321.

[2] S. Chen, Z. Guo, P. Liu, Z. Li, Advances in clean and low-carbon power generation planning, Comput. Chem. Eng. (2018), https://doi.org/10.1016/j.compchemeng.2018.02.012.

[3] J.H. Wee, A review on carbon dioxide capture and storage technology using coal fly ash, Appl. Energy (2013), https://doi.org/10.1016/j.apenergy.2013.01.062.

[4] Carbon Capture and Geological Storage | Climate Action." https://ec.europa.eu/clima/policies/innovation-fund/ccs_en (Accessed Sep. 12, 2019).

[5] E.S. Rubin, C. Chen, A.B. Rao, Cost and performance of fossil fuel power plants with CO2 capture and storage, Energy Pol. (2007), https://doi.org/10.1016/j.enpol.2007.03.009.

[6] Z. Ji, et al., Low-carbon power system dispatch incorporating carbon capture power plants, IEEE Trans. Power Syst. (2013), https://doi.org/10.1109/TPWRS.2013.2274176.

[7] P. Martens, E. Delarue, W. D'Haeseleer, A mixed integer linear programming model for a pulverized coal plant with post-combustion carbon capture, IEEE Trans. Power Syst. (2012), https://doi.org/10.1109/TPWRS.2011.2173506.

[8] A.S. Brouwer, M. van den Broek, A. Seebregts, A. Faaij, Operational flexibility and economics of power plants in future low-carbon power systems, Appl. Energy 156 (2015) 107—128.

[9] G.P. Hammond, S.S.O. Akwe, S. Williams, Techno-economic appraisal of fossil-fuelled power generation systems with carbon dioxide capture and storage, Energy 36 (2) (2011) 975—984.

[10] S. ReddyK, L. Panwar, B.K. Panigrahi, R. Kumar, Low carbon unit commitment (LCUC) with post carbon capture and storage (CCS) technology considering resource sensitivity, J. Clean. Prod. (2018), https://doi.org/10.1016/j.jclepro.2018.07.195.

[11] A.M. Abdilahi, M.W. Mustafa, S.Y. Abujarad, M. Mustapha, Harnessing flexibility potential of flexible carbon capture power plants for future low carbon power systems, Renew. Sustain. Energy Rev. 81 (2018) 3101–3110.

[12] A.S. Brouwer, M. van den Broek, W. Zappa, W.C. Turkenburg, A. Faaij, Least-cost options for integrating intermittent renewables in low-carbon power systems, Appl. Energy (2016), https://doi.org/10.1016/j.apenergy.2015.09.090.

[13] M. McPherson, N. Johnson, M. Strubegger, The role of electricity storage and hydrogen technologies in enabling global low-carbon energy transitions, Appl. Energy (2018), https://doi.org/10.1016/j.apenergy.2018.02.110.

[14] B. Lyseng, et al., System-level power-to-gas energy storage for high penetrations of variable renewables, Int. J. Hydrogen Energy (2018), https://doi.org/10.1016/j.ijhydene.2017.11.162.

[15] L. He, Z. Lu, J. Zhang, L. Geng, H. Zhao, X. Li, Low-carbon economic dispatch for electricity and natural gas systems considering carbon capture systems and power-to-gas, Appl. Energy (2018), https://doi.org/10.1016/j.apenergy.2018.04.119.

[16] X. Li, R. Zhang, L. Bai, G. Li, T. Jiang, H. Chen, Stochastic low-carbon scheduling with carbon capture power plants and coupon-based demand response, Appl. Energy (2018), https://doi.org/10.1016/j.apenergy.2017.08.119.

[17] D. Pudjianto, M. Aunedi, P. Djapic, G. Strbac, Whole-systems assessment of the value of energy storage in low-carbon electricity systems, IEEE Trans. Smart Grid (2014), https://doi.org/10.1109/TSG.2013.2282039.

[18] R. Zhang, et al., Adjustable robust power dispatch with combined wind-storage system and carbon capture power plants under low-carbon economy, Int. J. Electr. Power Energy Syst. (2019), https://doi.org/10.1016/j.ijepes.2019.05.079.

[19] M. Abbaspour, M. Satkin, B. Mohammadi-Ivatloo, F. Hoseinzadeh Lotfi, Y. Noorollahi, Optimal operation scheduling of wind power integrated with compressed air energy storage (CAES), Renew. Energy (2013), https://doi.org/10.1016/j.renene.2012.09.007.

[20] M.A. Mirzaei, A.S. Yazdankhah, B. Mohammadi-Ivatloo, M. Marzband, M. Shafie-khah, J.P.S. Catalão, Stochastic network-constrained co-optimization of energy and reserve products in renewable energy integrated power and gas networks with energy storage system, J. Clean. Prod. (2019), https://doi.org/10.1016/j.jclepro.2019.03.021.

[21] S. Shafiee, H. Zareipour, A.M. Knight, N. Amjady, B. Mohammadi-Ivatloo, Risk-constrained bidding and offering strategy for a merchant compressed air energy storage plant, IEEE Trans. Power Syst. (2017), https://doi.org/10.1109/TPWRS.2016.2565467.

[22] A. Dolatabadi, M. Jadidbonab, B. Mohammadi-Ivatloo, Short-term scheduling strategy for wind-based energy hub: a hybrid stochastic/IGDT approach, IEEE Trans. Sustain. Energy (2019), https://doi.org/10.1109/TSTE.2017.2788086.

[23] S. Nojavan, A. Akbari-Dibavar, K. Zare, Optimal energy management of compressed air energy storage in day-ahead and real-time energy markets, IET Gener., Transm. Distrib. 13 (16) (Jun. 2019) 3673–3679 [Online]. Available: https://digital-library.theiet.org/content/journals/10.1049/iet-gtd.2018.7022. (Accessed 23 August 2019).

[24] B. Mohammadi-Ivatloo, H. Zareipour, N. Amjady, M. Ehsan, Application of information-gap decision theory to risk-constrained self-scheduling of GenCos, IEEE Trans. Power Syst. (2013), https://doi.org/10.1109/TPWRS.2012.2212727.

[25] G. Cplex, The Solver Manuals. Gams/Cplex, 2014.

[26] A. Soroudi, Power System Optimization Modeling in GAMS, 2017.

Energy vehicles as means of energy storage: impacts on energy markets and infrastructure

Mahsa Bagheri Tookanlou[1], Mousa Marzband[1,2], Ameena Saad Al-Sumaiti[3], Somayeh Asadi[4]

[1]Faculty of Engineering and Environment, Department of Maths, Physics and Electrical Engineering, Northumbria University Newcastle, Newcastle upon Tyne, United Kingdom; [2]Center of Research Excellence in Renewable Energy and Power Systems, King Abdulaziz University, Jeddah, Saudi Arabia; [3]Advanced Power and Energy Center, Department of Electrical Engineering and Computer Science, Khalifa University, Abu Dhabi, United Arab Emirates; [4]Department of Architectural Engineering, Pennsylvania State University, State College, PA, United States

Nomenclature

$\overline{f}/f$ Maximum number of iterations/Number of iterations
$\overline{N}/N$ Maximum number/Number
$\overline{SOC}/\underline{SOC}$ Maximum/Minimum SOC
c Per-unit capacity cost
E Capacity (kWh)
C Cost ($)
D Battery degradation
L Location
P Power (kW)
R Revenue ($)
SOC State of charge of battery
Δt Time step (s)
$u/\overline{u}$ Number of chargers/available chargers
U Voltage (V)
$\underline{U}/\overline{U}$ Minimum/Maximum voltage of buses
W Waiting time
N^S, N^e, N^R Number of EVCSs, EVs, and trips
S_1, S_2 Acceleration coefficients

Energy Storage in Energy Markets
https://doi.org/10.1016/B978-0-12-820095-7.00016-9

r_1, r_2 Random numbers between 0 and 1
$\mathscr{B}$ Best individual solutions
$\mathscr{G}$ Best global solutions
u_0 Inertia weight
X Position
u_1, u_2 Initial and final value of the inertia weight
s_1^i, s_2^i Initial values of acceleration coefficients
s_1^f, s_2^f Final values of acceleration coefficients

Superscript

A Aggregator
B Battery
C Charger
CS Charging station
CY Cycling
D Departure
Des Destination
DG Degradation
EV EV
g Natural gas
M Market
N Nominal
P Operation
PV PV system
R Trip
Re Real
RQ Required
S Energy storage system
U Generation unit
initial Initial
OR EV's origin
? Charging/Discharging

Subscript

t Hours of a day
s Index of charging station
k Number of particles
j Index of chargers
e Index of EV
p.u Per-unit
r Index of trip

Greek symbols

$\overline{\pi}/\underline{\pi}$ Maximum/Minimum electricity price ($/kWh)
α Harmonic current
β Reliability coefficient of chargers
η Efficiency
J Overall correction coefficient of a CS
λ Simultaneity coefficient of the chargers
π Electricity price ($/kWh)
ρ Power factor
τ Heat value (kWh/m^3)
ζ υ

1. Introduction

Electric vehicles (EVs) have played a significant role in reducing greenhouse gas emissions into the environment [1–6]. However, large-scale EV charging demand has significant challenges for the operation of distribution network. The power system may encounter significant technical challenges [7–13] including voltage deviation, increasing power losses, and overloading of power lines [14–18]. Since several studies showed that the coordinated both charging and discharging operation of EVs can be advantageous for the distribution network, it is one of the indispensable challenges [19–24]. These challenges are due to managing the huge power required for charging the large number of EVs. Also, the problem becomes more complicated when there are multiple electric vehicle charging stations (EVCSs), which can be selected for charging and discharging operation because EVs should choose proper EVCSs to meet their driving requirements. Therefore, an optimal strategy for scheduling EVCSs and EVs as well as determining an economic strategy for choosing proper EVCSs by EVs in charging and discharging modes is needed. A scheme is required for charging and discharging EVs to minimize the cost of EVs and maximize the income of the EVCSs.

Recently, different studies have focused on charging and discharging operation of EVs. Optimal charging/discharging schedule of EVs is determined with [14,22,25–27] or without [28–32] considering its impact on the distribution network constraints. In Refs. [29,30], charging/discharging management of EVs has been done such that it satisfied the costumers' concerns. In Refs. [32,33], charging schedules of electric freight vehicles were modeled and analyzed that operated fixed delivery routes and performed several routes per day. A mathematical model that includes numerous attributes was presented. The features were related to the use of electric freight vehicles including a realistic process of charging, energy costs, battery aging, restrictions of electricity network, and facility-related demand charges. In Ref. [34], a multiobjective optimization is proposed for scheduling EV charging/discharging. Simultaneous optimization of electricity cost, battery degradation, grid net exchange, and CO_2 emissions has been performed. In Refs. [35–38], an economical and technical charging/discharging strategy that mainly focuses on finding proper charging stations by EV owners is developed. In Ref. [39], a strategy is proposed to find the EVCS based on the minimization of the travel time, waiting time, and cost of an EV in charging mode. Hosting capacity of the distribution network was evaluated in Refs. [14,40,41] based on a two-stage model considering technical constraints including bus voltage and line capacity. In Ref. [42], an optimization model was presented to determine EV charging at EVCSs with different types of chargers. Planning of EVCSs is obtained in Ref. [43] by proposing a mixed-integer linear programming (MILP) model. In Ref. [44], EV charging scheduling was studied. The objective is to minimize the cost of EV battery aging such that the features of EV battery charging are satisfied. In Ref. [45], a smart scheduling approach was presented for planning EVCSs and minimizing the total time of travel for each EV. Waiting times and overall travel times decrease significantly using the proposed approach. The review of the available literature indicates that studies related to both charging and discharging operation as well as EVCSs operation do not present optimal solutions for EVs and EVCSs simultaneously [14,29,30,39,40,42,43]. The optimal electricity prices in charging and discharging mode as well as combined charging and discharging operation with the minimum driving distance have not been determined such that the rewards of EVs and EVCSs are guaranteed [29,39].

This study aims to propose a holistic scheme for charging and discharging scheduling to meet rewards of EVs and EVCSs simultaneously. In this study, electricity prices for charging and discharging EVs are determined. Then, the minimum driving route and electricity prices for discharging and charging operation are considered to determine EVs charging/discharging mode. It is done by taking into account the benefits of EVs and EVCSs simultaneously as a bilevel optimization problem. It is assured that the technical constraints of EVCSs and also EVs' concerns are met at all times.

The rest of this chapter is organized as follows. Section 2 describes the overall picture of study. Section 3 presents the proposed bilevel optimization formulation. Case study and simulation results are presented and discussed in Section 4. Finally, in Section 5, conclusion is given.

2. Overall picture of study

While a wider utilization of EVs is encouraged, main obstacles prevent faster adoption of the new technology, such as coordinated/planned charging of EVs, coordinated/planned EV discharging to add a new revenue stream for the EVs, competitive prices for EVs and EVCSs during charging and discharging. As a result, advanced mechanisms are needed to guarantee the benefits of EVs and EVCSs.

In this study, EVCSs are charging stations responsible to provide electricity for EVs. This way, electricity from the distribution network will be delivered to EVs. EVs should choose their energy suppliers based on economic benefits. EVCSs can provide electricity for EVs from different resources including a generation unit, photovoltaic, and energy storage system (ESS). The generation unit that utilizes natural gas and the PV system produce electricity. EVCSs are able to sell excess power to the wholesale market if power produced by the generation unit and the PV system is more than the energy required for charging EV and ESS. Thus, we are assuming an aggregator whose business is to purchase energy from EVCSs and sell it in the market. However, if the electricity produced in EVCSs is less than the required energy for charging EV, the required electricity is provided by the distribution network. We are assuming that during a typical day, a number of EVs has two trips with different waiting times between the first and the second trip. At each hour, EVs plan their charging and discharging operation depending on the minimum driving route and economic analysis. Thus, EVs select proper EVCSs for charging or discharging operation. The driving routes for EVs to reach their destination in each trip and EVCSs are determined by network analyst toolbar of ArcGIS.

In this study, a scheduling system is responsible for collecting the required data from EVs and EVCSs. It is assumed that they communicate with scheduling system and send information to it for running the bilevel optimization problem. Then, the results of solving the bilevel optimization problem are sent to corresponding EVs and EVCSs. The data exchanged between EVs and EVCSs are mentioned in Table 6.1.

The data regarding the location of EVs and the initial state of charge (SOC) of EVs, minimum SOC of the EVs at their destination, and the waiting time between trips of the EVs are sent to the scheduling system for running the bilevel optimization problem. The driving route between the location and the destination of EV e, the location of EV e and the EVCS s, and the destination of EV e and the location of EVCS s are determined in the scheduling system by ArcGIS. Then, the driving distances corresponding to the routes are used to calculate the battery SOC required to drive each route.

TABLE 6.1 Input parameters and decision variables for EVs and EVCSs.

	Input	Output
EVs	$L_{t,e,r}^{OR}$, $L_{t,e,r}^{Des}$, $SOC_{t,e}^{initial}$, $\underline{SOC}_{t,e,r}^{Des}$, $W_{t,e,r}$	$P_{t,s,e,r}^{+}$, $P_{t,s,e,r}^{-}$, selected EVCS
EVCSs	L_s^{CS}, $u_{t,s}^{C}$, $E_{t,s,j}^{C}$	$\pi_{t,s}^{-}$, $\pi_{t,s}^{+}$, $P_{t,s}^{A}$, $P_{t,s}^{U}$, $P_{t,s}^{WM}$

The optimization problem for EVs and EVCSs is solved in a bilevel optimization problem to guarantee the reward of inner level (EVs) and upper level (EVCSs) simultaneously. The formulation of the two optimization problems and the optimization algorithm are described in Section 3. The first optimization problem will be solved in inner level and electricity purchased/sold from/to EVCSs in each trip for each hour will be determined. The optimal value of electricity traded between EVs and EVCSs will be sent to upper level and the optimization problem in this level will be solved to determine the power produced by conventional generation units of EVCSs, and the electricity purchased from the wholesale market, and power sold to the aggregator. Also, the electricity prices sold to EVs by EVCSs are determined in upper level.

3. Problem formulation

In this section, the objective functions and constraints for inner and upper level of the bilevel optimization problem are presented.

3.1 Optimization problem at inner level

The objective function and constraints in inner level are explained in this section.

3.1.1 Objective function in inner level

The total cost of EVs must be minimized at inner level, which is the difference between the cost of EVs including electricity bought from EVCSs and the revenue from selling electricity to EVCSs and also battery degradation cost during discharging period.

$$C^{EV} = \sum_{t=1}^{24}\sum_{s=1}^{N^s}\sum_{e=1}^{N^e}\sum_{r=1}^{N^R}C_{t,s,e,r}^{+EV} + \sum_{t=1}^{24}\sum_{k=1}^{N^e}\sum_{r=1}^{N^R}C_{t,e,r}^{DG} - \sum_{t=1}^{24}\sum_{s=1}^{N^s}\sum_{e=1}^{N^e}\sum_{r=1}^{N^R}R_{t,s,e,r}^{-EV} \qquad (6.1)$$

The cost of electricity purchased from EVCSs by EVs is obtained as the product of power required for charging EVs and the electricity price offered by EVCSs.

$$C_{t,s,e,r}^{+EV} = P_{t,s,e,r}^{+EV} \times \pi_{t,s}^{+} \qquad (6.2)$$

The battery degradation cost during discharging operation of EVs is obtained by Ref. [46]:

$$C_{t,e,r}^{DG} = c_{p.u}^{B} \times E_e^N \times \frac{D_{t,e,r}^C}{E_e^N - E_e^R} \tag{6.3}$$

$$E_e^R = 0.8 \times E_e^N \tag{6.4}$$

The revenue of EVs from selling electricity to EVCSs is obtained by

$$R_{t,s,e,r}^{-EV} = P_{t,s,e,r}^{-EV} \times \pi_{t,s}^- \tag{6.5}$$

3.1.2 Constraints in inner level

For EV e with charging or discharging mode, at the departure time from the EVCS, the SOC of EVs must not be less than the required SOC of EV.

$$SOC_{t,e,r}^D \geq SOC_{t,e,r}^{RQ} \tag{6.6}$$

For EV e, during the charging period, the SOC of batteries must not exceed a maximum value, and in discharging mode, the batteries must not be discharged completely, which means SOC must not be less than a minimum value.

$$\underline{SOC} \leq SOC_{t,e,r} \leq \overline{SOC} \tag{6.7}$$

The SOC of EV e after charging/discharging in each trip for each hour can be determined by

$$SOC_{t,e,r} = SOC_{t-1,e,r} + \frac{P_{t,e,r}^{+EV} \times \eta^{B+} \times \Delta t}{E_e^{EV}} - \frac{P_{t,e,r}^{-EV} \times \Delta t}{E_e^{EV} \times \eta^{B-}} \tag{6.8}$$

The SOC of EV e must be more than the minimum SOC at the final destination, which is obtained by

$$SOC_{t,e,r}^D \geq \underline{SOC}_{t,e,r}^{Des} \tag{6.9}$$

There is a limitation for electricity power purchased/sold from/to EVCSs by EVs.

$$0 \leq P_{t,e,r}^{+EV} \leq E_{t,s,j}^C \tag{6.10}$$

$$0 \leq P_{t,e,r}^{-EV} \leq E_{t,s,j}^C \tag{6.11}$$

For each hour, EV e must be only allocated in one of the discharging and charging operation.

$$P_{t,e,r}^{+\text{EV}} \times P_{t,e,r}^{-\text{EV}} = 0 \tag{6.12}$$

3.2 Optimization problem at upper level

The objective function and constraints in upper level are explained in this section.

3.2.1 Objective function in upper level

In this study, if power produced and energy stored in ESS are not enough to supply electricity for charging of EVs, the electricity is purchased from the wholesale market. Also, they purchase electricity from EVs during discharging period and sell it to the market. The objective function in this level consists of the net revenue of EVCSs, which is the difference between the revenue and the costs of EVCSs, which include the operation cost and the cost of purchasing electricity from EVs.

$$R^{\text{CS}} = \sum_{t=1}^{24} \sum_{s=1}^{N^S} R_{t,s}^A + \sum_{t=1}^{24} \sum_{s=1}^{N^S} \sum_{e=1}^{N^e} \sum_{r=1}^{N^R} R_{t,s,e,r}^{-\text{CS}} - \sum_{t=1}^{24} \sum_{s=1}^{N^{CS}} C_{t,s}^M - \sum_{t=1}^{24} \sum_{s=1}^{N^S} \sum_{e=1}^{N^e} \sum_{r=1}^{N^R} C_{t,s,e,r}^{+\text{CS}} + C^P \tag{6.13}$$

The revenue of selling electricity to the aggregator by EVCSs is given by

$$R_{t,s}^A = P_{t,s}^A \times \pi_{t,s}^A \tag{6.14}$$

The revenue of selling electricity to EVs by EVCSs is given by

$$R_{t,s,e,r}^{-\text{CS}} = P_{h,s,e,r}^{-\text{CS}} \times \pi_{t,s}^+ \tag{6.15}$$

The operation cost is obtained by:

$$C^P = C^{P,\text{CH}} + C^{P,\text{U}} \tag{6.16}$$

$$C^{P,\text{CH}} = \sum_{s=1}^{N^S} c_{p.u,s} \times \lambda_s \times J_s \sum_{j=1}^{N_s^C} \beta_{s,j} \times \alpha_{s,j} \times \frac{P_{s,j}^C}{\eta_{s,j}^C \times \rho_{s,j}^C} \tag{6.17}$$

$$C^{P,\text{U}} = \sum_{t=1}^{24} \sum_{s=1}^{N^S} \frac{P_{t,s}^U \times \pi_h^{\text{gas}}}{\eta_{t,s}^U \times \tau} \tag{6.18}$$

The cost of purchased electricity from the market by EVCSs is determined as product of electricity purchased from the market and the electricity price, as given by

$$C_{t,s}^{M} = P_{t,s}^{+CS} \times \pi_t^{M} \qquad (6.19)$$

The cost of purchased electricity from EVs by EVCSs is determined as product of power sold to EVCSs by EVs and the electricity price offered by EVCSs, as obtained by

$$C_{t,s,e,r}^{+CS} = P_{t,s,e,r}^{+CS} \times \pi_{t,s}^{-} \qquad (6.20)$$

3.2.2 Constraints in upper level

The power balance within an EVCS should be fulfilled at all times, which is achieved by:

$$P_{t,s}^{PV} + P_{t,s}^{U} \pm P_{t,s}^{S\pm} + \sum_{s=1}^{N^e} P_{t,s,e,r}^{+CS} + P_{t,s}^{M} = P_{t,s}^{A} + \sum_{e=1}^{N^e} P_{t,s,e,r}^{-CS} \qquad (6.21)$$

The number of used chargers in an EVCS must not be more than the number of available chargers in that station.

$$u_{t,s}^{C} \leq \overline{u}^{C} \qquad (6.22)$$

The SOC of the ESS for each EVCS is limited by the maximum and minimum value

$$\underline{SOC}^{S} \leq SOC_{t,s}^{S} \leq \overline{SOC}^{S} \qquad (6.23)$$

It is not possible to charge and discharge the ESS at the same time.

$$P_{t,s}^{+S} \times P_{t,s}^{-S} = 0 \qquad (6.24)$$

The electricity produced by a generation unit must be maintained within a lower and upper bound

$$\underline{P}^{U} \leq P_{t,s}^{U} \leq \overline{P}^{U} \qquad (6.25)$$

The electricity bought from the market by EVCS s is limited by

$$P_{t,s}^{M} \leq E_s^{CS} \qquad (6.26)$$

The electricity prices offered by EVCSs during charging and discharging are limited by a minimum and maximum values.

$$\underline{\pi}^{+} \leq \pi_{t,s}^{+} \leq \overline{\pi}^{+} \qquad (6.27)$$

$$\underline{\pi}^- \leq \pi^-_{t,s} \leq \overline{\pi}^- \tag{6.28}$$

The bus voltage must be within permissible range as:

$$\underline{U} \leq |U_{t,s}| \leq \overline{U} \tag{6.29}$$

3.3 Optimization algorithm

Solving the bilevel optimization problem starts from inner level. In inner level, a solution is obtained and used to find a solution for upper level. An evolutionary technique, called particle swarm optimization (PSO), is utilized in this study presented in Section 3. PSO algorithm is an evolutionary computation technique of particles inspired by social behavior of a flock of birds searching for food. Each bird's position is a solution (particle) for the optimization problem. The objective function of optimization problem determines the fitness value of each particle. The velocity and position of each particle are updated by the best individual solutions and the best global solutions in each iteration.

$$v_k(t+1) = \zeta[u_0 \times v_k(t) + S_1 \times r_1 \times (\mathscr{B}_k(t) - X_k(t)) + S_2 \times r_2 \times (\mathscr{G}_k(t) - X_k(t))] \tag{6.31}$$

$$X_k(t+1) = X_k(t) + v_k(t+1) \tag{6.32}$$

The balance between global and local search is dealt with u_0 whose value is obtained based on the global and local search. For each iteration, u_0 is determined as:

$$u_0 = (u_1 - u_2) \times \left(\frac{\overline{f} - f}{\overline{f}}\right) + u_2 \tag{6.33}$$

$$u_1 = 0.9, \quad u_2 = 0.4$$

S_1 and S_2 are considered as time varying in order to prevent convergence in the initial iterations. At first, S_1 is more than S_2. In the process of search space, S_1 rises and S_2 decreases as:

$$S_1 = \left(s_1^f - s_1^i\right)\frac{f}{\overline{f}} + S_1^i \tag{6.34}$$

$$S_2 = \left(s_2^f - s_2^i\right)\frac{f}{\overline{f}} + S_2^i \tag{6.35}$$

The coefficient of ζ is determined to improve convergence of PSO algorithm as:

$$\zeta = \frac{2}{\left|2 - \mu - \sqrt{\mu^2 - 4\mu}\right|} \tag{6.36}$$

$$\mu = S_1 + S_2 \tag{6.37}$$

$5 \times N^S \times 24$ decision variables are considered that correspond to the power produced by conventional generation unit, power bought from the market, and power sold to the aggregator, electricity prices for charging and discharging for 24 h ahead. In inner level, $2 \times 24 \times N^e$ decision variables are considered for the power sold/purchased to/from EVs.

4. Case study and simulation results

For simulation purposes and to visualize and validate the proposed model developed in Section 3, the test system and simulation results are presented in this section.

4.1 Test system

Six EVCSs are installed in San Diego where 400 EVs with 28 and 40 kWh battery capacity are charged and discharged. There are fast chargers with 50 kW capacity. Two trips are allocated for each EV during a day. EVs have different waiting times. EVCSs are connected to IEEE test system with 37 buses. The voltage of distribution network is 480 V and the permissible voltage amplitudes must be between 0.95 and 1.05. The electricity prices are extracted from the California Independent System Operator. Expenses regarding system maintenance, taxes, and ancillary services are considered by a coefficient of 4.5 for prices of the wholesale market to get the electricity prices purchased by the market by EVCSs. It is assumed that the electricity prices sold to EVCSs by EVs are 60%–85% less than what EVCSs pay to the market as:

$$\pi_{t,s}^- = \pi_t^M \times rand(0.6, 0.85) \tag{6.38}$$

The electricity price sold to the aggregator by EVCSs in discharging mode is 10% more than the electricity prices purchased from EVs as:

$$\pi_{t,s}^A = 1.1 \times \pi_{t,s}^- \tag{6.39}$$

4.2 Simulation results

In this section, simulation results will be discussed for the case study presented in Subsection 4.1. The optimal electricity prices offered by six EVCSs for all time are shown in Tables 6.2 and 6.3 for charging and discharging EVs, respectively. The revenue of EVCS3 and EVCSs 6 are the most and the least for a day, respectively, as highlighted in Tables 6.2 and 6.3. The

TABLE 6.2 Electricity prices during charging mode for all EVCSs and times.

	Electricity prices (cents/kWh)					
Time	1	2	3	4	5	6
$t = 1$	24.3	25.1	24.4	22.3	21.4	22.2
$t = 2$	24.2	22.9	22.6	22.2	21.3	22.1
$t = 3$	22.9	19.8	22.6	21.6	21.1	20.2
$t = 4$	22.7	19.7	20.3	21.5	21.1	20.2
$t = 5$	23.6	25.1	22.1	21.6	24.6	23.0
$t = 6$	23.6	25.2	24.7	23.7	26.3	26.5
$t = 7$	28.5	26.4	27.0	29.1	27.9	28.3
$t = 8$	27.0	23.8	25.1	26.2	23.6	23.7
$t = 9$	20.4	23.0	22.3	19.9	21.4	21.8
$t = 10$	20.4	18.5	19.6	19.9	18.9	17.2
$t = 11$	20.4	22.1	19.6	19.9	23.0	20.0
$t = 12$	22.3	22.2	21.2	23.5	23.1	20.4
$t = 13$	22.4	24.1	26.2	23.5	23.9	22.6
$t = 14$	27.6	27.2	26.2	24.8	27.3	22.6
$t = 15$	27.6	27.2	26.3	24.9	27.4	23.6
$t = 16$	27.7	27.8	28.4	26.5	27.4	27.4
$t = 17$	30.0	29.6	28.4	27.5	27.5	28.7
$t = 18$	30.0	30.4	29.8	30.7	31.2	30.4
$t = 19$	42.3	41.1	47.1	45.2	46.2	40.8
$t = 20$	42.2	40.6	42.0	42.7	38.2	39.7
$t = 21$	30.8	31.8	30.0	32.8	32.2	34.1
$t = 22$	28.9	29.2	27.7	28.5	25.8	30.4
$t = 23$	25.1	27.1	25.0	25.5	25.1	28.4
$t = 24$	25.1	23.8	23.6	25.5	24.4	28.0

number of EVs selecting each EVCS during charging and discharging mode is shown in Figs. 6.1 and 6.2.

In Table 6.4, the optimal value of the cost of EVs and revenue of EVCSs are reported for a case in which the objective function of each level is optimized individually. In addition, the optimal value of the cost of EVs and revenue of EVCSs that is obtained by solving the bilevel optimization problem are mentioned in Table 6.4. It can be seen that when the bilevel

TABLE 6.3 Electricity prices during discharging mode for all EVCSs and times.

Time	Electricity prices (cents/kWh)					
	1	2	3	4	5	6
t = 1	12.2	12.5	12.9	11.8	12.8	12.1
t = 2	11.8	11.4	11.3	11.7	11.7	11.5
t = 3	11.1	11.3	10.8	11.1	11.4	11.1
t = 4	11.0	10.2	10.5	10.7	11.2	10.4
t = 5	11.5	11.5	10.5	11.7	11.5	10.8
t = 6	13.0	12.5	12.6	12.2	11.9	12.5
t = 7	15.2	14.6	14.0	14.1	15.1	14.6
t = 8	13.4	13.5	13.5	13.3	12.7	13.6
t = 9	9.4	9.9	10.0	9.1	9.2	10.1
t = 10	9.1	9.2	8.8	9.0	9.2	8.5
t = 11	10.3	9.5	10.2	10.3	10.1	9.4
t = 12	11.1	11.1	10.2	10.6	10.9	10.3
t = 13	11.6	11.7	10.5	11.3	10.9	10.4
t = 14	13.5	12.7	12.4	12.9	12.8	12.5
t = 15	13.6	12.8	13.5	12.9	13.2	13.5
t = 16	13.6	13.8	13.7	13.6	13.4	13.8
t = 17	13.9	14.0	14.5	14.4	14.7	14.0
t = 18	17.8	18.1	18.3	17.7	17.9	17.7
t = 19	26.7	27.2	26.5	27.6	26.5	26.4
t = 20	24.0	24.2	24.6	24.3	24.2	23.8
t = 21	18.8	18.4	19.0	18.6	17.9	18.8
t = 22	16.0	14.9	15.0	15.7	15.7	15.8
t = 23	14.0	13.7	13.5	14.5	14.4	13.6
t = 24	13.1	12.9	12.2	11.9	13.1	12.6

optimization problem is considered for charging and discharging of EVs, in comparison with the case of individual optimization, the cost of EVs decreased from 1440.6 ($) to 1210.6 ($) and the revenue of EVCSs increased from 242.0 ($) to 318.4 ($). Therefore, there is a decrease in the costs of EVs by 19% and an increase in the revenue of EVCSs by 24%, respectively.

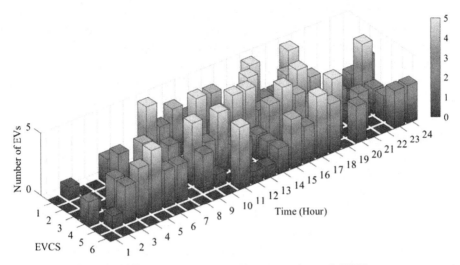

FIGURE 6.1 The number of EVs charged in each EVCS.

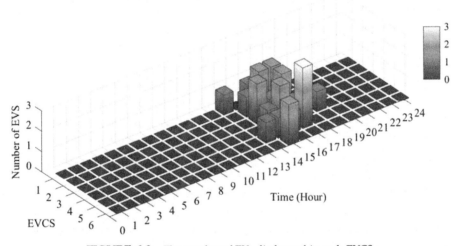

FIGURE 6.2 The number of EVs discharged in each EVCS.

TABLE 6.4 Simulation results for the bilevel optimization problem and individual problem.

Level	Type of optimization problem	Optimal value
Objective function of inner level	Bilevel problem	1210.6 ($)
	Individual problem	1440.6 ($)
Objective function of outer level	Bilevel problem	318.4 ($)
	Individual problem	242.0 ($)

5. Conclusions

In this study, a scheduling scheme for charging and discharging operation of EVs is presented, in which each EV finds optimal EVCSs, for charging and discharging. EVs plan their charging and discharging operation depending on the minimum driving route and economic analysis based on the prices offered by EVCSs. The benefits of EVs and EVCSs in EV charging/discharging are analyzed by a bilevel optimization problem. A scheduling system is considered to communicate with all EVs and EVCs, run the bilevel optimization problem, and send the results to them. Optimal electricity prices offered by EVCSs for charging/discharging EV are determined.

References

[1] B. Amirhosseini, S.H. Hosseini, Scheduling charging of hybrid-electric vehicles according to supply and demand based on particle swarm optimization, imperialist competitive and teaching-learning algorithms, Sustain. Cities Soc. 43 (2018) 339–349.

[2] S. Bellocchi, K. Klockner, M. Manno, M. Noussan, M. Vellini, On the role of electric vehicles towards low-carbon energy systems: Italy and Germany in comparison, Appl. Energy 255 (2019) 1–17.

[3] S. Kufeoglu, D.K.K. Hong, Emissions performance of electric vehicles: a case study from the United Kingdom, Appl. Energy 260 (2020) 1–15.

[4] B. Skugor, J. Deur, A bi-level optimisation framework for electric vehicle fleet charging management, Appl. Energy 184 (2016) 1332–1342.

[5] M. Tavakoli, F. Shokridehaki, M. Marzband, R. Godina, E. Pouresmaeil, A two stage hierarchical control approach for the optimal energy management in commercial building microgrids based on local wind power and pevs, Sustain. Cities Soc. 41 (2018) 332–340.

[6] V. Aryanpur, M.S. Atabaki, M. Marzband, P. Siano, K. Ghayoumi, An overview of energy planning in Iran and transition pathways towards sustainable electricity supply sector, Renew. Sustain. Energy Rev. 112 (2019) 58–74.

[7] Y. Zhao, X. He, Y. Yao, J. Huang, Plug-in electric vehicle charging management via a distributed neurodynamic algorithm, Appl. Soft Comput. 80 (2019) 557–566.

[8] K. Clement-Nyns, E. Haesen, J. Driesen, The impact of vehicle-to-grid on the distribution grid, Elec. Power Syst. Res. 81 (2011) 185–192.

[9] H.R. Gholinejad, A. Loni, J. Adabi, M. Marzband, A hierarchical energy management system for multiple home energy hubs in neighborhood grids, J. Build. Eng. 28 (2020) 101028.

[10] M. Marzband, E. Yousefnejad, A. Sumper, J.L. Domínguez-García, Real time experimental implementation of optimum energy management system in standalone microgrid by using multi-layer ant colony optimization, Int. J. Electr. Power Energy Syst. 75 (2016) 265–274.

[11] M. Marzband, M. Ghadimi, A. Sumper, J.L. Domínguez-García, Experimental validation of a real-time energy management system using multi-period gravitational search algorithm for microgrids in islanded mode, Appl. Energy 128 (2014) 164–174.

[12] H.J. Monfared, A. Ghasemi, A. Loni, M. Marzband, A hybrid price-based demand response program for the residential micro-grid, Energy 185 (2019) 274–285.

[13] M. Marzband, S.S. Ghazimirsaeid, H. Uppal, T. Fernando, A real-time evaluation of energy management systems for smart hybrid home microgrids, Elec. Power Syst. Res. 143 (2017) 624–633.

[14] J. Zhao, J. Wang, Z. Xu, C. Wang, C. Wan, C. Chen, Distribution network electric vehicle hosting capacity maximization: a chargeable region optimization model, IEEE Trans. Power Syst. 32 (2017) 4119–4130.

[15] M. Nazari-Heris, M.A. Mirzaei, B. Mohammadi-Ivatloo, M. Marzband, S. Asadi, Economic-environmental effect of power to gas technology in coupled electricity and gas systems with price-responsive shiftable loads, J. Clean. Prod. 244 (2020) 118769.

[16] M. Jadidbonab, B. Mohammadi-Ivatloo, M. Marzband, P. Siano, Short-term self-scheduling of virtual energy hub plant within thermal energy market, IEEE Trans. Ind. Electron. 68 (2020) 3124–3136.

[17] M.A. Mirzaei, A. Sadeghi-Yazdankhah, B. Mohammadi-Ivatloo, M. Marzband, M. Shafie-khah, J. ao P.S. Catalão, Integration of emerging resources in IGDT-based robust scheduling of combined power and natural gas systems considering flexible ramping products, Energy 189 (2019) 116195.

[18] M. Pourakbari-Kasmaei, M. Lehtonen, M. Fotuhi-Firuzabad, M. Marzband, J.R.S. Mantovani, Optimal power flow problem considering multiple-fuel options and disjoint operating zones: a solver-friendly MINLP model, Int. J. Electr. Power Energy Syst. 113 (2019) 45–55.

[19] P. Jain, A. Das, T. Jain, Aggregated electric vehicle resource modelling for regulation services commitment in power grid, Sustain. Cities Soc. 45 (2019) 439–450.

[20] L. Wang, S. Sharkh, A. Chipperfield, Optimal decentralized coordination of electric vehicles and renewable generators in a distribution network using A* search, Int. J. Electr. Power Energy Syst. 98 (2018) 474–487.

[21] M.R. Mozafar, M.H. Amini, M.H. Moradi, Innovative appraisement of smart grid operation considering large-scale integration of electric vehicles enabling V2G and G2V systems, Elec. Power Syst. Res. 154 (2018) 245–256.

[22] P. Richardson, D. Flynn, A. Keane, Optimal charging of electric vehicles in low-voltage distribution systems, IEEE Trans. Power Syst. 27 (2011) 268–279.

[23] M.A. Mirzaei, M. Hemmati, K. Zare, M. Abapour, B. Mohammadi-Ivatloo, M. Marzband, A. Anvari-Moghaddam, A novel hybrid two-stage framework for flexible bidding strategy of reconfigurable micro-grid in day-ahead and real-time markets, Int. J. Electr. Power Energy Syst. 123 (2020) 106293.

[24] M.S. Jonban, L. Romeral, A. Akbarimajd, Z. Ali, S.S. Ghazimirsaeid, M. Marzband, G. Putrus, Autonomous energy management system with self-healing capabilities for green buildings (microgrids), J. Build. Eng. (2020) 101604.

[25] C. Cao, B. Chen, Generalized nash equilibrium problem based electric vehicle charging management in distribution networks, Int. J. Energy Res. 42 (2018) 4584–4596.

[26] K. Zhang, Y. Mao, S. Leng, Y. He, S. Maharjan, S. Gjessing, Y. Zhang, D.H.K. Tsang, Optimal charging schemes for electric vehicles in smart grid: a contract theoretic approach, IEEE Trans. Intell. Trans. Syst. 19 (2018) 3046–3058.

[27] N. Nasiri, A. Sadeghi Yazdankhah, M.A. Mirzaei, A. Loni, B. Mohammadi-Ivatloo, K. Zare, M. Marzband, A bi-level market-clearing for coordinated regional-local multicarrier systems in presence of energy storage technologies, Sustain. Cities Soc. 63 (2020) 102439.

[28] J. Jannati, D. Nazarpour, Optimal performance of electric vehicles parking lot considering environmental issue, J. Clean. Prod. 206 (2019) 1073–1088.

[29] W. Su, M.-Y. Chow, Performance evaluation of an EDA-based large-scale plug-in hybrid electric vehicle charging algorithm, IEEE Trans. Smart Grid 3 (2011) 308–315.

[30] M. Honarmand, A. Zakariazadeh, S. Jadid, Optimal scheduling of electric vehicles in an intelligent parking lot considering vehicle-to-grid concept and battery condition, Energy 65 (2014) 572–579.

[31] A. Baziar, M.R. Akbarizadeh, A. Hajizadeh, M. Marzband, R. Bo, A robust integrated approach for optimal management of power networks encompassing wind power plants, IEEE Trans. Ind. Appl. (2020), 1–1.

[32] M.A. Mirzaei, M. Hemmati, K. Zare, B. Mohammadi-Ivatloo, M. Abapour, M. Marzband, A. Farzamnia, Two-stage robust-stochastic electricity market clearing considering mobile energy storage in rail transportation, IEEE Access 8 (2020) 121780–121794.

[33] S. Pelletier, O. Jabali, G. Laporte, Charge scheduling for electric freight vehicles, Trans. Res. Part B Methodol. 115 (2018) 246–269.

[34] R. Das, Y. Wang, G. Putrus, R. Kotter, M. Marzband, B. Herteleer, J. Warmerdam, Multi-objective techno-economic-environmental optimisation of electric vehicle for energy services, Appl. Energy 257 (2020) 113965.

[35] M.B. Tookanlou, M. Marzband, J. Kyyrä, A. Al Sumaiti, K.A. Hosani, Charging/discharging strategy for electric vehicles based on bi-level programming problem: San francisco case study, in: 2020 IEEE 14th International Conference on Compatibility, Power Electronics and Power Engineering (CPE-POWERENG) vol. 1, IEEE, 2020, pp. 24–29.

[36] M.B. Tookanlou, M. Marzband, A. Al Sumaiti, A. Mazza, Cost-benefit analysis for multiple agents considering an electric vehicle charging/discharging strategy and grid integration, in: 2020 IEEE 20th Mediterranean Electro-technical Conference (MELECON), IEEE, 2020, pp. 19–24.

[37] H. Ganjeh Ganjehlou, H. Niaei, A. Jafari, D.O. Aroko, M. Marzband, T. Fernando, A novel techno-economic multi-level optimization in home-microgrids with coalition formation capability, Sustain. Cities Soc. 60 (2020) 102241.

[38] M.A. Mirzaei, M. Nazari-Heris, K. Zare, B. Mohammadi-Ivatloo, M. Marzband, S. Asadi, A. Anvari-Moghaddam, Evaluating the impact of multi-carrier energy storage systems in optimal operation of integrated electricity, gas and district heating networks, Appl. Therm. Eng. 176 (2020) 115413.

[39] Z. Moghaddam, I. Ahmad, D. Habibi, Q.V. Phung, Smart charging strategy for electric vehicle charging stations, IEEE Trans. Trans. Electr. 4 (2017) 76−88.

[40] F. Manríquez, E. Sauma, J. Aguado, S. de la Torre, J. Contreras, The impact of electric vehicle charging schemes in power system expansion planning, Appl. Energy 262 (2020) 1−15.

[41] M.A. Mirzaei, M. Nazari-Heris, B. Mohammadi-Ivatloo, K. Zare, M. Marzband, A. Anvari-Moghaddam, A novel hybrid framework for co-optimization of power and natural gas networks integrated with emerging technologies, IEEE Syst. J. (2020) 1−11.

[42] L. Luo, W. Gu, S. Zhou, H. Huang, S. Gao, J. Han, Z. Wu, X. Dou, Optimal planning of electric vehicle charging stations comprising multi-types of charging facilities, Appl. Energy 226 (2018) 1087−1099.

[43] X. Wang, M. Shahidehpour, C. Jiang, Z. Li, Coordinated planning strategy for electric vehicle charging stations and coupled traffic-electric networks, IEEE Trans. Power Syst. 34 (2019) 268−279.

[44] Z. Wei, Y. Li, L. Cai, Electric vehicle charging scheme for a park-and-charge system considering battery degradation costs, IEEE Trans. Intell. Veh. 3 (2018) 361−373.

[45] V. del Razo, H.-A. Jacobsen, Smart charging schedules for highway travel with electric vehicles, IEEE Trans. Trans. Electr. 2 (2016) 160−173.

[46] A. Ahmadian, M. Sedghi, B. Mohammadi-ivatloo, A. Elkamel, M.A. Golkar, M. Fowler, Cost-benefit analysis of V2G implementation in distribution networks considering pevs battery degradation, IEEE Trans. Sustain. Energy 9 (2017) 961−970.

CHAPTER

7

Application of electric vehicles as mobile energy storage systems in the deregulated active distribution networks

Meisam Ansari[1], Ali Parizad[1], Hamid Reza Baghee[2], Gevork B. Gharehpetian[2]

[1]Electrical and Computer Engineering, Southern Illinois University, Carbondale, IL, United States; [2]Department of Electrical Engineering, Amirkabir University of Technology, Tehran, Iran

1. Introduction

Several decades ago, the electrical industry experienced the first round of restructuring. During that period, the generation and transmission sections profoundly changed. The outcome of that revolution was the deregulated energy markets and private companies that own different industrial energy parts. After passing the transient time, the second round of restructuring is happening on the distribution side. One of the restructuring signs in the distribution systems is the birth of active microgrids (MGs). Consider a distribution system with several MGs and a massive number of active elements such as electric vehicles (EVs), distributed generations (DGs), and demand response (DR) capable loads. The distribution system operator (DSO) will have trouble to manage such a system due to independent actions from the aforementioned entities. It may cause overloading in the system when a large group of the consumers demands electricity at once, and the DGs are not willing or able to support it. Without a market solution that can manage these active participants, the DSO must maintain too much line capacity to keep the system at an acceptable security level during the operation. However, it is not a cost-effective and secure way to rely only on the reserved capacity of distribution lines in the long-term operation due to the growing number of residential loads. If the MGs are managed by independent microgrid operators (MGOs) with enough privileges, the DSO can ask the MGOs to adjust their demand whenever it was needed.

The MGOs can then provide enough incentive for the active constituents in their territory to motivate them to change their demand/production as the DSO requires. The framework that can support this process should be iterative and bidirectional.

In this chapter, the main aim is to provide a complete model for all engaged entities (DSO, MGO, EV, DER, aggregators) and propose a holistic market-based framework that can facilitate their interaction.

2. A modern distribution system: structure and mathematical modeling

The new structure of distribution systems is supposed to be more flexible and adapted to the deregulated environments. In the modern distribution systems, all the end users or the intermediator entities can act fluently and the user rights are very significant. Each user has this opportunity to act as a seller or buyer in this system fairly. To have such a system, a proper framework is needed. In the rest of this part, the different aspects of such a system are explained [4].

2.1 Main entities definition

Fig. 7.1 indicates the different layers in the new structure of a modern distribution system. As is inferred from the figure, there are four hierarchical layers in this structure as:

- Distribution company (Disco) layer: In the new structure, each distribution company is divided into two different entities as: (1) the distribution market operator (DMO), which is responsible for the commercial settlements, and (2) the DSO, which takes care of the operation and management of the system.
- Microgrid (MG) layer: Each part of the grid can be owned and managed by private companies are called microgrid operators (MGOs). Several MGOs can communicate with a similar Disco. Also, each MGO can supervise several MGs on the same grid. According to the given privilege to the MGOs, an MGO can act as a monitoring entity or a local DSO.
- Aggregator's layer: Since there are a massive number of end users in a distribution system, the presence of the aggregators can reduce the transactions between end users and MGOs. Generally, three types of aggregators can be defined in a modern distribution system: (1) distributed generation aggregators (DGAG), (2) electrical vehicle aggregator (EVAG), and (3) demand response aggregator (DRAG). All the aggregators have a contract with the end users for each service.

Fig. 7.2 illustrates the interaction between entities in the new structure. The SOs are in contact with the MGOs. Also, the MGOs interact with the AGGs. This interaction, especially in abnormal conditions such as overloading the system, can be beneficial. The end-layer owners can have a contract with the MGOs to serve a part of their firm load. Simultaneously, the owners can contract with AGGs to provide cheaper energy for their nonfirm load or be a part of the demand response program [5].

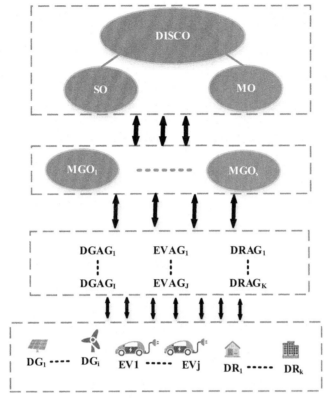

FIGURE 7.1 Different entity layers in a modern distribution system.

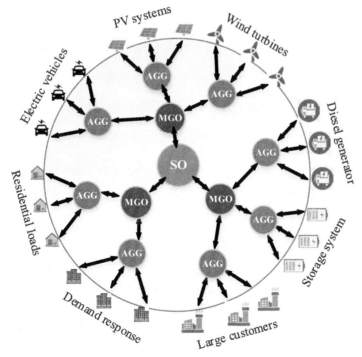

FIGURE 7.2 Communication in a modern distribution system.

2.2 Aggregators mathematical modeling

2.2.1 EVAG optimization model

The main objective of an EVAG is to minimize the cost of purchasing power from the grid for its clients. Eqs. (7.1)–(7.7) describe the optimization problem. In these equations, j implies to the jth EVAG; P^{EVAG}, ρ^{MG}, P^{EVBB}, and E^{EVBB} represent the offered power by EVAG, the electricity price, the exchanged power, and the available stored energy regarding the battery bank. The integer variable X is used to consider wth EV owner if it is plugged in. The customer requests for $E_w^{EVBB,sch}$ amount of charge and supposes to have it at t_{end}. Moreover, $w\epsilon W\{j\}$ means the wth EV owner has a contract with the jth EVAG [6].

$$\text{min: } \sum_{t=t_0}^{24} \sum_{w\in W\{j\}} P_{w,t}^{EVBB} \rho_{s,w,t}^{MG} \tag{7.1}$$

s.t.

$$P_{j,t}^{EVAG} = \sum_{w\in W\{j\}} P_{w,t}^{EVBB} \tag{7.2}$$

$$-P_{w,t}^{EVBB} \leq P_w^{EVBB,disch} X_{w,t} \tag{7.3}$$

$$P_{w,t}^{EVBB} \leq P_w^{EVBB,ch} X_{w,t} \tag{7.4}$$

$$E_{w,t}^{EVBB} = E_{w,t-1}^{EVBB} - P_{w,t}^{EVBB} \tag{7.5}$$

$$\begin{aligned} E_{w,t}^{EVBB} &= E_w^{EVBB,sch}, \text{ if } \quad t = t_{end} \\ E_w^{min} &\leq E_{w,t}^{EVBB} \leq E_w^{max} \end{aligned} \tag{7.6}$$

$$X_{w,t} = \begin{cases} 1 & if \quad t_{st} \leq t \leq t_{end} \\ 0 & otherwise \end{cases} \tag{7.7}$$

2.2.2 DGAG optimization model

The DG owners in the system are the sellers of electrical energy. The role of the aggregators, in this case, is to manage their battery charging/discharging and the injected power to the grid. Eqs. (7.8)–(7.14) describe the objective function for the ith DGAG at time t, which is included in the sth MG territory. In these equations, $v\epsilon V\{i\}$ means the vth DG owner has a contract with the ith DGAG. Also, P^{DG}, P^{DGBB}, P^{DGin}, E^{DGBB}, and E_0 represent the total power, the battery bank power, the produced power by DG, the available energy at the next Δt time, and the remained energy at the end of the day, respectively.

$$\text{max } \sum_{t=t_0}^{24} \sum_{v\in V\{i\}} P_{v,t}^{DG} \rho_{s,v,t}^{MG} \tag{7.8}$$

s.t.

$$P_{v,t}^{DG} = P_{v,t}^{DGin} + P_{v,t}^{DGBB} \tag{7.9}$$

$$P_{v,t}^{DGBB} \leq P_v^{DGBB,disch} \tag{7.10}$$

$$-P_{v,t}^{DGBB} \leq P_v^{DGBB,ch} \tag{7.11}$$

$$E_{v,t}^{DGBB} = E_{v,t-1}^{DGBB} - P_{v,t}^{DGBB} \tag{7.12}$$

$$E_v^{DGBB,min} \leq E_{v,t}^{DGBB} \leq E_v^{DGBB,max} \tag{7.13}$$

$$E_{v,24}^{DGBB} \geq E_0 \tag{7.14}$$

2.2.3 DRAG optimization model

By aggregating the demand response service, a DRAG can play a significant role in the market. Therefore, the main objective for a typical DRAG is to maximize the benefits regarding the demand response service for its clients. Eqs. (7.15) and (7.16) represent the optimization problem for the kth DRAG where P^{cur} and ρ^{DR} stand for the power (kW) and the price for each curtailed step. Also, X^{DR} is a binary variable, which is 1 if the corresponding DR step is selected and otherwise is 0. Moreover, $u \varepsilon U\{k\}$ represents that the uth DR owner has a contract with the kth DR aggregator to reduce the maximum $E_l^{cur,max}$ energy per day.

$$\max rev_k = \sum_{t=t_0}^{T} \sum_{u \in U\{k\}} \sum_{l=1}^{4} P_{u,l,t}^{cur} X_{u,l,t}^{DR} \rho_{u,l,t}^{DR} \tag{7.15}$$

s.t.

$$\sum_{t=t_0}^{T} \sum_{u \in U\{k\}} \sum_{l=1}^{4} P_{u,l,t}^{cur} X_{u,l,t}^{DR} \leq E_l^{cur,max} \tag{7.16}$$

After solving the optimization problem, the DRAG sends the following equivalent values to the MGO where P_k^{DRAG} and ρ_k^{DRAG} indicate the aggregated DR power and corresponding price, respectively.

$$P_k^{DRAG} = \sum_{u \in U\{k\}} \sum_{l=1}^{4} Pst_{u,l}^{DR} X_{u,l}^{DR} \tag{7.17}$$

$$\rho_k^{DRAG} = \frac{rev_k}{P_k^{DRAG}} \tag{7.18}$$

2.3 MGO modeling

Generally, an MGO tries to maximize the usage of its system. Therefore, the load curtailment is not the first option for an MGO in an abnormal condition such as congestion in its system. Eq. (7.19) represents the objective function for the sth MGO. In this equation f_1 is the total cost of electricity that is bought from the grid or the DGs, and f_2 is the rescheduling cost consisting of DR and EV action in an abnormal situation. If an MGO finds that the considered schedule for the next hour puts the system into risk and rescheduling by DRs and/or EVs is needed, then f_1 and f_2 should be considered in the optimization problem. Otherwise only f_1 should be taken into account.

$$\min: \begin{cases} f_1 = P_s^{GR} \rho_s^{GR} \\ f_2 = \sum_{k \in K\{s\}} P_k^{DRAG} \rho_k^{DRAG} \end{cases} \tag{7.19}$$

In Eq. (7.19), P_s^{GR} represents total active power purchased from the grid by sth MGO with the price of ρ_s^{GR}. The active and reactive power, which is demanded by sth MG, can be formulated as Eqs. (7.20) and (7.21), respectively. PL, PEV, PDG, and P^{loss} represent the active load after DR implementation, EV demand, DG production, and active power loss in each MG, respectively. Also, QL, QDG, and Q^{loss} stand for reactive demand, reactive power associated with DGs, and reactive loss in each MG.

$$P_s^{GR} = \sum_{ph \in \{a,b,c\}} \sum_{n=1}^{N} \left[PL_n^{*,ph} + PEV_n^{ph} - PDG_n^{ph} \right] + P_s^{loss} \tag{7.20}$$

$$Q_s^{GR} = \sum_{ph} \sum_{n=1}^{N} \left[QL_n^{*,ph} - QDG_n^{ph} \right] + Q_s^{loss} \tag{7.21}$$

Eq. (7.22) describes the maximum power that MG can get from the grid where $P_s^{MG,max}$ and $P_s^{MG,ord}$ represent the transformer capacity and the maximum power, which the sth MG can get from the grid to prevent the congestion in the upstream network. $P_s^{MG,ord}$ is determined by DSO if any congestion happens in the system.

$$P_s^{MG} \leq \begin{cases} P_s^{MG,max} & \text{in norml condition} \\ P_s^{MG,ord} & \text{in congestion condition} \end{cases} \tag{7.22}$$

If an MGO acts as a local DSO, the operation and management of the system in its supervision are its responsibility. As a result, the MGO should consider all the unbalanced power flow (ULF) constraints in its optimization problem [7]. Therefore, the equations in Section 2.4 should be added to Eqs. (7.19)–(7.22).

2.4 Unbalanced load flow: d-q formulation

The mathematical optimization methods such as Lagrange cannot easily handle the conventional formulation for unbalanced load flow (ULF) because the real and imaginary

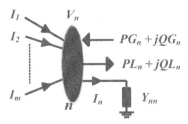

FIGURE 7.3 The input/output current and power in a typical node.

variables are mixed. Therefore, a decomposed model for ULF would be useful in an optimization problem with ULF constraints. In this section, a d-q decomposed formulation is proposed for unbalanced systems. Fig. 7.3 illustrates a typical bus of a system. The active and reactive power injected into the nth bus can be calculated as Eqs. (7.23) and (7.24). Also, Eqs. (7.25) and (7.26) formulate the complementary equations referring to the KCL. In these equations, the variables with d index stand for the real part, and the variables with q index stand for the imaginary part of each variable. For example, the voltage is shown in complex form as $V = V^d + jV^q$ [3].

$$PG_n - PL_n = V_n^d \left(I_{nn}^d - I_n^d \right) + V_n^q \left(I_{nn}^q - I_n^q \right) \tag{7.23}$$

$$QG_n - QL_n = V_n^q \left(I_{nn}^d - I_n^d \right) - V_n^d \left(I_{nn}^q - I_n^q \right) \tag{7.24}$$

$$I_{nn} = \sum_{\substack{m=1 \\ m \neq n}}^{N} I_{mn} \tag{7.25}$$

$$I_n = V_n Y_{nn} \tag{7.26}$$

PG_n and QG_n stand for active and reactive power generation in the nth node. Also, PL_n and QL_n represent the consumed active and reactive power at the same node. Eqs. (7.23) and (7.24) show the calculation of these parameters where PL_n^{sch}, QL_n^{sch}, and $QL_{u,l}^{cur}$ represent the active and reactive scheduled power before any curtailment and the reactive curtailed power, respectively. The curtailment of power is done under the demand response contract between the loads and aggregators.

$$PL_n = \left(PL_n^{sch} - \sum_l P_{u,l}^{cur} \right) + PEV_n, u \in n \tag{7.27}$$

$$QL_n = \left(QL_n^{sch} - \sum_l Q_{u,l}^{cur} \right), u \in n \tag{7.28}$$

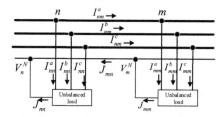

FIGURE 7.4 The relation between the currents for two connected nodes in an unbalanced system.

Fig. 7.4 shows the connections for two nodes with unbalanced loads. As is evident, the summation of the input currents in each node is not necessarily equal to zero. Therefore, the remained current goes back to the slack node through the null path. According to KVL, Eqs. (7.29) and (7.30) show the voltage drop associated with the two nodes in Fig. 7.4. Also, Eqs. (7.31) and (7.32) can be written based on KCL in node n. In these equations, Z^{ph} and Z^N stand for the impedance of phase wires and neutral wires, respectively. Since the network is radial, therefore the number of branches is N−1. As a result, in Eq. (7.31), only N−1 variables (J_{mn}) are available, and these linear equations can be solved.

$$V_m^{ph} = V_n^{ph} - Z_{nm}^{ph} I_{nm}, \; n \neq m, \; n \neq slack \tag{7.29}$$

$$V_n^N = V_m^M - Z_{mn}^N J_{mn}, \; n \neq m, \; n \neq slack \tag{7.30}$$

$$J_{nn} + \sum_{n=1, n\neq m}^{N} J_{mn} = 0, \; n \neq slack \tag{7.31}$$

$$J_{nn} = I_{nn}^a + I_{nn}^b + I_{nn}^c \tag{7.32}$$

Eqs. (7.29)–(7.32) are in complex format. After decomposition, the d-q format of these equations can be written as Eqs. (7.33)–(7.40). In these equations, R and X are the resistance and reactance associated with the system. All the equations are written according to phase a. Therefore, 120 degrees difference phases should be considered to project the values from the phasor domain to the d-q domain. Fig. 7.5 demonstrates the projection process. Thus Eq. (7.32) is represented as Eqs. (7.39) and (7.40) after projection where $\phi^a = 0$, $\phi^b = \frac{-2\pi}{3}$, and $\phi^c = \frac{2\pi}{3}$.

$$V_n^{ph,d} - V_m^{ph,d} + R_{mn}^{ph} I_{mn}^d - X_{mn}^{ph} I_{mn}^q, \; n \neq \{m, slack\} \tag{7.33}$$

$$V_n^{ph,q} - V_m^{ph,q} + R_{mn}^{ph} I_{mn}^q + X_{mn}^{ph} I_{mn}^d, \; n \neq \{m, slack\} \tag{7.34}$$

$$V_n^{N,d} - V_m^{N,d} + R_{mn}^N J_{mn}^d - X_{mn}^N J_{mn}^q, \; n \neq \{m, slack\} \tag{7.35}$$

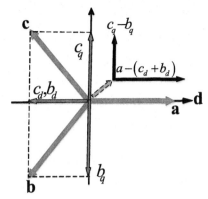

FIGURE 7.5   Conversion from phasor to d-q domain.

$$V_n^{N,q} - V_m^{N,q} + R_{mn}^N J_{mn}^q + X_{mn}^N J_{mn}^d, \; n \neq \{m, slack\} \qquad (7.36)$$

$$J_{nn}^d + \sum_{m=1}^{N} J_{mn}^d = 0, \; n \neq m \qquad (7.37)$$

$$J_{nn}^q + \sum_{m=1}^{N} J_{mn}^q = 0, \; n \neq m \qquad (7.38)$$

$$\begin{bmatrix} J_{nn}^d \\ J_{nn}^q \end{bmatrix} = \sum_{\substack{m=1 \\ m \neq n}}^{N} A(\phi^a) \begin{bmatrix} I_{mn}^{d,a} \\ I_{mn}^{q,a} \end{bmatrix} + A(\phi^b) \begin{bmatrix} I_{mn}^{d,b} \\ I_{mn}^{q,b} \end{bmatrix} + A(\phi^c) \begin{bmatrix} I_{mn}^{d,b} \\ I_{mn}^{q,c} \end{bmatrix} \qquad (7.39)$$

$$A(\phi) = \begin{bmatrix} \cos \phi & -\sin \phi \\ \sin \phi & \cos \phi \end{bmatrix} \qquad (7.40)$$

2.5 The DSO decision-making

In this framework, the management of MGs is left to MGOs. Therefore, the DSO only checks the load flow equations, and if there is any congestion in the system, it takes the necessary actions. The congestion happens when the power flowing through a branch is more than the maximum allowed. Suppose the DSO received all the data from the aggregators and smart meters and wants to schedule the system for the next hour. After solving the ULF problem, three situations might happen for each branch is the system. Eq. (7.41) formulates all these situations, where P, P^{sch}, and P^{mrg} represent the actual power, the scheduled power, and the considered margin for the branch.

$$
\begin{cases}
\text{if} & P - P^{sch} \leq 0 & Flag = -1 \\
\text{if} & 0 \leq P - P^{sch} \leq P^{mrg} & Flag = 0 \\
\text{if} & P^{sch} \geq P^{mrg} & Flag = 1
\end{cases}
\tag{7.41}
$$

Flag equal to -1, 0, and 1 means the branch is not congested, partially congested, and overloaded. According to the output flag, the DSO should act as follows:

- Flag$=-1$: Since there is no congestion in the system, the DSO accepts the schedule and confirms it for implementation.
- Flag$=0$: The system is not overloaded, but the demand is more than the expected value. As a result, there is not enough margin in the system. Without any action, the DSO should operate the system with a low-level risk of congestion. To eliminate the risk, the DSO can directly ask a group of EVAGs under the congested areas to reduce their demand at the operation time. The optimization problem that should be solved by DSO at this step is formulated in Eqs. (7.42) and (7.43) where Δf_j^{EVAG}, ΔP_j^{EVAG}, and X_j stand for the cost incremental, demand reduction, and a binary variable to determine if the wth EVAG participates or not. Also, $j \in B\{l\}$ means the EVAGs that have clients down the congested branch l.

$$
\min : \sum_{j \in B\{l\}} \Delta f_j^{EVAG} X_j
\tag{7.42}
$$

$$
s.t.
$$

$$
\sum_{j \in B\{l\}} \Delta P_j^{EVAG} X_j \geq P_l^{mrg} - \left(P_l - P_l^{sch} \right)
\tag{7.43}
$$

- Flag$=1$: The branch is overloaded. Therefore, the system is in a high-level risk of congestion (HLRC). In this case, the DSO must limit the MGs' demand and relive the congested parts. The first step after detecting HLRC is to identify the MGs under the congested area. Then the DSO provides a list of the MGOs that should reduce their demand to suppress the congestion in the upstream network. The share of the demand reduction for each MG under the congested area is calculated by DSO, according to Eq. (7.44) where ΔP_s is the share of sth MG, which is connected to the congested part of the system and has a demand for ΔP_s. Also, ΔP_{cong} is the total mismatched power that should be reduced by all MGs under the congested branch.

$$
\Delta P_s = \frac{P_s}{\sum_{s \in D} P_s} \Delta P_{cong}
\tag{7.44}
$$

3. The operation framework

3.1 Main flowchart

Fig. 7.6 illustrates an adaptive framework that can facilitate cooperation between all entities. In this framework, each owner has a contract with an aggregator, and the aggregators compete under the MGOs' supervision. There are six significant steps in the proposed framework as follows:

Step 1: In this step, all aggregators solve their optimization problem independently considering the electricity price $(\rho^{\tau MG})$, which is determined by MGOs, and send their outputs to corresponding MGOs.

Step 2: MGOs use Eqs. (7.19)–(7.40) to find their needed power from the grid (P^{MG}) and send it to the DSO. The DSO needs these data to calculate the power flow and check the steady-state constraints of the system.

Step 3: The DSO uses Eqs. (7.23)–(7.40) to solve the ULF. Then, the DSO usesEq. (7.41) to check the line's capacity and, according to the situation, proceed with the next steps.

Step 4: If the DSO identifies any LLRC in the system, this step should be taken into account. In this step, DSO uses Eqs. (7.42) and (7.43) and asks the EVAGs to reduce their demand at the congested time.

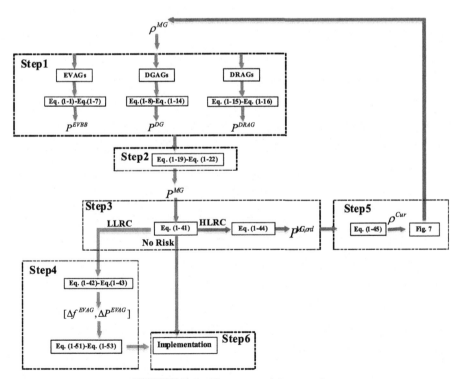

FIGURE 7.6   The proposed framework.

Step 5: If the system is at an HLRC, the DSO uses Eq. (7.44) to limit the MGs' demand. After determining the maximum allowable load by the DSO, the MGOs use the DT (Section 3.2) process to reduce the reliance on the grid.

Step 6: The DSO implements the final obtained schedule in this step.

3.2 Dynamic tariff mechanism

If the system is at HLRC for the next operation time, the DSO should use a DT mechanism to motivate the aggregators to change their schedule and alleviate the congested lines [8]. The DLMP in this framework is considered to cover all electricity costs entirely and, at the same time, to create enough motivation for the aggregators to participate in the congestion prevention process. Fig. 7.7 shows the elements of DLMP in the process of money transfer in the market for congestion prevention. (The DSO and MGOs' rate is not considered in this part.) As is shown, the initial electricity price is the wholesale market rate (ρ^{WM}), and the rest of the elements are added to reflect the power loss cost and congestion cost in the system.

ρgr_s^{Loss} and $\rho_{s,n}^{Loss}$ represent the price associated with power loss in the DSO's network and MGO's system, respectively. Also, ρgr_s^{Cur} is the equivalent curtailment rate and is calculated as Eq. (7.45).

$$\rho_s^{Cur} = \frac{\sum\limits_{k \in K\{s\}} P_k^{DRAG} \rho_k^{DRAG}}{P_s^{GR}} \tag{7.45}$$

Fig. 7.8 illustrates the input and output cash flow related to each MG. The revenues and costs are defined using Eqs. (7.46)–(7.50). In this framework, all received money by the microgrid financial center (MGFC) is equal to the paid money to the DSO and MGs' entities. As a result, there is no residual money within the process.

In Eqs. (7.46)–(7.50), R^{EV} and R^L are the cash amounts collected by the MGO removed from the corresponding EVs and other loads, respectively. Also C^{GR}, C^{DG}, and C^{DR} are the cash amounts that MGO should pay to the SO, DGs, and curtailed loads, respectively.

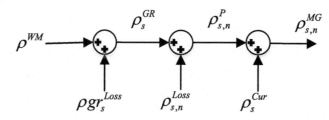

FIGURE 7.7   Electricity rate calculation process.

FIGURE 7.8   Cash flow process.

$$R^{EV} = \sum_n \sum_{ph} PEV_n^{ph} \rho_{s,n}^{MG} \tag{7.46}$$

$$R^L = \sum_n \sum_{ph} \left(PL_n^{sch,ph} - \sum_l P_{u,l}^{cur,ph} \right) \rho_{s,n}^{MG} \tag{7.47}$$

$$C^{GR} = P_s^{GR} \rho_s^{GR} \tag{7.48}$$

$$C^{DG} = \sum_n \sum_{ph} PDG_n^{ph} \rho_{s,n}^{MG} \tag{7.49}$$

$$C^{DR} = \sum_{k \in K\{s\}} P_k^{DRAG} \rho_k^{DRAG} \tag{7.50}$$

3.3 EVAGs' activity in an LLRC

At an LLRC situation, the EVAGs should send an offer with the amount of power reduction and cost incremental due to rescheduling for the next hour. Suppose the jth EVAG is willing to reduce its demand by $P_{j,t0}^{EVAG}$ where t_0 is the next hour. Eqs. (7.51)–(7.53) show the optimization problem in this situation. Eq. (7.51) calculates the minimum cost incremental if the aggregator wants to reduce its demand by $P_{j,t0}^{EVAG}$ for the next time. Also, Eq. (7.53) guarantees there is no curtailment in the rescheduling process.

$$\text{min: } \Delta f_j^{EVAG} = \sum_{t=t_0}^{24} \sum_{w \in W\{j\}} \Delta P_{w,t}^{EVBB} \rho_{s,w,t}^{MG} \tag{7.51}$$

s.t.

$$\Delta P_{j,t0}^{EVBB} = \sum_{w \in W\{j\}} \Delta P_{w,t0}^{EVBB} \tag{7.52}$$

$$\sum_{t=t_0}^{24} \sum_{w \in W\{j\}} \Delta P_{w,t}^{EVBB} = 0 \tag{7.53}$$

4. EVAGs' activity in an HLRC

If the system falls into a high-level risk situation, then the MGOs use an iterative process to motivate the aggregators to change the risk level. The proposed dynamic tariff method is introduced in Section 3.2. After each round of DLMP modification, the EVAGs use Eqs. (7.1)–(7.7) and resolve their optimization problem, as well as the other aggregators.

5. The link between MATLAB and GAMS

5.1 General view

One of the most powerful software that can solve the complicated optimization problems is the General Algebraic Modeling Language (GAMS). GAMS uses several powerful solvers to handle the linear, nonlinear, and mixed-integer problems. The MATLAB's flexibility, combined with the GAMS's power in handling complicated optimizations, makes a powerful tool that can optimize large-scale cases in a meaningful time.

Fig. 7.9 illustrates the way that the MATLAB and GAMS can communicate [1]. As is shown, the GAMS optimization code is used as a MATLAB's built-in function in this scheme. There are three steps to create this platform:

- First, the input variables that the GAMS needs for optimization should be defined in the MATLAB. The variables should be generated in the "*.gdx" format, which can be supported by GAMS.
- The second step is to execute the GAMS scripts from MATLAB.
- The third step is to extract the results from the GAMS's outputs and bring them into the MATLAB.

This platform provides this ability to call GAMS scripts from MATLAB multiple times during an optimization process.

5.2 Define GAMS in MATLAB's path list

The first step to link GAMS and MATLAB is to define the GAMS installation directory as one of the known paths in MATLAB. To do this, in MATLAB home, the "set path" option should be selected, and the GAMS directory should be added using the "add" button. Fig. 7.10 shows the process. Without this step, the necessary commands to define the inputs, execute the GAMS files, and extract the results don't work in MATLAB.

5.3 MATLAB commands to generate gdx file for input parameters

In this structure, the GAMS optimization files should take the input parameters and sets from the MATLAB. Therefore, in the second step, the proper files should be generated in MATLAB.

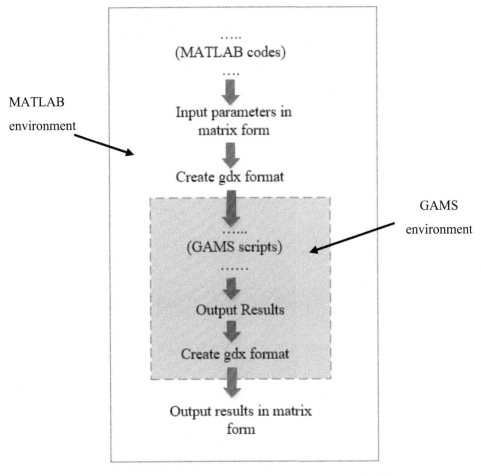

FIGURE 7.9 The interaction between MATLAB and GAMS.

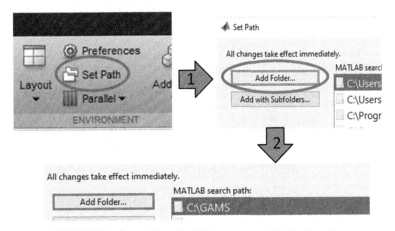

FIGURE 7.10 Adding GAMS directory to MATLAB paths.

5.3.1 Generated gdx files in MATLAB

The template to define the sets and tables in gdx format is different. To define the sets in gdx format, the following template should be used in MATLAB script:

[*the set name in MATLAB*].name='[*the used set name in gams*]'

[*the set name in MATLAB*].uels={'[*value 1*]','[*value 2*]',...};

wgdx('[*the gdx file name*]', [*the set name in MATLAB*]);

For example, following is the MATLAB code to define a set to show the time in GAMS. The set's name in GAMS is "t" and the values are {1,2}. Also, the file is saved in "t_file.gdx".

```
t_set.name='t';
t_set.uels={'1','2'};
wgdx('t_file ', t_set)
```

The tables should be inserted as parameters into the GAMS. A general format to define an array in gdx format is as follows:

iwgdx('[*the gdx file name*]','[*MATLAB Array1*]', '[*MATLAB Array2*]',...)

According to the format, it is possible to save several arrays with different dimensions in a single gdx file. As an example, the matrixes R and X are saved into the gdx file named "Impedance" as follows.

```
R=[ 0   0.02; 0.02   0];
X=[ 0   0.1; 0.1   0];
iwgdx('Impedance','R','X')
```

5.3.2 Read gdx files from GAMS

After using "iwgdx" and "wgdx" commands to generate gdx files from tables and sets, the following format is used in GAMS to read the generated gdx files.

Reading gdx file associated with sets:

$GDXIN [*gdx file with full drectory*]

$LOAD [*set name*]

$GDXIN

Example GAMS commands:

$GDXIN C:\User\Desktop\ t_file.gdx

$LOAD t

$GDXIN

Reading gdx file associated with arrays:

$GDXIN [*gdx file with full drectory*]

$LOADIDX [*arrays' names*]

$GDXIN

Example:

$GDXIN C:\User\Desktop\Impedance.gdx

$LOADIDX R   X

$GDXIN

5.4 Saving the output GAMS results in gdx format

The GAMS and MATLAB can support the gdx format. Therefore, all the exchanged data should be under this format. After solving the optimization problem by GAMS, the following command is needed to save the preferred results in gdx format. In the next step, the MATLAB can extract the values from these gdx files.

execute_UnloadIdx '[*gdx file name with full directory*]' [*GAMS result to be saved*]

In this step, the GAMS result should be in the parameter format. Therefore, the results obtained by GAMS should be saved in proper parameters before using the command, as mentioned earlier. Following is an example of GAMS commands to save the magnitude of voltage after a ULF calculation in a file named "MagV.gdx". Vd and Vq are the GAMS optimization variables for real and imaginary parts of the voltage.

```
Parameter V(n) ;

V(n)=sqrt(V_d.l(n)*V_d.l(n)+ V_q.l(n)*V_q.l(n));

execute_UnloadIdx 'C:\User\Desktop\ MagV'  V
```

5.5 Calling GAMs model from MATLAB

When all the input sets and parameters were provided, and the corresponding commands to generate gdx from the results were added to the gams file, the following template can be used to execute the GAMS file from MATLAB:

gams('[*gams file name*]')

Here is an example of running a GAMS file named "loadflow.gms" from MATLAB:

```
gams('loadflow') ;
```

5.6 Extracting results from generated gdx files

The following command in MATLAB can be used to extract the optimization results from the GAMS's generated gdx files.

irgdx '[*generated gdx file by GAMS*]'

Here is an example:

```
irgdx 'MagV.gdx';
```

5.7 A complete example

Here is a complete example of an ULF problem associated with a two-bus test system. GAMS commands for defining sets, parameters, and variables:

```
SETS
n    bus numbers
ph phase sequence
;
$GDXIN C:\Users\Desktop\n
$LOAD n
$GDXIN
;
$GDXIN C:\Users\Desktop\ph
$LOAD ph
$GDXIN
;
alias (n,m)
;
parameters
R(n,m)
Rn(n,m)
X(n,m)
Xn(n,m)
SP(n,PH)
SQ(n,PH)
;
$GDXIN C:\Users\Desktop\InputD
$LOADIDX  R Rn X Xn SP SQ
$GDXIN
;
variables
V_x(ph,n)
V_y(ph,n)
I_x(ph,n,m)
I_y(ph,n,m)
In_x(ph,n)
In_y(ph,n)
J_x(n,m)
J_y(n,m)
Ploss
Qloss
;
```

GAMS commands for modeling ULF equations and setting the output gdx:

```
eq1(ph).. V_x(ph,'1')=e=1.05$(ord(ph) ne 4)+0$(ord(ph)=4);
eq01(ph).. V_y(ph,'1')=e=0;
PLoss1.. Ploss=e= sum((n,m,ph)$((ord(ph) ne 4)and(ord(n) ne ord(m))),V_x(ph,n)*I_x(ph,n,m)+V_y(ph,n)*I_y(ph,n,m))
        +sum(n,V_x('4',n)*J_x(n,n)+V_y('4',n)*J_y(n,n));
QLoss1.. Qloss=e= sum((n,m,ph)$((ord(ph) ne 4)and(ord(n) ne ord(m))),V_y(ph,n)*I_x(ph,n,m)-V_x(ph,n)*I_y(ph,n,m))
        +sum(n,V_y('4',n)*J_x(n,n)-V_x('4',n)*J_y(n,n));
eq2(n,ph)$((ord(ph) ne 4)and(ord(n)>1)).. -SP(n,ph)-(V_x(ph,n)*I_x(ph,n,n)+V_y(ph,n)*I_y(ph,n,n))=e=0;
eq3(n,ph)$((ord(ph) ne 4)and(ord(n)>1)).. -SQ(n,ph)-(V_y(ph,n)*I_x(ph,n,n)-V_x(ph,n)*I_y(ph,n,n))=e=0;
eq4(ph,n).. I_x(ph,n,n)=e=-sum(m$(ord(n) ne ord(m)),I_x(ph,m,n));
eq5(ph,n).. I_y(ph,n,n)=e=-sum(m$(ord(n) ne ord(m)),I_y(ph,m,n));
eq6(ph,n,m).. V_x(ph,m)=e=V_x(ph,n)-(R(n,m)*I_x(ph,n,m)-X(n,m)*I_y(ph,n,m));
eq7(ph,n,m).. V_y(ph,m)=e=V_y(ph,n)-(R(n,m)*I_y(ph,n,m)+X(n,m)*I_x(ph,n,m));
eq8(n,m).. V_x('4',n)=e=V_x('4',m)-(Rn(m,n)*J_x(m,n)-Xn(m,n)*J_y(m,n));
eq9(n,m).. V_y('4',n)=e=V_y('4',m)-(Rn(m,n)*J_y(m,n)+Xn(m,n)*J_x(m,n));
eq10(n).. sum(m,J_x(n,m))=e=0;
eq11(n).. sum(m,J_y(n,m))=e=0;
eq12(n).. -J_x(n,n)=e=sum(m$(ord(m) ne ord(n)),I_x('1',m,n)-I_x('2',m,n)*0.5+0.866*I_y('2',m,n)-0.5*I_x('3',m,n)-0.866*I_y('3',m,n)  );
eq13(n).. -J_y(n,n)=e=sum(m$(ord(m) ne ord(n)),I_y('1',m,n)-I_x('2',m,n)*0.866-0.5*I_y('2',m,n)+0.866*I_x('3',m,n)-0.5*I_y('3',m,n)  );
model ULF /all/
option     optca=1e-10,optcr=1e-10,nlp=CONOPT;
solve ULF minimizing Ploss using nlp
;
parameter V(ph,n)
;
V(ph,n)=sqrt(V_x.l(ph,n)*V_x.l(ph,n)+ V_y.l(ph,n)*V_y.l(ph,n));
execute_UnloadIdx 'C:\Users\Desktop\V' V
```

MATLAB commands:

```
t_file.name='t';
t_file.uels={'1','2'};
ph.name='ph';
ph.uels={'1','2','3','4'};
R= [ 0     0.02;0.02      0];
Rn=[ 0     0.01; 0.01     0];
X= [ 0     0.025;0.025    0];
Xn=[ 0     0.015;0.015    0];
%     a     b     c    n
SP=[ 0     0     0    0
     1    0.5   0.7  0 ];
%     a     b     c    n
SQ=[ 0     0     0    0
     0.4  0.15  0.2  0 ];
% All arrays
iwgdx('InputData','R','Rn','X','Xn','SP','SQ')
% sets
wgdx('n',n)
wgdx('ph',ph)
gams('loadflow')
irgdx 'V.gdx'
V
```

6. Simulation results

The modified IEEE 13-bus unbalanced test system [2] is chosen for the numerical study to evaluate the proposed framework. Four MGs with active EVs are added to this system to change it to a modern distribution system. Fig. 7.11 exhibits a single diagram of the main system and the MGs, where MG3 is presented with more details as an example. Each EV icon in the diagram stands for parking with EV charging ability.

Table 7.1 shows the data associated with three different EV types that are modeled. According to Table 7.1, the first EV type can be charged up to 16 kWh by a maximum of 4.5 kW/h charging/discharging ramp. The owners of this type can also plug their EVs to the grid from 1 a.m. to 7 a.m. The aggregators should schedule how to fully charge the vehicles during this interval, knowing that the initial charge is 25%. The second and third EV types have the maximum capacity of 10 and 25 kWh with the ramp of 25 and 4 kW/h, respectively. Their desired charging time is from 10 a.m. to 2 p.m. and 3–8 p.m., respectively. The initial charge is considered 25% for these types as well.

To facilitate engaging the EVs in the system operation, an EVAG is considered for each MG. Table 7.2 shows the number of EVs within each MG.

The hourly aggregated load and the grid electricity price are shown in Fig. 7.12A and B, respectively. The values of the electrical load are without aggregating the EVAGs' demands. To avoid complexity, no DGs and DRs are considered in this study.

According to the framework (Fig. 7.6), all the EVAGs solve the optimization problem formulated in Eqs. (7.1)–(7.7) and send their aggregated demand to the corresponding MGO. Fig. 7.13 illustrates the aggregators' hourly demand at the initial step of the

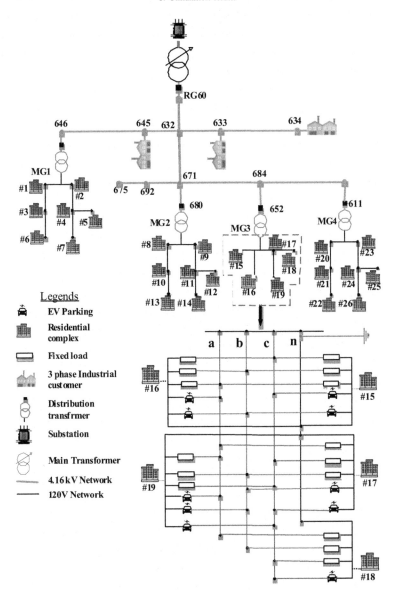

FIGURE 7.11 Modified IEEE 13-bus unbalanced system.

framework. The MGOs should aggregate these amounts to their local load and solve the corresponding optimization problem to find out their hourly demand from the grid. The results will be shared with the DSO for evaluation.

The DSO identifies two congestion risk during the next day's operation. This is found by comparing the line flows with the maximum values (see Section 2.3). Fig. 7.14 specifies the identified congestion risks on the single diagram. As is inferred, branch 623.671 will have

TABLE 7.1   Types of modeled EVs.

Types	E^{max}	E^{min}	P^{ch}_{EVBB}	P^{disch}_{EVBB}	t_{st}	t_{end}	SOC_0
1	16	5.4	4.5	4.5	1	7	%30
2	10	3.6	4	4	10	14	%40
3	25	6.6	2.75	2.5	15	20	%50

TABLE 7.2   EVs' population associated with each MG.

MG	EV type			Total
	Type1	**Type2**	**Type3**	
MG1	125	211	150	486
MG2	210	455	152	817
MG3	355	302	260	917
MG4	170	180	105	455
Total	860	1148	667	2675

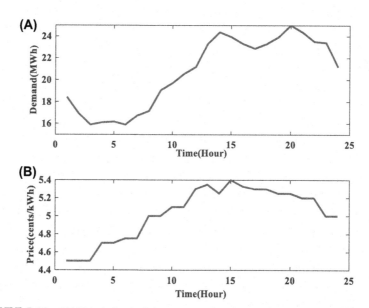

FIGURE 7.12   (A) Hourly load of the system, (B) Hourly electricity price of the system.

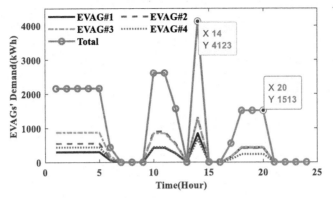

FIGURE 7.13   Aggregators demand in the initial schedule.

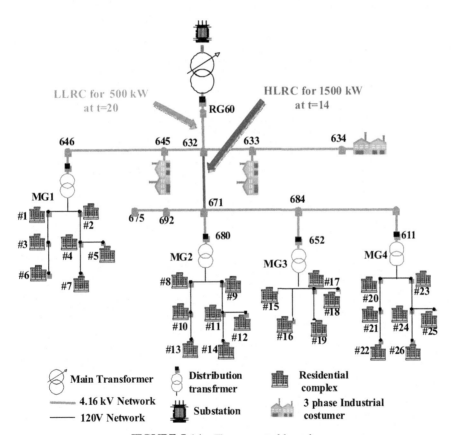

FIGURE 7.14   The congested branches.

an HLRC situation at 2 p.m., and the branch RG60-632 will experience an LLRC at 8 p.m. To avoid the HLR, the DSO should reduce the downstream load by 1500 kW at 2 p.m. Also, 500 kW reduction is needed to suppress the LLRC at 8 p.m.

Fig. 7.15 illustrates the risks of the hourly demand and price curve. As is evident, the HLRC happens when the electricity price is lower than the price at the previous and next hour. The MGOs, with the collaboration of EVAGs, can relive this issue by adjusting the price.

After taking action by the DSO and MGOs, the EVAGs are motivated to reschedule their demand. According to the proposed DT method in Section 3.2, the price at all MGs is adjusted by the MGOs to relive the HLRC in branch 632.671. Fig. 7.16 shows the electricity prices on all buses before and after the DT process. There is an increase in the electricity price for MG2, MG3, and MG4, located down the HLRC branch. This increase motivates the EVAGs in these MGs to reduce their demand. As a result, the burden on branch 632.671 is reduced. At the same time, the price for MG1 is decreased to provide a cheaper charge for the EVs. It gives this chance to that group of owners in the other MGS that need to charge their vehicle at 2 p.m. to drive and use a charge station in MG1.

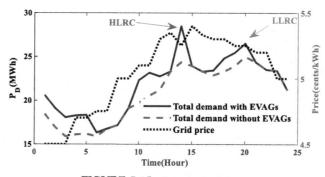

FIGURE 7.15   Initial scheduling.

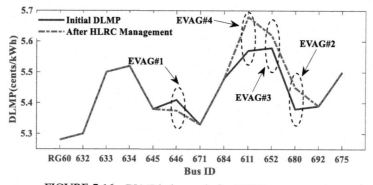

FIGURE 7.16   DLMP before and after HLRC management.

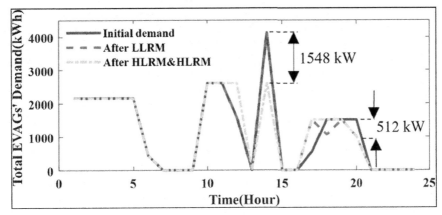

FIGURE 7.17 EVAGs' demand after LLRC and HLRC management.

After this step, the DSO should take care of the LLRC at 8 p.m. It can be done by following the process in Section 3.4. Fig. 7.17 shows the total EVAGs' demand after each round of rescheduling. The outcome of two rescheduling rounds is reducing the demand by 1548 kW at 2 p.m. and 512 kW at 8 p.m. This reduction is enough to resolve the risk issues for the next day's operation.

References

[1] GDXMRW Manual, n.d. [Online]. Available: https://www.gams.com/latest/docs/T_GDXMRW.html.

[2] IEEE PES AMPS DSAS Test Feeder Working Group, n.d. [Online] https://site.ieee.org/pes-testfeeders/resources/.

[3] A. Saleh, The formulation of a power flow using d-q reference frame components—part I: Balanced 3φ systems, IEEE Trans. Ind. Appl. 52 (5) (2016) 3682–3693.

[4] S.A.A. Kazmi, et al., Smart distribution networks: A review of modern distribution concepts from a planning perspective, Energies (2017) 501–548, https://doi.org/10.3390/en10040501.

[5] C. Zhu, et al., A real-time battery thermal management strategy for connected and automated hybrid electric vehicles (CAHEVs) based on iterative dynamic programming, IEEE Trans. Vehicular Techol. 67 (9) (2018) 8077–8084.

[6] D. Said, H.T. Mouftah, A novel electric vehicles charging/discharging management protocol based on queuing model, IEEE Trans. Intel. Vehicles 5 (1) (2020) 100–111, https://doi.org/10.1109/TIV.2019.2955370.

[7] Y. Liao, et al., Unbalanced multi-phase distribution grid topology estimation and bus phase identification, IET Smart Grid 2 (4) (2018) 557–570, https://doi.org/10.1049/iet-stg.2018.0291.

[8] S. Huang, et al., Distributed optimization-based dynamic tariff for congestion management in distribution networks, IEEE Trans. Smart Grid 10 (1) (2019) 184–192, https://doi.org/10.1109/TSG.2017.2735998.

Economic analysis of energy storage systems in multicarrier microgrids

Amin Mansour-Saatloo[1], Masoud Agabalaye-Rahvar[1],
Mohammad Amin Mirzaei[1], Behnam Mohammadi-Ivatloo[1,2],
Kazem Zare[1]

[1]Faculty of Electrical and Computer Engineering, University of Tabriz, Tabriz, East Azerbaijan, Iran; [2]Department of Energy Technology, Aalborg University, Aalborg, Denmark

Nomenclature

Index:

i Thermal units
j Index of minimum up-time and down-time running from 1 to max $\{T_i^{On}, T_i^{OFF}\}$
s Scenario index
t Time interval

Constants:

DR Allowed shiftable electrical load (%)
R_i^{up}, R_i^{down} Thermal units ramp-up and ramp-down (kW)
$EL_{t,s}$ Electrical load at time t and scenario s (kW)
λ_t^E Electricity price at time t (¢/kWh)
λ_t^G Gas price at time t (¢/MBtu)
λ_t^H Heat price at time t (¢/kWth)
η^{P2H}, η^{H2P} HSS charging and discharging efficiencies (%)
$C^{chr, HSS}$ Hydrogen price at time t (¢/kWh)
HB^-/HB^+ Min/Max generated heat by boiler (kWt)
P_i^-/P_i^+ Min/Max generated power by thermal units (kW)
T_i^{ON}, T_i^{OFF} Minimum up/down-time of power plant i
$A_{t,s}^{-, ESS}/A_{t,s}^{+, ESS}$ Min/Max electrical storage
$A_{t,s}^{-, TSS}/A_{t,s}^{+, TSS}$ Min/Max thermal storage
$A_{t,s}^{-, HSS}/A_{t,s}^{+, HSS}$ Min/Max hydrogen storage
$Pc^{-, ESS}/Pc^{+, ESS}$ Min/Max charge of electrical storage
$Pc^{-, TSS}/Pc^{+, TSS}$ Min/Max charge of thermal storage

$Pc^{-,\,HSS}/Pc^{+,\,HSS}$ Min/Max charge of hydrogen storage
$Pd^{-,\,ESS}/Pd^{+,\,ESS}$ Min/Max discharge of electrical storage
$Pd^{-,\,TSS}/Pd^{+,\,TSS}$ Min/Max discharge of thermal storage
$Pd^{-,\,HSS}/Pd^{+,\,HSS}$ Min/Max discharge of hydrogen storage
sug_i, sud_i Required gas at the start-up and shutdown period for thermal units (MBtu)
$\eta^{chr,\,TSS}$, $\eta^{dischr,\,TSS}$ TSS charging and discharging efficiencies (%)
HR_i/HR^B Thermal units/Boiler heat rates (MBtu/kWh)

Variables:

$Pc_{t,s}^{HSS}/Pd_{t,s}^{HSS}$ HSS charging/discharging at time t and scenario s
$SUC_{i,t,s}/SDC_{i,t,s}$ Start-up and shutdown cost of power plant i
$GB_{t,s}$ Boiler gas consumption at time t and scenario s (MBtu)
$A_{t,s}^{ESS}$ Charge level of ESS at time t and scenario s (kWth)
$A_{t,s}^{TSS}$ Charge level of TSS at time t and scenario s (kWth)
$A_{t,s}^{HSS}$ Charge level of HSS at time t and scenario s (kWth)
$G_{i,t,s}$ Thermal units gas consumption at time t and scenario s (MBtu)
$Pc_{t,s}^{ESS}/Pd_{t,s}^{ESS}$ ESS charging/discharging at time t and scenario s
$Pc_{t,s}^{TSS}/Pd_{t,s}^{TSS}$ TSS charging/discharging at time t and scenario s
$HB_{t,s}$ Generated heat via boiler at time t and scenario s (kWt)
$H_{t,s}^{CHP}$ Generated heat via CHP unit at time t and scenario s (kWt)
$P_{i,t,s}$ Generated power via thermal units at time t and scenario s (kW)
$P_{t,s}^{G,\,buy}$ Purchased gas from the main grid at time t and scenario s (MBtu)
$P_{t,s}^{E,\,buy}$ Purchased electrical energy from the main grid at time t and scenario s (MBtu)
$P_{t,s}^{H,\,buy}$ Purchased thermal energy from the main grid at time t and scenario s (MBtu)
$P_{t,s}^{E,\,buy}$ Sold thermal energy to the main grid at time t and scenario s (MBtu)
$P_{t,s}^{E,\,sell}$ Sold electrical energy to the main grid at time t and scenario s (MBtu)
$EL_{t,s}^{DR}$ The amount of electrical demand at time t and scenario s after DR program (kW)
$DR_{t,s}^{crt}$ The amount of curtailed electrical load at time t and scenario s (kW)
$DR_{t,s}^{shft}$ The amount of shifted electrical load at time t and scenario s (kW)
$I_{i,t,s}$ On/off status of thermal units at time t and scenario s
$I_{t,s}^B$ On/off status of boiler at time t and scenario s
$Ic_{t,s}^{ESS}/Id_{t,s}^{ESS}$ Charging/discharging mode of ESS at time t and scenario s
$Ic_{t,s}^{TSS}/Id_{t,s}^{TSS}$ Charging/discharging mode of TSS at time t and scenario s
$Ic_{t,s}^{TSS}/Id_{t,s}^{TSS}$ Charging/discharging mode of TSS at time t and scenario s

1. Introduction

In the past years, the great acceleration of integration distributed generations (DGs), especially renewable energy sources (RESs), up to 60% until 2050 [1], into the power system brings a new concept of microsystem, which today is known as microgrid (MG). An MG is a system that delivers energy services to end users and itself can be a consumer or producer from the dependent system operator (DSO) point of view. A survey on MG energy management, including optimization techniques, can be found in Ref. [2]. As soon as the emergence of MG, because of the multiple demands of consumers, the concept of MG developed into multicarrier microgrid (MCMG). MCMG is an infrastructure that integrated different energy carriers, e.g., electricity, gas, and heat to operate together [3].

MCMGs are constructed to meet the various demands of consumers with high-efficiency, more reliability, and less environmental emissions. So, the optimal planning of MCMG has been executed in Ref. [4] in which the operating, maintenance, and investment costs are effects on determining the best component and size of equipment in the regarded period. After completing the design and planning of interdependent infrastructures, the optimal management of MCMGs has appeared to reduce the total operating costs as well as maintain the consumers' welfare. In Ref. [5], the optimal energy management approach for MCMG with considering various technical and economic constraints between electrical and natural gas networks was proposed based on mixed-integer linear programming (MILP) formulation. Another optimal energy management between electricity and thermal systems by applying demand response (DR) program was presented in Ref. [6], which is solved using particle swarm optimization (PSO) method. For both islanded and grid-connected MCMGs, a coordinated optimal energy dispatch was proposed in Ref. [7] to reduce the net operation cost of the system as well as improve the dispatch flexibility of power and heat/cool for the day-ahead energy market. Ref. [8] was established as a two-stage energy scheduling framework for MCMGs with an MILP and Stackelberg game theory solution methodologies in these two stages, respectively. Authors in Ref. [9] introduced two-stage stochastic energy and reserve scheduling structure by taking into account the uncertainties of wind and solar power generation, electrical and thermal demands besides deploying DRP with the aim of reducing total expected cost and improving security factors. Also, for a network-constrained MCMG with different uncertain parameters, probabilistic optimal energy flow along with modern time-based DRP was proposed in Ref. [10]. Long-term planning for optimal sizing of islanded MCMG was established in Ref. [11] in which the robust metaheuristic optimization approach was deployed to solve the proposed problem by indicating the capability of optimal configuration in terms of economic, efficiency, and environmental factors. By increasing electric vehicles in the MGs and the obligation of integration between EVs and RES, a cooperative optimization algorithm was presented in Ref. [12] to achieve the economic dispatch and capacity allotment in the regional MCMG. A novel energy management structure-based dynamic game theory approach was reported in Ref. [13] to reduce the operating cost of the distribution network operator in which the distribution network contains multimicrogrids.

Due to environmental issues, researchers try to turn clean energy carriers. Hydrogen is one of those clean energy carriers, which had significant attention in recent years. Hydrogen production options include gray hydrogen generated from fossil fuels, blue hydrogen generated from carbon captured fossil fuels, and green hydrogen generated from renewables. Green hydrogen production is going to grow rapidly in the near future. In Germany, Amprion (operator of the transmission system) and OGE (operator of the gas network) have designed a project that will be online until 2023 for 100 MW electrolyzer and dedicated hydrogen pipeline [14]. Another project in Fukushima, Japan, has been done by Toshiba up to 10 MW electrolyzer capacity to produce 9000 tonnes of hydrogen from renewable yearly [15]. Furthermore, the hydrogen facility has been embedded into the power system researches in the near past. A security-constraint stochastic model has been proposed in Ref. [16], where the hydrogen storage system (HSS) and DR program were considered. In Ref. [17], a risk-based stochastic model was presented for cryogenic energy storage in microgrids. In addition to the HSS, multiple energy storage systems such as electrical storage system (ESS) and thermal storage system (TSS) have considerable impacts on energy markets. An optimal operation model for the residential energy system, including electrical and thermal storage systems, was proposed in Ref. [18] to minimize the EH operation cost. A two-stage robust cooptimization model for EH was carried out in Ref. [19], where a precise model for the economical storage system was considered. Optimal scheduling of EH, including energy storage integrated with electric vehicles, was presented in Ref. [20]. Robust day-ahead scheduling for a smart microenergy system using power-to-hydrogen technology was carried out in Ref. [21].

According to the aforementioned scholars and to the best of the authors' knowledge, there is a research gap in considering multienergy storage systems (MESSs), specifically HSSs, into the multicarrier infrastructure in order to participate in the multienergy markets. Utilization of MESSs can bring many economic benefits to EH operators by saving energy at low price periods and delivering at high price periods. So, this chapter focuses on this issue along with considering the uncertainty of wind power, electrical and thermal loads through a scenario-based optimization. To do so, three different storage systems i.e., ESS, TSS, and HSS, are utilized. In addition, demand-side management is considered using an incentive-based DR program. Using the DR program, end users can shift their loads from high energy price hours to the other periods, which, as a consequence, EH operator costs reduce.

The rest of the chapter is organized as follows: Section 2 provides a precise description of the system, including HSS technology. The mathematical model of the system, along with scenario generation and reduction details, is presented in Section 3. The next section provides simulation results and discussions. Finally, Section 5 concludes the chapter.

2. System description

Fig. 8.1 shows the proposed MCMG schematically, which is composed of CHP unit, boiler, gas turbine (GT), wind turbine (WT), ESS, TSS, and HSS with electrical energy, thermal energy, and gas at input terminals and electrical energy and thermal energy at the output terminal. The proposed MCMG is equipped with MESSs to participate in a multienergy market in the way of storing energy at low energy price periods, then delivers stored energy at high

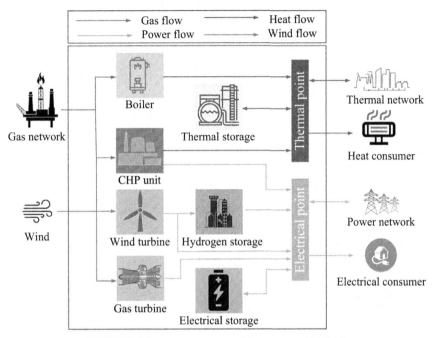

FIGURE 8.1 Basic structure of the proposed MCMG.

price periods if required. Fig. 8.2 shows the HSS structure with more details. Electrical energy from WT is used to water electrolysis. Electrolysis of water reveals hydrogen gas and oxygen gas through the following chemical reaction: $H_2O \rightarrow H_2 + \frac{1}{2}O_2$. This is the brief of power-to-hydrogen (P2H) technology. Then produced hydrogen can be stored in a tank without any degradation. At the required time, the inverse of the mentioned reaction in a hydrogen fuel cell can convert the hydrogen back to power as follows: $H_2 + \frac{1}{2}O_2 \rightarrow H_2O$, which indicates hydrogen-to-power (H2P) technology [22]. In addition, an incentive-based DR program is

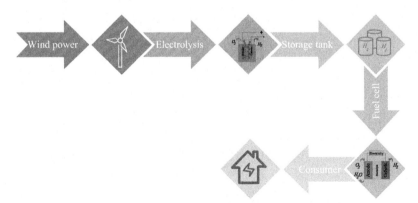

FIGURE 8.2 Basic structure of the hydrogen storage.

applied to increase the flexibility of the system. In doing so, end users would be incentivized to have a role in the scheduling program by curtailing their loads at on-peak hours and shift their loads at off-peak hours. So, end users' load patterns can be controlled in both technical and economical manner. Finally, to alleviate uncertainties of wind power, electrical and thermal loads, stochastic programming takes into account by producing a high number of scenarios. Because of computational complexity, it is hard to deal with a high number of scenarios in the optimization problem. To this end, a proper mathematical algorithm, namely fast forward selection [23], is employed to reduce the number of scenarios.

3. Scenario-based problem formulation

This section covers the mathematical modeling of all considered technologies along with objective function and scenario generation and reduction part.

3.1 Scenario generation and reduction

In order to model uncertainties of wind and loads in scenario-based problems, different probability density functions (PDFs) are used. Wind power depends on wind speed at every moment, so wind speed must be considered to scenario generation. Usually, Weibull PDF was used in the literature to wind speed scenario generation, and in this work, Rayleigh PDF, as represented in Eq. (8.1), a specific case of Weibull PDF, is used. In addition, wind turbine output power is a piecewise function of wind speed, which is given in Eq. (8.2).

$$PDF(v) = \left(\frac{v}{c^2}\right)\exp\left[-\left(\frac{v^2}{2c^2}\right)\right] \tag{8.1}$$

$$P^{wind}(v) = \begin{cases} 0 & v \leq v^c_{cut-in} \quad \text{or} \quad v \geq v^c_{cut-out} \\[2ex] \dfrac{v - v^c_{cut-in}}{v^r - v^c_{cut-in}}P^r_{wind} & v^c_{cut-in} \leq v \leq v^r \\[2ex] P^r_{wind} & \text{otherwise} \end{cases} \tag{8.2}$$

Furthermore, normal PDF is used to model the uncertainty of both electrical and thermal loads. The following equation gives the normal PDF:

$$PDF(d) = \frac{1}{\sqrt{2\pi\sigma_d^2}} * \exp\left[-\frac{\left(d^{EL/TH} - \mu_d\right)^2}{2\sigma_d^2}\right] \tag{8.3}$$

A high number of scenarios by applying Monte Carlo simulation can be generated. Then, a fast forward selection algorithm can be used to decrease the number of scenarios.

3.2 Multienergy storage systems

This subsection provides all constraints that are required for mathematical modeling of considered MESSs, i.e., ESS, TSS, and HSS. Energy storage systems are considered to be charged at off-peak hours and discharged at on-peak hours if required. But each storage system depends on its technical characteristics, has limitation to charge and discharge. These limitations are considered using Eq. (8.4) for charging rates and Eq. (8.5) for discharging rates. In the case of HSS, charging means converting power to hydrogen and discharging means converting hydrogen to power. In addition, it does not make sense that a storage system charges and discharges simultaneously, so relations of Eq. (8.6) are considered to prevent this from happening. Depending on the amount of charging and discharging at any time slot, the level of stored energy increases and decreases, respectively. The level of stored energy in a storage system at time t known is as the state of charge (SoC) and can be calculated using Eq. (8.7). Moreover, SoC must keep in an acceptable range for any storage system, which satisfies by Eq. (8.8). Finally, equations of Eq. (8.9) indicate that at the end of the scheduling time, horizon level of charge must be the same as the initial state.

$$
\begin{cases}
Pc_{t,s}^{-,\,ESS} Ic_{t,s}^{ESS} \leq Pc_{t,s}^{ESS} \leq Pc_{t,s}^{+} Ic_{t,s}^{ESS} \\
Pc_{t,s}^{-,\,TSS} Ic_{t,s}^{TSS} \leq Pc_{t,s}^{TSS} \leq Pc_{t,s}^{+} Ic_{t,s}^{TSS} \\
Pc_{t,s}^{-,\,HSS} Ic_{t,s}^{HSS} \leq Pc_{t,s}^{HSS} \leq Pc_{t,s}^{+,\,HSS} Ic_{t,s}^{HSS}
\end{cases}
\tag{8.4}
$$

$$
\begin{cases}
Pd_{t,s}^{-,ESS} Id_{t,s}^{ESS} \leq Pd_{t,s}^{ESS} \leq Pd_{t,s}^{+,ESS} Id_{t,s}^{ESS} \\
Pd_{t,s}^{-,TSS} Id_{t,s}^{TSS} \leq Pd_{t,s}^{TSS} \leq Pd_{t,s}^{+,TSS} Id_{t,s}^{TSS} \\
Pd_{t,s}^{-,HSS} Id_{t,s}^{HSS} \leq Pd_{t,s}^{HSS} \leq Pd_{t,s}^{+,HSS} Id_{t,s}^{HSS}
\end{cases}
\tag{8.5}
$$

$$
\begin{cases}
Ic_{t,s}^{ESS} + Id_{t,s}^{ESS} \leq 1 \\
Ic_{t,s}^{TSS} + Id_{t,s}^{TSS} \leq 1 \\
Ic_{t,s}^{HSS} + Id_{t,s}^{HSS} \leq 1
\end{cases}
\tag{8.6}
$$

$$
\begin{cases}
A_{t,s}^{ESS} = A_{t-1,s}^{ESS} + \left(\eta^{chr} \times Pc_{t,s}^{ESS} - \dfrac{Pd_{t,s}^{ESS}}{\eta^{dischr}} \right) \times \Delta t \\[4mm]
A_{t,s}^{TSS} = A_{t-1,s}^{TSS} + \left(\eta^{chr} \times Pc_{t,s}^{TSS} - \dfrac{Pd_{t,s}^{TSS}}{\eta^{dischr}} \right) \times \Delta t \\[4mm]
A_{t,s}^{HSS} = A_{t-1,s}^{HSS} + \left(\eta^{chr} \times Pc_{t,s}^{HSS} - \dfrac{Pd_{t,s}^{HSS}}{\eta^{dischr}} \right) \times \Delta t
\end{cases}
\tag{8.7}
$$

$$\begin{cases} A_{t,s}^{-,ESS} \leq A_{t,s}^{ESS} \leq A_{t,s}^{+,ESS} \\ A_{t,s}^{-,TSS} \leq A_{t,s}^{TSS} \leq A_{t,s}^{+,TSS} \\ A_{t,s}^{-,HSS} \leq A_{t,s}^{HSS} \leq A_{t,s}^{+,HSS} \end{cases} \tag{8.8}$$

$$\begin{cases} A_{(t=0),\ s}^{ESS} = A_{(t=0),\ s}^{ESS} \\ A_{(t=0),\ s}^{TSS} = A_{(t=0),\ s}^{TSS} \\ A_{(t=0),\ s}^{HSS} = A_{(t=0),\ s}^{HSS} \end{cases} \tag{8.9}$$

3.3 Boiler

A boiler can supply thermal demand using natural gas as its fuel. Eq. (8.10) indicates the amount of natural gas that is used by the boiler to produce thermal energy. In addition, the generation of thermal energy via the boiler is limited by Eq. (8.11).

$$GB_{t,\ s} = HB_{t,\ s} \times HRB \tag{8.10}$$

$$HB^- \times I_{t,\ s}^B \leq HB_{t,\ s} \leq HB^+ \times I_{t,\ s}^B \tag{8.11}$$

3.4 Demand response program

DR helps end users to participate in the scheduling program by shifting their flexible loads from on-peak hours to off-peak hours. Relation Eq. (8.12) indicates the amount of load that can be curtailed and relation Eq. (8.13) indicates the amount of load that can be shifted. The total amount of curtailed loads must be equal to total shifted loads, which are considered using Eq. (8.14). Finally, the net electrical load after curtailing and shifting loads is given by Eq. (8.15).

$$0 \leq DR_{t,s}^{crt} \leq DR \times EL_{t,s} \tag{8.12}$$

$$0 \leq DR_{t,s}^{shft} \leq DR \times EL_{t,s} \tag{8.13}$$

$$\sum_{t=t_m}^{NT_m} DR_{t,\ s}^{crt} = \sum_{t=t_m}^{NT_m} DR_{t,\ s}^{shft} \tag{8.14}$$

$$EL_{t,\ s}^{DR} = EL_{t,\ s} + DR_{t,\ s}^{shft} - DR_{t,\ s}^{crt} \tag{8.15}$$

3.5 Thermal units

GT and CHP units are considered as thermal units. Producing electrical and thermal units is dependent on each other in the CHP unit. This characteristic of the CHP unit is known as FOR characteristic that is shown in Fig. 8.3. To model this region for CHP unit, relations Eqs. (8.16)–(8.19) are taken into account. In addition, Eq. (8.20) is considered for both of the thermal units to limit their generated power.

$$P_{t, s}^{CHP} - P_A^{CHP} - \frac{P_A^{CHP} - P_B^{CHP}}{H_A^{CHP} - H_B^{CHP}} \times \left(H_{t, s}^{CHP} - H_A^{CHP}\right) \leq 0 \tag{8.16}$$

$$P_{t, s}^{CHP} - P_B^{CHP} - \frac{P_B^{CHP} - P_C^{CHP}}{H_B^{CHP} - H_C^{CHP}} \times \left(H_{t, s}^{CHP} - H_B^{CHP}\right) \geq -\left(1 - I_{t, s}^{CHP}\right) \times M \tag{8.17}$$

$$P_{t, s}^{CHP} - P_C^{CHP} - \frac{P_C^{CHP} - P_D^{CHP}}{H_C^{CHP} - H_D^{CHP}} \times \left(H_{t, s}^{CHP} - H_C^{CHP}\right) \geq -\left(1 - I_{t, s}^{CHP}\right) \times M \tag{8.18}$$

$$0 \leq H_{t, s}^{CHP} \leq H_B^{CHP} \times I_{t, s}^{CHP} \tag{8.19}$$

$$P_i^- I_{i, t, s} \leq P_{i, t, s} \leq P_i^+ I_{i, t, s} \tag{8.20}$$

3.5.1 Ramp rate limits

Ramp rate limits must be considered because thermal units cannot change their output power more than a certain amount. These constraints are as follows:

$$\Delta P_{i, t, s} = P_{i, t, s} - P_i^- \times I_{i, t, s} \tag{8.21}$$

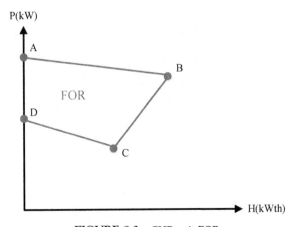

FIGURE 8.3 CHP unit FOR.

$$\Delta P_{i,\,t-1,\,s} - \Delta P_{i,\,t,\,s} \leq R_i^{up} \tag{8.22}$$

$$\Delta P_{i,\,t,\,s} - \Delta P_{i,\,t-1,\,s} \leq R_i^{down} \tag{8.23}$$

3.5.2 Minimum up/down-time

A thermal unit has a limitation to restart after shutting down, and it must stay at OFF state for a certain hour, and it must stay ON after start-up for a certain hour. These statements are given through Eqs. (8.24)–(8.27):

$$I_{i,\,t,\,s} - I_{i,\,t-1,\,s} \leq I_{i,\,t+TU_{i,\,j},\,s} \tag{8.24}$$

$$TU_{i,\,j} = \begin{cases} j; & j \leq T_i^{ON} \\ 0; & j \geq T_i^{ON} \end{cases} \tag{8.25}$$

$$I_{i,t-1,s} - I_{i,t,s} + I_{i,t+TU_{i,j},s} \leq 1 \tag{8.26}$$

$$TD_{i,\,j} = \begin{cases} j; & j \leq T_i^{OFF} \\ 0; & j \geq T_i^{OFF} \end{cases} \tag{8.27}$$

3.5.3 Start-up and shutdown costs

The start-up and shutdown costs are modeled as the fuel cost (i.e., natural gas) that thermal units consumed at start-up and shutdown time.

$$SU_{i,\,t,\,s} \geq sug_i(I_{i,\,t,\,s} - I_{i,\,t-1,\,s}) \tag{8.28}$$

$$SD_{i,\,t,\,s} \geq sdg_i(I_{i,\,t-1,\,s} - I_{i,\,t,\,s}) \tag{8.29}$$

$$SU_{i,t,s} \geq 0 \tag{8.30}$$

$$SD_{i,t,s} \geq 0 \tag{8.31}$$

$$G_{i,\,t,\,s} = P_{i,\,t,\,s} \times HR_i + SU_{i,\,t,\,s} + SD_{i,\,t,\,s} \tag{8.32}$$

3.6 Multienergy balancing

MCMG operator must meet multiple demands using included technologies. Electrical demand must be satisfied by operating ESS, HSS, CHP unit, GT, and interaction with the main grid. Thermal demand must be satisfied by operating TSS, CHP unit, boiler, and interaction with the main grid. Finally, components gas demand must be satisfied only by interaction with the gas network. Relations Eqs. (8.33)–(8.35) indicate these statements.

$$P_{t,s}^{E,\,buy} - P_{t,s}^{E,\,sell} + P_{i,t,s} + P_{t,s}^{wind} - Pc_{t,s}^{HSS} + Pd_{t,s}^{HSS} - Pc_{t,s}^{ESS} + Pd_{t,s}^{ESS} = EL_{t,s}^{DR} \tag{8.33}$$

$$P_{t,s}^{H,\,buy} - P_{t,s}^{H,\,sell} + Pd_{t,s}^{TSS} - Pc_{t,s}^{TSS} + H_{t,s}^{CHP} + HB_{t,s} = HL_{t,s}^{DR} \tag{8.34}$$

$$P_{t,s}^{G,\,buy} = GB_{t,s} + G_{i,t,s} \tag{8.35}$$

3.7 Objective function

The objective of the MCMG operator is minimizing total operation cost. In the relation of Eq. (8.36), the first and second terms indicate the cost of exchanged electrical and thermal energy with the main grid, respectively. The third term is the cost of purchasing natural gas from the gas network and the last term is the HSS charging cost.

$$\text{Cost} = Min \sum_{s=1}^{N_s} \pi_s \sum_{t=1}^{N_t} \begin{bmatrix} \lambda_t^E \left(P_{t,s}^{E,\,buy} - P_{t,s}^{E,\,sell} \right) + \lambda_t^H \left(P_{t,s}^{H,\,buy} - P_{t,s}^{H,\,sell} \right) \\ + \lambda_t^G P_{t,s}^{G,\,buy} + C^{chr,\,HSS} Pc_{t,s}^{HSS} \end{bmatrix} \tag{8.36}$$

4. Results and discussion

The proposed model is carried out on a MCMG as depicted in Fig. 8.1. The overall employed methodology is reviewed in Fig. 8.4. CHP technical data, including FOR characteristics, are obtained from Ref. [3]. MESSs' data are provided in Table 8.1 and other components of technoeconomic data, including wind power details, are obtained from Ref. [24]. In addition, forecasted prices, along with forecasted thermal demand and electrical demand, are illustrated in Figs. 8.5 and 8.6, respectively. It is assumed that natural gas is purchased from the market with 2¢/MBtu rate for all of the time slots. Scenario generation and reduction were performed using the Monte Carlo simulation and fast forward selection algorithm. To this end, 1000 scenarios were generated for wind power using Rayleigh PDF, electrical and thermal load using normal PDF, and then scenario sets reduced to 10 ones. Finally, to solve the proposed scenario-based mixed-integer linear problem, the CPLEX solver is selected. CPLEX is a solver that usually uses the branch and cut method and provides an optimal solution [25].

The stochastic problem of Eqs. (8.4)–(8.36) is solved and the optimal solution is achieved. The expected operation cost of considered MCMG is $1096.96. Table 8.2 figures out the total operation cost of each scenario. After this, all reported results are provided for a specific scenario of four. Exchanged power and thermal unit power dispatch are depicted in Fig. 8.7. It can be seen that the optimal situation for MCMG operator between t = 1, t = 8, and

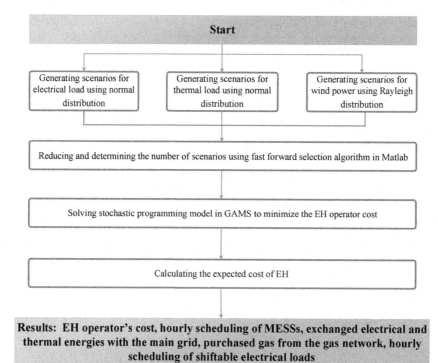

FIGURE 8.4   The review of employed methodology.

TABLE 8.1   Technoeconomic data for MESSs.

Thermal storage			Electrical storage			Hydrogen storage		
Parameter	Unit	Value	Parameter	Unit	Value	Parameter	Unit	Value
$\eta^{chr,\,hs}$	%	90	$\eta^{chr,ES}$	%	96	$C^{chr,HSS}$	¢/kWh	0.1
$\eta^{dischr,hs}$	%	90	$\eta^{dischr,ES}$	%	96	$\eta^{chr,HSS}$	%	80
$A^{+,TSS}$	kWth	500	$A^{+,ESS}$	kWh	100	$\eta^{dischr,HSS}$	%	70
$A^{-,TSS}$	kWth	0	$A^{-,ESS}$	kWh	5	$A^{+,HSS}$	kWh	100
$Pc^{+,TSS}$	kWth	50	$Pc^{+,ESS}$	kWh	25	$A^{-,HSS}$	kWh	20
$Pc^{-,TSS}$	kWth	0	$Pc^{-,ESS}$	kWh	5	$Pc^{+,HSS}$	kW	30
$Pd^{+,TSS}$	kWth	50	$Pd^{+,ESS}$	kWh	25	$Pc^{-,HSS}$	kW	10
$Pd^{-,TSS}$	kWth	0	$Pd^{-,ESS}$	kWh	5	$Pd^{+,HSS}$	kW	30
						$Pd^{-,HSS}$	kW	10

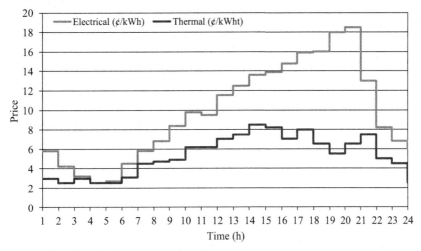

FIGURE 8.5 Forecasted electrical and thermal price.

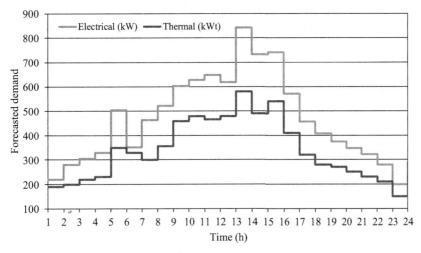

FIGURE 8.6 Forecasted electrical and thermal demand.

t = 24 is purchasing the whole power from the market to meet demand because at these hours electricity price is low so, the cost of operating thermal units is more than purchasing power from the market. But after t = 9 until t = 23, CHP unit starts to be online since electricity price rises at these hours. In addition, at peak electricity price hours, i.e., t = 19 to t = 21, GT produces the maximum power as it can. In doing so, the MCMG operator can benefit from selling surplus power to the main grid during peak electricity price hours. It should be noted that during hours t = 9 to t = 11 although electricity price is not very high, the CHP unit is online. This is because of thermal demand that forces the CHP unit to be online and because of CHP unit FOR region, some power is produced along with producing thermal energy.

TABLE 8.2 Total operation cost of each scenario.

Scenario	Probability	Total operation cost ($)
1	0.08701	1166.808
2	0.07449	1163.185
3	0.10354	1174.820
4	0.11853	1028.772
5	0.09949	1128.877
6	0.16094	1015.409
7	0.04634	1172.052
8	0.22379	1076.709
9	0.05307	1091.895
10	0.03092	1101.370

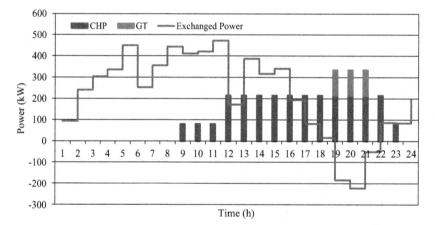

FIGURE 8.7 Exchanged power and dispatch of thermal units.

Fig. 8.8 shows the exchanged thermal energy along with the thermal dispatch of the CHP unit and boiler. If the figure checked out, it can be seen that the CHP unit goes to be online from t = 9 with an increase at thermal demand. After then, the boiler joins the CHP unit to produce thermal energy at time t = 13 with a considerable increase in demand and also an increase in thermal energy price. After the hour t = 16, thermal demand decreases but the price of thermal energy is still at high amounts. So, the MCMG operator can benefit from selling surplus thermal energy to the main grid after that hour, which happened during t = 17, 18, 21, 22. Moreover, the boiler is not online at t = 19, 20. The reason is at these hours, a decrease in both thermal demand and thermal energy price can be observable; so, the optimal decision for MCMG is purchasing thermal energy from the market instead of keeping the boiler online. In addition to the electrical and thermal energy purchasing, the MCMG

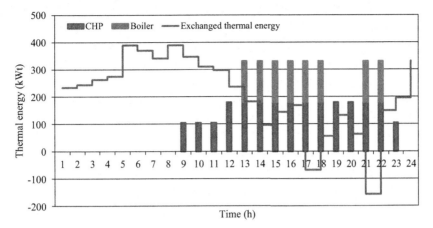

FIGURE 8.8   Exchanged thermal energy along with thermal dispatch of CHP unit and boiler.

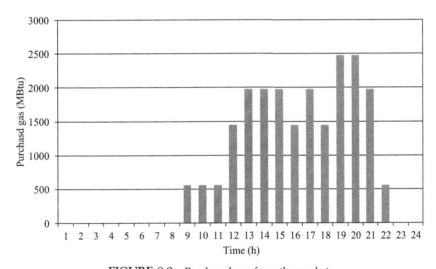

FIGURE 8.9   Purchased gas from the market.

operator must purchase natural gas to meet components fuel. Fig. 8.9 illustrates the amount of purchased gas from the network. As can be seen from the figure, gas is purchased exactly at hours that CHP unit, GT, and boiler are online, i.e., t = 9 to t = 22.

SoC for MESSs are depicted in Fig. 8.10. ESS and HSS show same performance during the scheduling time horizon and both of them applied in charge mode at initial hours in which electricity price is low, then discharges at peak price hours (t = 17 to t = 20). Furthermore, TSS charges at initial hours too because of low thermal energy price, then applied in discharge mode with increases in both demand and price of thermal energy. Finally, the DR effect on electrical demand is shown in Fig. 8.11. As can be seen in the figure, DR incentivize end users to shift their load from peak on-hours (t = 12 to t = 22) to off-peak hours (t = 1 to t = 11 and t = 23, 24). The economic benefit of both MESSs and DR is analyzed in Table 8.3. Using these technologies causes decrease in total operation cost up to $63.

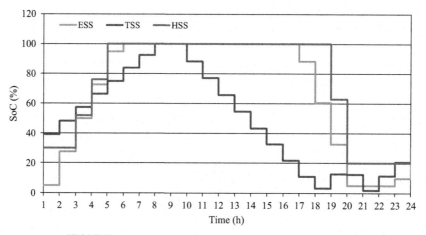

FIGURE 8.10   Multienergy storage systems' state of charge.

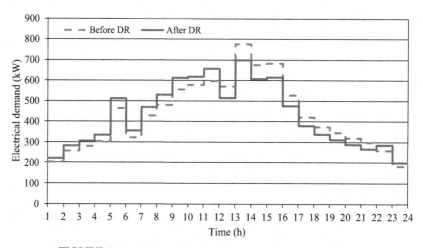

FIGURE 8.11   Effect of demand response on electrical demand.

TABLE 8.3   Economic analysis of MESSs and DR.

Technology Cost ($)	Without any storage and DR	With ESS	With ESS and TSS	With ESS, TSS, and HSS	With ESS, TSS, HSS, and DR
Exchanged electrical energy	432.8973	421.9045	42190.45	414.8645	375.0106
Exchanged thermal energy	235.2814	235.2814	221.4972	221.4972	221.4972
Purchased gas	423.9507	423.9507	423.9507	423.9507	423.9507
Total operation cost	1092.1294	1081.1366	1067.352	1060.3125	1028.7721

5. Conclusion

This chapter focused on the economic analysis of MESSs in the multienergy markets. To this end, an MCMG is equipped with three types of energy storage systems, i.e., the ESS, TSS, and HSS. HSS is a novel technology in which power from a wind turbine at low electricity price hours can be converted to hydrogen (P2H), a clean energy carrier, and vice versa (H2P) at high electricity price hours. In doing so, the MCMG operator has benefited nearly 0.65%. In addition, the utilization of ESS and TSS has caused total operation cost to decrease by 2.35%. Incentive-based DR program has been used to demand-side management, which causes a decrease up to 3% in total operation cost. Different uncertainties, including wind power intermittency, electrical and thermal demand fluctuations, are alleviated using scenario-based optimization. To do so, Monte Carlo simulation has been utilized to scenario generation based on PDFs. Then fast forward selection algorithm has been applied to scenario reduction.

References

[1] R.C. Statistics, International Renewable Energy Agency (IRENA), 2016.
[2] S. Saidi, L. Kattan, P. Jayasinghe, P. Hettiaratchi, J. Taron, Integrated infrastructure systems—a review, Sustain. Cities Soc. 36 (2018) 1—11.
[3] M. Nazari-Heris, B. Mohammadi-Ivatloo, G.B. Gharehpetian, M. Shahidehpour, Robust short-term scheduling of integrated heat and power microgrids, IEEE Syst. J. 13 (3) (2018) 3295—3303.
[4] V. Amir, S. Jadid, M. Ehsan, Optimal design of a multi-carrier microgrid (MCMG) considering net zero emission, Energies 10 (12) (2017) 2109.
[5] T. Shekari, A. Gholami, F. Aminifar, Optimal energy management in multi-carrier microgrids: an MILP approach, J. Mod. Power Syst. Clean Energy 7 (4) (2019) 876—886.
[6] S.M. Moghaddas-Tafreshi, S. Mohseni, M.E. Karami, S. Kelly, Optimal energy management of a grid-connected multiple energy carrier micro-grid, Appl. Therm. Eng. 152 (2019) 796—806.
[7] Z. Li, Y. Xu, Optimal coordinated energy dispatch of a multi-energy microgrid in grid-connected and islanded modes, Appl. Energy 210 (2018) 974—986.
[8] T. Rui, G. Li, Q. Wang, C. Hu, W. Shen, B. Xu, Hierarchical optimization method for energy scheduling of multiple microgrids, Appl. Sci. 9 (4) (2019) 624.
[9] M.H. Shams, M. Shahabi, M.E. Khodayar, Stochastic day-ahead scheduling of multiple energy carrier microgrids with demand response, Energy 155 (2018) 326—338.
[10] V. Amir, S. Jadid, M. Ehsan, Probabilistic optimal power dispatch in multi-carrier networked microgrids under uncertainties, Energies 10 (11) (2017) 1770.
[11] A. Lorestani, G.B. Gharehpetian, M.H. Nazari, Optimal sizing and techno-economic analysis of energy-and cost-efficient standalone multi-carrier microgrid, Energy 178 (2019) 751—764 (C).
[12] J. Chen, C. Chen, S. Duan, Cooperative optimization of electric vehicles and renewable energy resources in a regional multi-microgrid system, Appl. Sci. 9 (11) (2019) 2267.
[13] X. Tong, C. Hu, C. Zheng, T. Rui, B. Wang, W. Shen, Energy market management for distribution network with a multi-microgrid system: a dynamic game approach, Appl. Sci. 9 (24) (2019) 5436.
[14] Energate, Shell: Large Electrolyzer Goes Online in 2020 (PEM Electrolysis), Neue Markte, 2018 [Online]. Available, www.energate-messenger.de/news/180256/shell-grosselektrolyseur-geht-2020-ans-netz?media=print.
[15] E. Ohira, Hydrogen Cluster in Japan - Regional Activities to Promote Hydrogen, 2019.
[16] M.A. Mirzaei, A.S. Yazdankhah, B. Mohammadi-Ivatloo, Stochastic security-constrained operation of wind and hydrogen energy storage systems integrated with price-based demand response, Int. J. Hydrogen Energy 44 (27) (2019) 14217—14227.
[17] F. Kalavani, B. Mohammadi-Ivatloo, K. Zare, Optimal stochastic scheduling of cryogenic energy storage with wind power in the presence of a demand response program, Renew. Energy 130 (2019) 268—280.

[18] M.H. Barmayoon, M. Fotuhi-Firuzabad, A. Rajabi-Ghahnavieh, M. Moeini-Aghtaie, Energy storage in renewable-based residential energy hubs, IET Gener. Transm. Distrib. 10 (13) (2016) 3127—3134.

[19] C. Chen, H. Sun, X. Shen, Y. Guo, Q. Guo, T. Xia, Two-stage robust planning-operation co-optimization of energy hub considering precise energy storage economic model, Appl. Energy 252 (2019) 113372.

[20] Y. Luo, X. Zhang, D. Yang, Q. Sun, Emission trading based optimal scheduling strategy of energy hub with energy storage and integrated electric vehicles, J. Mod. Power Syst. Clean Energy 8 (2) (2020) 267—275.

[21] A. Mansour-Saatloo, M. Agabalaye-Rahvar, M.A. Mirzaei, B. Mohammadi-Ivatloo, K. Zare, Robust scheduling of hydrogen based smart micro energy hub with integrated demand response, J. Clean. Prod. (2020) 122041.

[22] S. Niaz, T. Manzoor, A.H. Pandith, Hydrogen storage: materials, methods and perspectives, Renew. Sustain. Energy Rev. 50 (2015) 457—469.

[23] A.J. Conejo, M. Carrión, J.M. Morales, Decision Making Under Uncertainty in Electricity Markets, vol. 1, Springer, 2010.

[24] M.J. Vahid-Pakdel, S. Nojavan, B. Mohammadi-Ivatloo, K. Zare, Stochastic optimization of energy hub operation with consideration of thermal energy market and demand response, Energy Convers. Manag. 145 (2017) 117—128.

[25] L. Sun, W. Liu, B. Xu, T. Chai, The scheduling of steel-making and continuous casting process using branch and cut method via CPLEX optimization, in: 5th International Conference on Computer Sciences and Convergence Information Technology, 2010, pp. 716—721.

Optimal resilient scheduling of multicarrier energy distribution system considering energy storages and plug-in electric hybrid vehicles contribution scenarios

Mehrdad Setayesh Nazar[1], *Alireza Heidari*[2]

[1]Faculty of Electrical Engineering, Shahid Beheshti University, Tehran, Iran;
[2]School of Electrical Engineering and Telecommunications (EE&T), The University of New South Wales (UNSW), Sydney, NSW, Australia

Nomenclature

Discharge Power discharge
ECDF Electrical customer damage function
ENSC Energy not supplied costs
H Heating power
HCDF Heating customer damage function
NEL_Critical Number of critical electrical load
NEL_Dispatchable Number of dispatchable electrical load
NESSAS Number of ESSA contribution scenarios
NEXSS Number of external shocks scenarios
NHL_Critical Number of critical heating load
NHL_Dispatchable Number of dispatchable heating load
NINSS Number of internal shocks scenarios
NPHEVAS Number of PHEVA contribution scenarios
NPVS Number of photovoltaic electricity generation scenarios
NWTS Number of wind turbine electricity generation scenarios
revenue Revenue of energy sold to customers
V Voltage of electrical node

Energy Storage in Energy Markets
https://doi.org/10.1016/B978-0-12-820095-7.00018-2

W Weighting factor

C_{Op} Operational costs

C^{PHEVAs} PHEVAs contribution costs

C^{ESSAs} ESSAs contribution costs

$C^{Purchase}$ Energy purchased from upward market costs

P Active electric power

$P^{Load}_{Critical}$ Critical electrical load

$P^{Load}_{Dispatchable}$ Dispatchable electrical load

$H^{Load}_{Critical}$ Critical heating load

$H^{Load}_{Dispatchable}$ Dispatchable heating load

$Energy(t)$ Energy of PHEVA or ESSA facility

ν^{Charge} Charge limitation ratio

Charge Power charge

$\nu^{Discharge}$ Discharge limitation ratio

ϕ Decision variable for commitment

1. Introduction

A multicarrier energy distribution system (MCEDS) should tolerate the external shock that may be imposed to its system, continue to deliver energy to the customers through its energy networks, recover from the shock condition, and resume to new stable conditions [1]. The shock sources may be categorized into the external and internal shocks. The external shocks are natural or cyber—physical attacks to the power system. The internal shocks are the severe contingencies of the energy system and should be considered in the security assessment of the system [1]. Multiple resilient operational paradigms of electric systems and/or natural gas systems are presented in the recent literature.

Ref. [2] proposed a two-stage optimization framework to quantify the resiliency of the electric distribution system and determine the optimal formation of microgrids. The algorithm controlled the external shock impacts and used the analytical hierarchical process (AHP) to define a resiliency index and reconfigure the system. Ref. [3] presented a two-layer metaheuristics algorithm to coordinate the distributed energy resources (DERs) of a system using graph theory. The optimal formation of microgrids was determined in the introduced method and the effectiveness of the method was assessed for the 33-bus IEEE test system.

Ref. [4] introduced a probabilistic analyzing tool for weather-driven forecasting algorithm that was integrated into a vulnerability assessment method. The proposed method utilized a risk-based corrective switching for the restoration of the most important loads of power system.

Ref. [5] proposed a three-step procedure to enhance the resilience of an electric distribution system. At the first step, the hardening preparation procedure was considered. In the second step, the corrective switching actions and microgrids formation for increasing system resiliency were performed and in the third stage, the service restoration procedure was carried out.

Ref. [6] proposed an algorithm for prioritizing of load shedding of critical loads of islanded microgrids following an external shock occurrence. The procedure utilized a mixed-integer nonlinear programming (MINLP) optimization algorithm to minimize the aggregated critical load shedding. The demand response programs and plug-in hybrid electric vehicles (PHEVs) are modeled in the proposed method.

Ref. [7] introduced a day-ahead operational planning algorithm of a distribution system that minimized the costs and maximized utility sustainability. The algorithm utilized a switching procedure to sectionalize the system in external shock conditions and the effectiveness of the proposed method was assessed for the 633-bus IEEE test system.

Ref. [8] evaluated the modularity idea to quantify the resiliency level of an electric distribution system. The proposed method utilized the graph theory and dependency-based indices to form the microgrids in the external shock and worst-case contingencies conditions.

Ref. [9] proposed a two-stage algorithm for the day-ahead optimal DERs scheduling and formation of microgrids in sever system contingencies. The first stage minimized costs of DERs operation; meanwhile, the second stage reduced the interrupted load of the system.

Ref. [10] introduced the model of interaction between electric and natural gas systems in external shock conditions. The model considered precondition and postcondition of both systems for external shock occurrence to enhance the resiliency of the electric system. A two-stage robust optimization framework was presented to assess the performance of the electric system in the worst-case shock conditions.

The ESSAs and PHEVAs commitment strategies in external shock conditions is a crucial issue in operational paradigms of RMCEDS that can completely change the energy resources controllability based on the fact that these resources can be utilized to mitigate the impacts of shocks. Thus, the multicarrier energy distribution system operator (MCEDSO) must exactly determine the ESSAs and PHEVAs contribution scenarios for the internal and external shock conditions of day-ahead operation scheduling.

An MCEDS can utilize distributed generations (DGs), combined heat and power (CHP) units, thermal energy storages (TESs), boilers, intermittent power generation (IPG) facilities, and demand response programs (DRPs). The MCEDSO can determine the optimal commitment of ESSAs and PHEVAs facilities in its optimal day-ahead resource scheduling.

Fig. 9.1 depicts the system configuration and its electrical and heating systems.

The optimal resilient day-ahead scheduling (ORDAS) of MCEDS consists of optimization of the commitment of system resources, ESSAs, and PHEVAs facilities that depends on the energy system uncertainties, resiliency criteria, and cost—benefit analysis.

This book chapter is about the ORDAS algorithm that considers the ESSAs/PHEVAs contribution uncertainties and shock scenarios.

2. Problem modeling and formulation

The ORDAS has different sources of uncertainties that are the solar and photovoltaic power generations, the location and intensity of shocks, and the ESSAs and PHEVAs contribution scenarios.

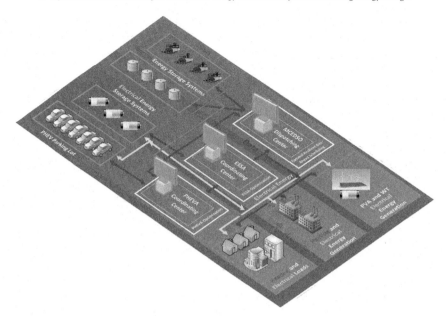

FIGURE 9.1 Schematic diagram of a multicarrier energy distribution system.

The MCEDSO forecasts its load and the upward market price using the proposed method of Ref. [11]. Then, at the first stage of ORDAS, the MCEDSO optimizes its DERs scheduling considering the different sources of uncertainties for the day-ahead horizon and mitigates the impact of the estimated external and internal shocks. Then, the MCEDSO updates its databases of load, wholesale market price, weather, and the imposed external and internal shocks data in real-time operation procedure. The system operator utilizes the second stage of ORDAS to perform the on-event corrective actions. Finally, at the third stage, the MCEDSO evaluates the effectiveness of performed procedures and implements the postevent corrective actions.

The six sources of problem uncertainties can be modeled by stochastic programming and the probability distribution functions for each stochastic process should be calculated. Then, the distribution functions are discretized and the scenario-driven method is utilized for modeling of the stochastic process. Thus, the objective functions of ORDAS can be written as a function of the probability of random variables and the expected value of the objective functions can be optimized. In this book chapter, the autoregressive integrated moving average (ARIMA) models are utilized for intermittent electricity generation. For other sources of uncertainties, the Monte Carlo stochastic process simulation procedures are used to model the location and intensity of the external shocks [12]. The internal shocks location and intensity are estimated by the historical database and modeled by the proposed model of Ref. [13]. The scenario generation and reduction procedure is utilized and the reduced number of stochastic process are delivered to the ORDAS.

2.1 First-stage problem formulation

At the first stage of ORDAS, the MCEDSO minimizes the expected costs of operation and energy not supplied for the day-ahead horizon. The first-stage problem determines the optimal MCEDS's DERs commitment schedule for the next 24-hour horizon. The problem evaluates the adequacy of system resources for the internal and external shock conditions. The objective function of the first-stage problem is introduced as follows:

$$
\text{Min } \mathbb{Z} =
\begin{pmatrix}
\left(C_{Op}^{DG} \cdot \phi^{DG} + C_{Op}^{TES} \cdot \phi^{TES} + C_{Op}^{Boiler} \cdot \phi^{Boiler} + C_{Op}^{CHP} \cdot \phi^{CHP} + C_{Op}^{DRP} \cdot \phi^{DRP} \right. \\
\left. \sum_{NPVS} prob \cdot C_{Op}^{PV} + \sum_{NWTS} prob \cdot C_{Op}^{WT} \right) \\
+ \sum_{NPHEVAS} prob \cdot C^{PHEVAs} \cdot \phi^{PHEVAs} \\
+ \sum_{NESSAS} prob \cdot C^{ESSAs} \cdot \phi^{ESSAs} \\
+ C^{Purchase} + \sum_{NEXSS} prob \cdot ENSC \\
+ \sum_{NINSS} prob \cdot ENSC - revenue
\end{pmatrix}
\tag{9.1}
$$

The objective function of the first-stage problem consists of the following parameters: (1) MCEDS DGs, TESs, boilers, CHPs, PVs, WTs, DRPs operation costs; (2) the costs of purchased energy from PHEVAs; (3) the costs of purchased energy from ESSAs; (4) the costs of purchased energy from the upward wholesale market; (5) the ENSCs of external shocks; (6) the ENSCs of internal shocks; and (7) the revenue of electrical and heating energy sold to customers.

The WT and PV electrical generation facilities are equipped with the ESSs and their electricity generation can be dispatched by the MCEDSO. The device loading constraints and electrical and thermal load flow constraints are constrained in the first-stage problem [14].

The electrical power balance constraint of MCEDS can be written as (9.2):

$$
\sum P^{CHP} + \sum P^{DG} \mp \sum P^{ESSAs} - \sum P^{Loss} - \sum P^{Load}
$$

$$
\mp \sum P^{PHEVAs} + \sum P^{WT} + \sum P^{PV} \mp \sum P^{DRP} + \sum P^{Upward} = 0
\tag{9.2}
$$

The thermal power balance constraint can be written as (9.3):

$$
\sum H^{CHP} + \sum H^{Boiler} - \sum H^{Loss} \pm \sum H^{TES} - \sum H^{Load} + \sum H^{Flow} = 0
\tag{9.3}
$$

The MCEDS electrical and thermal loads consist of critical and dispatchable loads and the dispatchable loads can be controlled by the MCEDSO. Thus, the DRP constraints for the MCEDS can be written as [14]:

$$P^{\text{Load}} = P^{\text{Load}}_{\text{Critical}} + P^{\text{Load}}_{\text{Dispatchable}} \tag{9.4}$$

$$H^{\text{Load}} = H^{\text{Load}}_{\text{Critical}} + H^{\text{Load}}_{\text{Dispatchable}} \tag{9.5}$$

$$\Delta P^{\text{Load Min}}_{\text{Dispatchable}} \leq P^{DRP} \leq \Delta P^{\text{Load Min}}_{\text{Dispatchable}} \tag{9.6}$$

$$\Delta H^{\text{Load Min}}_{\text{Dispatchable}} \leq H^{DRP} \leq \Delta H^{\text{Load Max}}_{\text{Dispatchable}} \tag{9.7}$$

The ESSAs, TESs, PHEVAs constraints are maximum capacity, charge and discharge constraints. Further, the mass balance constraints of the heating system and boiler constraints are considered that are available in Ref. [14].

The charge/discharge of PHEVAs is modeled as stochastic parameters. Thus, the ESSAs and PHEVAs contribution scenarios can be presented by probability distributions and it is assumed that the PHEVAs and ESSAs have two smart charge/discharge modes. The location of ESSs and parking lots of PHEVAs are fixed and the state of charge of their energy storage facilities can be modeled as a stochastic process. Further, the PHEVAs and ESSAs are independently dispatched. According to the aforementioned assumptions, historical data can be used to compute the probability density function [15].

The energy balance of PHEVAs and ESSAs, their battery energy limits and discharge, and charge rates can be formulated as the following equations:

$$\text{Energy}_X(t) = \text{Energy}_X(t-1) + v_X^{\text{Charge}} \cdot \text{Charge}^X(t) \cdot \Delta t$$

$$-\frac{1}{v_X^{\text{Discharge}}} \cdot \text{Dischargee}^X \cdot \Delta t \quad \forall X \in \{PHEVAs, ESSAs\} \tag{9.8}$$

$$\text{Energy}_X^{\min} \leq \text{Energy}_X \leq \text{Energy}_X^{\max} \quad \forall X \in \{PHEVAs, ESSAs\} \tag{9.9}$$

$$0 \leq \text{Charge}^X \leq \text{Charge}^{X \ \text{Max}} \quad \forall X \in \{PHEVAs, ESSAs\} \tag{9.10}$$

$$0 \leq \text{Disharge}^X \leq \text{Disharge}^{X \ \text{Max}} \quad \forall X \in \{PHEVAs, ESSAs\} \tag{9.11}$$

2.2 Second-stage problem formulation

The external and internal shocks may change the control variables of the MCEDS. The second stage of ORDAS considers the impacts of the shocks on the system and performs the on-event corrective actions in the optimization procedure.

The proposed objective function of the second-stage problem minimizes the penalty costs of electrical and heating loads and the deviations of bus voltages from their normal values. The proposed formulation of the second-stage problem can be written as (9.11):

$$\text{Min } \Omega = W_1 \cdot \sum_{NEL\_Critical} P_{Critical}^{Load} \cdot ECDF + W_2 \cdot \sum_{NHL\_Critical} H_{Critical}^{Load} \cdot HCDF + W_3 \cdot \sum_{NEB} \left| V - V^{Normal} \right|$$

(9.11)

The components of Eq. (11) are the penalty costs of interrupted critical electrical loads (first sentence), the penalty cost of interrupted critical heating loads (second sentence), and the deviation of voltages from third normal values (third sentence).

The constraints of this problem can be categorized into the following groups:

The upward and downward ramps of energy generation facilities, the AC load flow of the electrical system, the state of charge of ESSAs and PHEVAs, the minimum and maximum value of operating parameters of energy systems, and the energy balance of heating system [14].

2.3 Third-stage problem formulation

The third-stage problem performs the postevent corrective actions to restore the energy services of dispatchable heating and electrical loads. The objective function of the third-stage problem can be introduced as follows:

$$\text{Min } \Gamma = \Omega + W_4 \cdot \sum_{NEL\_Dispatchable} P_{Dispatchable}^{Load} \cdot ECDF_{Dispatchable}$$
$$+ W_5 \cdot \sum_{NHL\_Dispatchable} H_{Dispatchable}^{Load} \cdot HCDF_{Dispatchable}$$

(9.12)

The components of Eq. (9.12) are the second-stage objective (first sentence), the penalty cost of interrupted critical electrical loads (second sentence), and the penalty cost of interrupted critical heating loads (third sentence). The constraints of this problem are the same as the second-stage problem.

3. Solution algorithm

The first-stage problem has multiple nonlinear discrete decision variables and a nonlinear solver is utilized to solve the first-stage problem. The second- and third-stage problems are linearized and solved by the linear programming algorithms. Fig. 9.2 presents the optimization algorithm procedure.

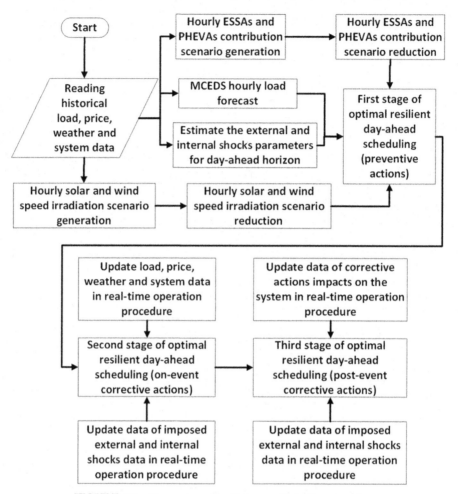

FIGURE 9.2 The proposed multistage optimization procedure.

At first, the ARIMA model forecasts the intermittent electricity generation of MCEDS solar and photovoltaic power generations. Further, the Monte Carlo procedure simulates the external shocks location and intensity [12,13]. The historical database is utilized to estimate the parameters of the internal shocks using the custom reliability assessment methods.

The MATLAB software is utilized to generate ARIMA and Monte Carlo stochastic process and the scenario generation. Then, the scenario reduction procedures are implemented using the GAMS/SCENRED library.

The first-stage problem is optimized for the generated scenarios using the DICOPT solver of GAMS package. Then, the second- and third-stage problems are optimized using the CPLEX solver.

4. Numerical results

The introduced model was implemented on the 33-bus IEEE test system and an industrial district energy distribution system. Figs. 9.3 and 9.4 depict the 33-bus and industrial town distribution system, respectively.

The 33-bus IEEE test system data is presented in Refs. [16]. Further, the wind turbine and solar panel data are available in [14].

4.1 33-bus test system

Fig. 9.5 presents the forecasted heating and electrical loads. Further, Fig. 9.6 depicts the forecasted upward network electricity price of the 33-bus IEEE test system. The MU stands for Monetary Unit.

Table 9.1 shows the optimization input data for the 33-bus IEEE test system.

Table 9.2 presents the external shocks categories.

Figs. 9.7 and 9.8 present different reduced scenarios of the day-ahead forecasted electricity generation of the 33-bus system PV and WT facilities, respectively. The maximum and minimum values of PV electricity generation were 159.8 and 0.00 kW, respectively. Further, the maximum and minimum values of WT electricity generation were 151.32 and 5.69 kW, respectively.

Fig. 9.9 depicts the scenarios of the day-ahead forecasted electricity injection/withdrawal of the 33-bus IEEE test system ESSAs and PHEVAs.

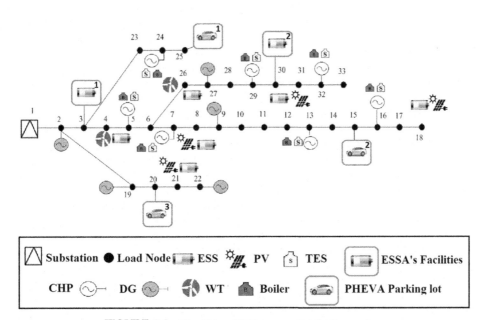

FIGURE 9.3 The 33-bus IEEE energy distribution system.

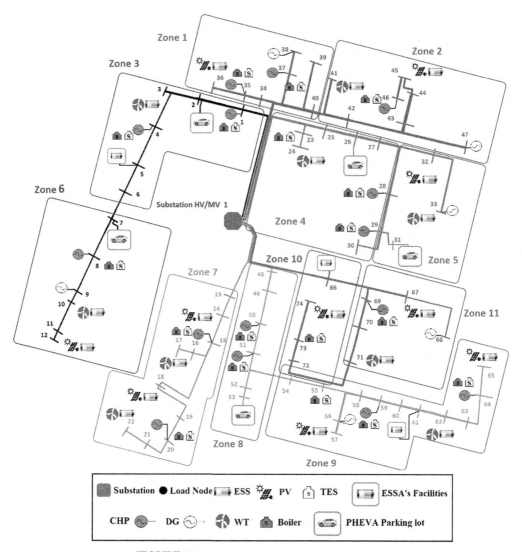

FIGURE 9.4   The industrial district energy system.

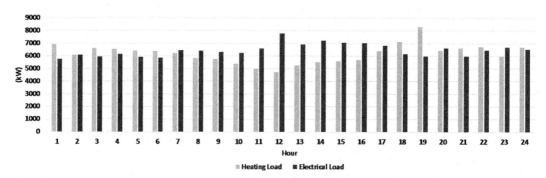

FIGURE 9.5   The forecasted heating and electrical loads of the 33-bus IEEE energy system.

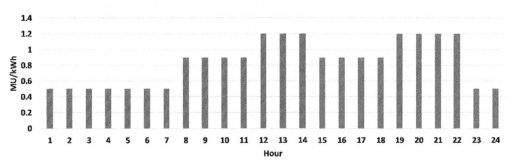

FIGURE 9.6 The forecasted upward network electricity price of the 33-bus IEEE test system.

TABLE 9.1 The optimization input data for the 33-bus IEEE test system.

System parameter	Value
Number of solar irradiation scenarios	500
Number of wind turbine power generation scenarios	500
Number of PHEVAs contribution scenarios	500
Number of ESSAs contribution scenarios	500
Number of solar irradiation reduced scenarios	10
Number of wind turbine power generation reduced scenarios	10
Number of PHEVAs contribution reduced scenarios	10
Number of ESSAs contribution reduced scenarios	10
W	1
V^{Normal}	1 pu

TABLE 9.2 The external shocks categories.

External shock category	Shocks
Extreme	Triple-line outage
	Triple DG outage
	Combination of above shocks
Expected	Double-line outage
	Double DG outage
	Single-line and double DG outage
	Combination of above shocks
Routine	Single-line and DG outage
	Double-line outage

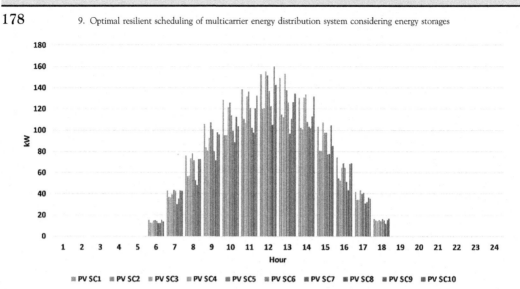

FIGURE 9.7   The reduced scenarios of the day-ahead forecasted electricity generation of the 33-bus IEEE test system PV facilities.

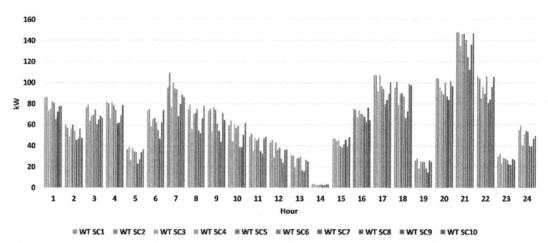

FIGURE 9.8   The reduced scenarios of the day-ahead forecasted electricity generation of the 33-bus IEEE test system WT facilities.

Fig. 9.10 shows the estimated values of CHPs and DGs electricity generation for the first stage of ORDAS procedure. As shown in Fig. 9.10 the CHPs were fully committed and the DGs were tracking the electrical load.

Fig. 9.11 presents the estimated values of heating storage charge and discharge for the day-ahead horizon and for the first-stage problem optimization.

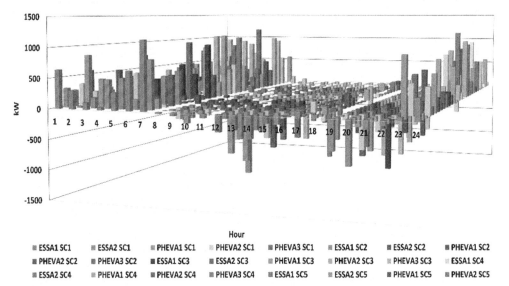

FIGURE 9.9 The reduced scenarios of the day-ahead forecasted electricity injection/withdrawal of the 33-bus IEEE test system ESSAs and PHEVAs.

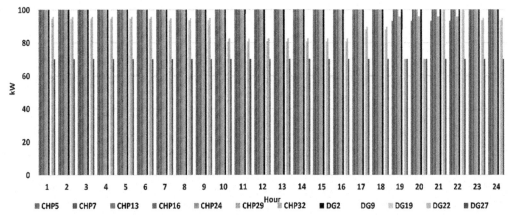

FIGURE 9.10 The estimated values of CHPs and DGs electricity generation for the first stage of ORDAS procedure.

At the second stage of ORDAS, the MCEDSO received the online data of shocks impacts on its system, updated its database, and optimized the proposed second stage of ORDAS. Then, the MCEDSO performed the on-event corrective actions to reduce the impacts of the imposed shocks.

Fig. 9.12 presents the second-stage optimization outputs for the electricity injection/withdrawal of the 33-bus system ESSAs and PHEVAs for the worst-case scenario of external shock. Fig. 9.13 presents the CHPs and DGs electricity generation for the on-event corrective

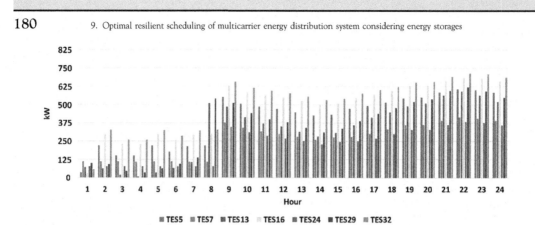

FIGURE 9.11 The estimated values of heating storage charge and discharge for the day-ahead horizon and for the first-stage problem optimization.

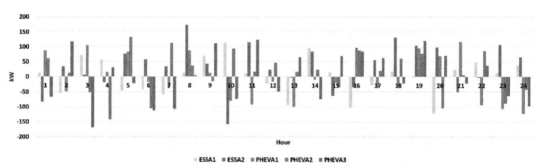

FIGURE 9.12 The electricity injection/withdrawal of the 33-bus system ESSAs and PHEVAs for the on-event corrective actions for the worst-case scenario of external shock.

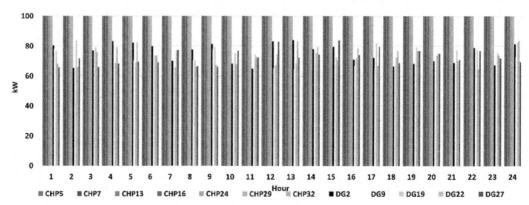

FIGURE 9.13 The CHPs and DGs electricity generation for the on-event corrective actions for the worst-case scenario of external shock.

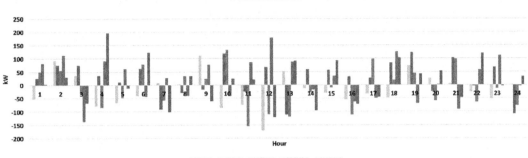

FIGURE 9.14 The electricity injection/withdrawal of the 33-bus system ESSAs and PHEVAs for the third stage of ORDAS.

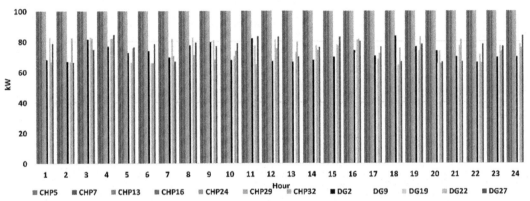

FIGURE 9.15 The CHPs and DGs electricity generation of the third-stage problem.

actions for the worst-case scenario of the 33-bus system external shock. The second-stage optimization procedure restored the entire critical load of the system for the worst-case external shock scenario.

At the third stage, the MCEDSO performed the postevent corrective actions to restore the electricity of dispatchable loads. Fig. 9.14 presents the electricity injection/withdrawal of ESSAs and PHEVAs for the third stage of ORDAS. Fig. 9.15 presents the CHPs and DGs electricity generation of the third-stage problem. All of the dispatchable loads were restored in the third-stage optimization process.

4.2 Industrial district energy system

Fig. 9.16 shows the forecasted heating and electrical loads of the industrial district energy system. Fig. 9.17 depicts the estimated day-ahead upward network electricity price. The data of system facilities is presented in Ref. [14].

Table 9.3 presents the input data for the industrial district energy system optimization process. Figs. 9.18 and 9.19 show the day-ahead forecasted electricity generation PV and WT facilities, respectively. The maximum and minimum values of WT electricity generation were

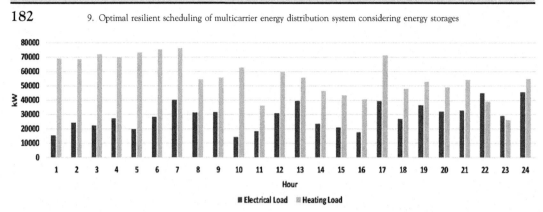

FIGURE 9.16   The forecasted heating and electrical loads of the industrial district energy system.

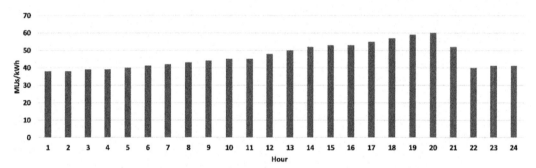

FIGURE 9.17   The estimated day-ahead upward network electricity price.

TABLE 9.3   The optimization input data for the industrial district energy system.

System parameter	Value
Number of solar irradiation scenarios	1000
Number of wind turbine power generation scenarios	1000
Number of PHEVAs contribution scenarios	1500
Number of ESSAs contribution scenarios	1500
Number of solar irradiation reduced scenarios	10
Number of wind turbine power generation reduced scenarios	10
Number of PHEVAs contribution reduced scenarios	10
Number of ESSAs contribution reduced scenarios	10
W	1
V^{Normal}	1 pu

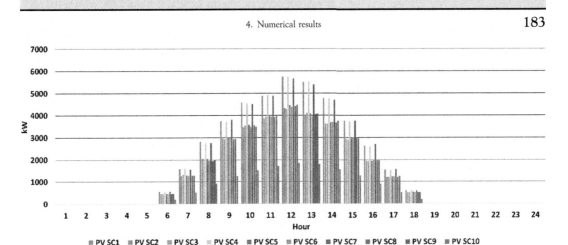

FIGURE 9.18 The reduced scenarios of the day-ahead forecasted electricity generation of the industrial district energy system PV facilities.

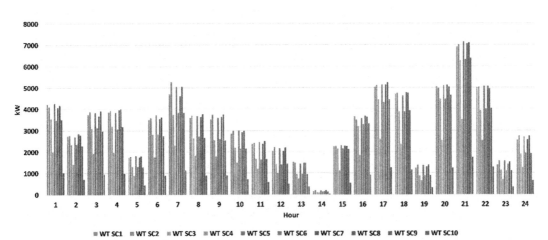

FIGURE 9.19 The reduced scenarios of the day-ahead forecasted electricity generation of the industrial district energy system WT facilities.

7151.136 and 34.90 kW, respectively. Further, the maximum and minimum values of PV electricity generation were 5474.01 and 0 kW, respectively.

Fig. 9.20 shows the scenarios of the day-ahead forecasted electricity injection/withdrawal of the industrial district energy system ESSAs and PHEVAs.

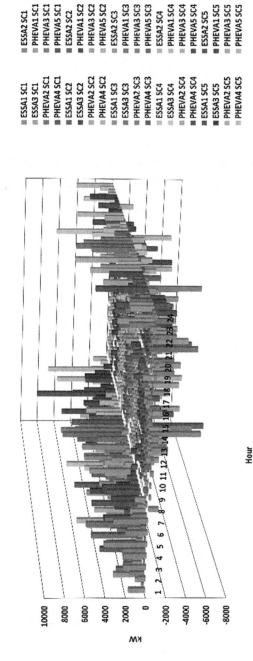

FIGURE 9.20   The scenarios of the day-ahead forecasted electricity injection/withdrawal of the industrial district energy system ESSAs and PHEVAs.

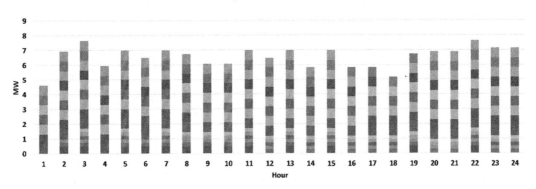

FIGURE 9.21   The estimated values stacked column of electricity generation of CHPs for the first stage of ORDAS procedure.

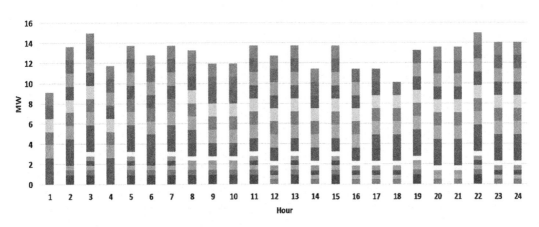

FIGURE 9.22   The estimated values stacked column of heat generation of CHPs for the first stage of ORDAS procedure.

Figs. 9.21 and 9.22 show the estimated values stacked column of electricity and heat generation of CHPs for the first stage of ORDAS procedure, respectively. The maximum values of electricity and heat generation of CHPs were 7.64 and 15.00 MW, respectively.

Fig. 9.23 presents the estimated day-ahead heating load, aggregated CHPs' heat generation, TESs charge and discharge, and boilers heat generation for the industrial district energy system. The boilers were the main source of heat generation and the TESs are optimally dispatched to supply the heating loads.

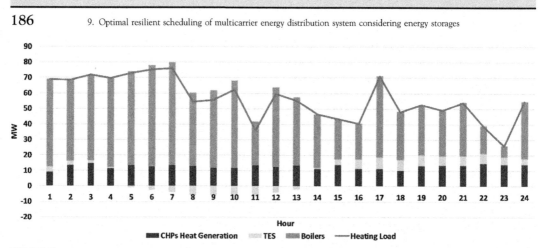

FIGURE 9.23 The estimated day-ahead heating load, aggregated CHPs' heat generation, TESs charge and discharge, and boilers heat generation for the industrial district energy system.

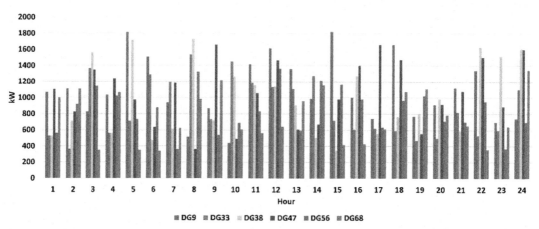

FIGURE 9.24 The estimated optimal dispatch of DGs for the first stage of ORDAS problem.

Fig. 9.24 presents the estimated optimal dispatch of DGs for the first stage of ORDAS problem.

Fig. 9.25 presents the on-event corrective dispatch of the second-stage problem. The second-stage problem determined the optimal values of electricity injection/withdrawal of

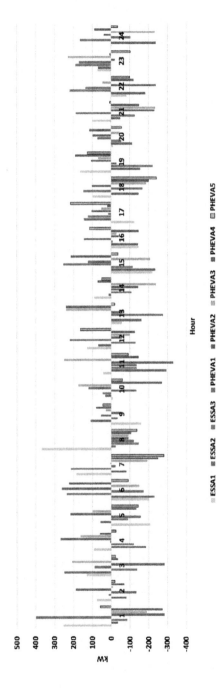

FIGURE 9.25 The on-event corrective dispatch of the second-stage problem that determines the optimal values of electricity injection/withdrawal of ESSAs and PHEVAs for the worst-case scenario of external shock.

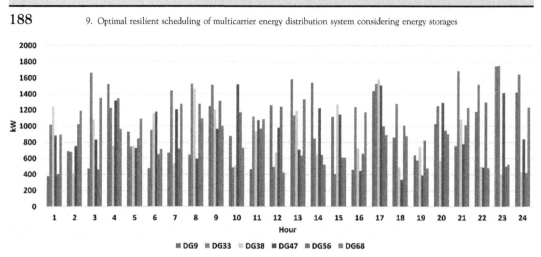

FIGURE 9.26   The DGs electricity generation for the worst-case scenario of the industrial district energy system.

ESSAs and PHEVAs for the worst-case scenario of external shock. Fig. 9.26 presents the DGs electricity generation for the worst-case scenario of the industrial district energy system. The second-stage optimization procedure restored the entire critical load of the system for the worst-case external shock scenario.

Finally, at the third stage of ORDAS, the postevent corrective actions were performed to restore the dispatchable loads. Fig. 9.27 depicts the electricity injection/withdrawal of ESSAs and PHEVAs for the third stage of ORDAS. Fig. 9.28 shows the DGs electricity generation of the third-stage problem. All of the dispatchable loads were restored in the third-stage optimization procedure.

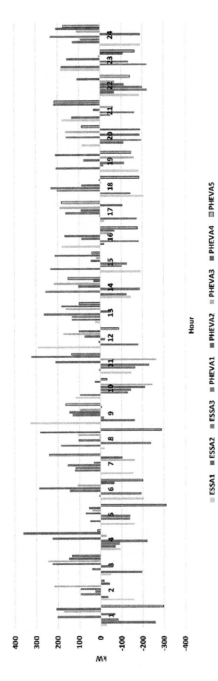

FIGURE 9.27 The electricity injection/withdrawal of ESSAs and PHEVAs for the third stage of ORDAS.

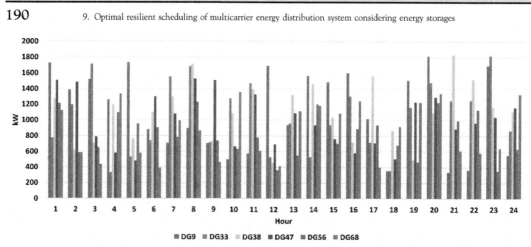

FIGURE 9.28   The DGs electricity generation of the third-stage problem.

5. Conclusions

A day-ahead multistage optimization procedure for scheduling of energy distribution system was reviewed in the present chapter. The proposed framework considered PHEVAs and ESSAs contribution scenarios in the preventive and corrective actions. The uncertainties of intermittent electricity generation, the shocks parameters, and ESSAs and PHEVAs contributions were considered in the introduced model. The first-stage problem optimized the commitment of energy resources of the system and performed preventive actions to reduce the impacts of shocks. At the second stage, the on-event corrective actions were carried out to minimize the critical load shedding. Finally, at the third stage, the postevent corrective actions were implemented to restore the dispatchable loads. The introduced algorithm was successfully assessed for the 33-bus and an industrial district energy system.

References

[1] W. Kröger, E. Zio, Vulnerable Systems, Springer, 2011.
[2] S. Chanda, A.K. Srivastava, Defining and enabling resiliency of electric distribution systems with multiple microgrids, IEEE Trans. Smart Grid 7 (2016) 2859–2868.
[3] M. Zadsar, M.R. Haghifam, S.M.M. Larimi, Approach for self-healing resilient operation of active distribution network with microgrid, IET GTD 11 (2017) 4633–4643.
[4] P. Dehghanian, B. Zhang, T. Dokic, M. Kezunovic, Predictive risk analytics for weather-resilient operation of electric power systems, IEEE Trans. Sustainable Energy 10 (2019) 3–15.
[5] A. Hussain, V.H. Bui, H.M. Kim, Microgrids as a resilience resource and strategies used by microgrids for enhancing resilience, Appl. Energy 240 (2019) 56–72.
[6] B. Balasubramaniam, P. Saraf, R. Hadidi, E.B. Makram, Energy management system for enhanced resiliency of microgrids during islanded operation, Elec. Power Syst. Res. 137 (2016) 133–141.
[7] V. Hosseinnezhad, M. Rafiee, M. Ahmadian, P. Siano, A comprehensive framework for optimal day-ahead operational planning of self-healing smart distribution systems, Int. J. Electr. Power Energy Syst. 99 (2018) 28–44.
[8] S. Mousavizadeh, T.G. Bolandi, M.R. Haghifam, M. Moghimi, J. Lu, Resiliency analysis of electric distribution networks: a new approach based on modularity concept, Int. J. Electr. Power Energy Syst. 117 (2020) 105669.
[9] M. Zadsar, S. Sebtahmadi, M. Kazemi, S.M.M. Larimi, M.R. Haghifam, Two stage risk based decision making for operation of smart grid by optimal dynamic multi-microgrid, Int. J. Electr. Power Energy Syst. 118 (2020) 105791.

[10] A.R. Sayed, C. Wang, T. Bia, Resilient operational strategies for power systems considering the interactions with natural gas systems, Appl. Energy 241 (2019) 548−566.

[11] M.S. Nazar, A.E. Fard, A. Heidari, M. Shafie-khah, J.P.S. Catalão, Hybrid model using three-stage algorithm for simultaneous load and price forecasting, Elec. Power Syst. Res. 165 (2018) 214−228.

[12] Y. Wang, C. Chen, J. Wang, R. Baldick, Research on resilience of power systems under natural disasters—a review, IEEE Trans. Power Syst. 31 (2016) 1604−1613.

[13] M.S. Nazar, A. Heidari, Multi-stage resilient distribution system expansion planning considering non-utility gas-fired distributed generation, power system resilience, in: N.M. Tabatabaei, et al. (Eds.), Power Systems Resilience, Springer, 2018, pp. 192−222.

[14] F. Varasteh, M.S. Nazar, A. Heidari, M. Shafie-khah, H.P.S. Catalão, Distributed energy resource and network expansion planning of a CCHP based active microgrid considering demand response programs, Energy 172 (2019) 79−105.

[15] A. Bostan, M.S. Nazar, Optimal scheduling of distribution systems considering multiple downward energy hubs and demand response programs, Energy 116349 (2020).

[16] W. El-Khattam, Y.G. Hegazy, M.M.A. Salama, An integrated distributed generation optimization model for distributed system planning, IEEE Trans. Power Syst. 20 (2005) 1158−1165.

Optimal participation of electric vehicles aggregator in energy and flexible ramping markets

Faezeh Jalilian[1], Amin Mansour-Saatloo[1],
Mohammad Amin Mirzaei[1], Behnam Mohammadi-Ivatloo[1,2],
Kazem Zare[1]

[1]Faculty of Electrical and Computer Engineering, University of Tabriz, Tabriz, East Azerbaijan, Iran;
[2]Department of Energy Technology, Aalborg University, Aalborg, Denmark

1. Introduction

Nowadays, greenhouse emissions due to mass motorization and increasing urbanization are a major challenge. Transportation electrification as a significant opportunity to alleviate the issues posed by transportation fleets is heeded by researchers. Electric vehicles (EVs) have a considerable role in the realization of transportation electrification, with advantages of reducing fossil fuel consumption, climate change mitigation, and increasing power grid flexibility. Indeed, from the power system operator point of view, the EV battery is an active load that can participate in the energy market and can be utilized as a distributed energy storage if effectively managed. In addition, the EV battery, due to its fast ramping capability, can participate in the ancillary services market.

Flexible ramping product (FRP) market is an ancillary services market that has been introduced by CAISO [1] and MISO [2] to provide more flexibility in the electricity market. FRP includes both upward ramping products (URPs) and downward ramping products (DRPs), where ramping constraints of the system can be satisfied. Many researches have investigated the impact of FRP market from economic advantages to the operators. A comprehensive review of FRP impact on operational flexibility of the power system can be found in Ref. [3]. An optimal bidding strategy for energy storage systems in energy and FRP markets has been presented in Ref. [4]. Security-constraint stochastic scheduling for integrated wind power and

Energy Storage in Energy Markets
https://doi.org/10.1016/B978-0-12-820095-7.00001-7

compressed air energy storage (CAES) system has been proposed in Ref. [5]. So, an EV aggregator as a mediator between EV owner and grid can bid on the day-ahead energy and FRP markets to maximize its profit [6]. However, one of the most important challenges that an EV aggregator faces is the uncertainty of different parameters, e.g., electricity market price, driving patterns of EV owners.

So far, many papers have proposed different models for scheduling of EV aggregators considering uncertainties. Day-ahead scheduling of EV aggregators to participate in the energy and regulation market considering the uncertainty of the electricity market and EV fleet patterns has been conducted in Ref. [7]. A stochastic model to deal with the energy market, reserve market, and availability of vehicles has been introduced in Ref. [8]. In order to integrate renewable energy sources and EVs, dynamic programming has been proposed in Ref. [9], where EVs can be charged with different electricity prices. The information gap decision-making (IGDT) approach has been presented in Ref. [10] to deal with electricity market uncertainty for EV aggregator day-ahead charging and discharging schedule. A bi-level methodology has been applied to the scheduling of EV aggregators in the day-ahead market in Ref. [11], where the first level aims to minimize aggregator cost, and market clearing is done in the second level. A real-time model for scheduling of EVs charging in the unregulated market has been proposed in Ref. [12], where the robust approach was applied to the optimization problem to tackle with uncertainty. In addition, the robust approach has been utilized in Ref. [13] to optimal scheduling of EV aggregators to participate in the electricity market. A stochastic multiobjective optimization model for optimal scheduling of large-scale EV fleets in the distribution system has been presented in Ref. [14]. In Ref. [15], a two-stage stochastic linear programming for real-time scheduling of the EV charging process has been presented. Authors in Ref. [16] have proposed a hybrid stochastic/IGDT model for EVs aggregator to participate in the energy market. In Ref. [17,18], risk-based models have been proposed for the optimal bidding strategy of EV aggregators to participate in the energy market.

Dealing with uncertainties is the most important challenge to optimal scheduling of EV aggregator. According to the reviewed papers and best of the found knowledge, there is no focus on the optimal scheduling of EV aggregators to participate in both energy and FRP markets considering uncertainties. So, this chapter aims to provide a model for the optimal bidding strategy of EV aggregators in energy and FRP markets. To this end, a scenario-based stochastic linear programming optimization approach is employed to face uncertainties associated with market price and EV owners' driving patterns.

The rest of the chapter is organized as follows: Section 3 provides the scenario generation and reduction algorithms along with problem formulation, including objective function and all related constraints. Section 4 introduces a case study and numerical results to verify the effectiveness of the proposed model. Finally, section 5 concludes the chapter.

2. Problem formulation

In this section, a stochastic structure is designed for solving the optimal scheduling problem of EVs aggregator. The EVs aggregator implements the stochastic optimization to explain the optimal programs of EVs' charging and discharging times. Any EV should be charged or discharged due to the diverse values of the primary state of charge (SOC), arrival time (AT), and exit time (ET) relevant to the driving pattern. To modeling the mentioned uncertainties, a stochastic program based on the scenario is considered. Furthermore, there are different types of EVs in the energy market and flexible ramping product (FRP) market, and each type uses special features for the EVs' battery, which includes it is capacity and charge and discharge rate. In this study, five types of EV with various capacities of batteries and charge and discharge rates have been used.

2.1 Objective function

The purpose of the proposed model is to maximize EV aggregator profit according to the forecasted electricity prices, which is defined as (10.1). EV aggregator income will be obtained by charging vehicles and sales of power in the energy market and FRP market. The EV aggregator allocates part of its capacity that it cannot use to the FRP market and thus can make more profit. Furthermore, the cost of EV aggregator is regarded as battery degeneration and electricity preparation from the energy market and FRP market.

The first term in Eq. (10.1) express to aggregator buying $P_{r,t}^{c,s}$ at times when electricity prices are low and sells the value $P_{r,t}^{d,s}$ power at times when electricity prices are high in the energy market and FRP market. The second term indicates the power sold by the aggregator to EV owners. The third term indicates the flexible ramp up reserve (FRUR) and flexible ramp down reserve (FRDR) provided by aggregator in the FRP market. The fourth term indicates the cost of discharging EV.

$$
\max \Upsilon = \max \sum_{t=1}^{T} \begin{bmatrix} \lambda_t \cdot \sum_{r=1}^{N_r} NV \cdot \rho_r \sum_{s=1}^{N_s} \rho_s \cdot \left(P_{r,t}^{d,s} - P_{r,t}^{c,s} \right) \\ + \sum_{r=1}^{N_r} NV \cdot \rho_r \sum_{s=1}^{N_s} \rho_s \cdot \left(\lambda_{co} \cdot P_{r,t}^{s,co} + lfrp_t \cdot FRU_{r,t}^{d,s} \right. \\ \left. + lfrp_t \cdot FRD_{r,t}^{c,s} - Deg \cdot FRU_{r,t}^{d,s} - Deg \cdot P_{r,t}^{d,s} \right) \end{bmatrix} \tag{10.1}
$$

Where $P_{r,t}^{d,s} / P_{r,t}^{c,s}$ describes discharge/charge power of EV r at time t and scenario s, respectively. The amount of power exchanged between the energy market and the EVs aggregator is traded through the predicted energy market price λ_t, on the other hand, $lfrp_t$ parameter indicates the price of the FRP market. Furthermore, $FRU_{r,t}^{d,s}$ is upward reserve in the discharge mode EV r at time t and scenario s, $FRD_{r,t}^{c,s}$ is descending reserve in charge mode EV r at time t and scenario s. The quantity of sold power to every vehicle, $p_{r,t}^{s,co}$, is defined at time t and scenario s, with the contractual price λ_{co} among EV owners and aggregator. NV and ρ_r describe the whole number of combined EVs and the percent of electric vehicles that have the same type, respectively. The occurrence possibility of any scenario relevant to SOC, AT, and ET is determined by ρ_s. On the other hand, because the vehicle to grid (V2G) application accelerates battery destruction, EV owners are reluctant to allow the EV aggregator to act battery in V2G mode. The word *Deg* indicates the amount of return cost to EV owners due to the cost of battery destruction owed to V2G discharge. Also, the amount of *Deg* is obtained through degradation cost of V2G reported in Refs. [19].

2.2 Problem constraints

Eq. (10.2) updates the amount of energy stored in any EV r at time t and scenario s.

$$EB_{r,t}^s = EB_{r,t-1}^s + \mu_c . P_{r,t}^{c,s} - 1/\mu_d . P_{r,t}^{d,s} \tag{10.2}$$

In Eq. (3.2), μ_c and μ_d indicate the charge/discharge efficiency of EV.

Eq. (10.3), to express the stored energy in the EV r at time t and scenario s, $EB_{r,t}^s$, is limited between the maximum and minimum amount of battery capacity.

$$\underline{EB_r} + FRU_{r,t}^{d,s} \leq EB_{r,t}^s \leq \overline{EB_r} - FRD_{r,t}^{c,s} \tag{10.3}$$

As indicated by $\overline{EB_r}$ and $\underline{EB_r}$ parameters, respectively.

Eqs. (10.4) and (10.5) indicate that $P_{r,t}^{c,s}$ and $P_{r,t}^{d,s}$ are limited between the maximum and minimum amount of their capacity.

$$\underline{P_r^c} \cdot \beta_{r,t}^{c,s} \leq P_{r,t}^{c,s} \leq \overline{P_r^c} \cdot \beta_{r,t}^{c,s} - FRD_{r,t}^{c,s} \quad \forall t \in [AT, ET] \tag{10.4}$$

$$\underline{P_r^d} \cdot \beta_{r,t}^{d,s} \leq P_{r,t}^{d,s} \leq \overline{P_r^d} \cdot \beta_{r,t}^{d,s} - FRU_{r,t}^{d,s} \quad \forall t \in [AT, ET] \tag{10.5}$$

Which in this equations, $\overline{P_r^c} \big/ \underline{P_r^c}$ are maximum/minimum charge power rates and $\overline{P_r^d} \big/ \underline{P_r^d}$ are maximum/minimum discharge power rates.

Also, the effects of the flexible ramp market, including FRDC and FRUD at time t and scenario s, are shown in (10.3–10.5) constraints.

Eq. (10.6) expresses that the battery of each vehicle cannot be charged and discharged at the same time due to technical limitations.

$$\beta_{r,t}^{c,s} + \beta_{r,t}^{d,s} \leq 1 \quad \forall t [AT, ET] \tag{10.6}$$

Binary variables $\beta_{r,t}^{c,s}$ and $\beta_{r,t}^{d,s}$ are utilized to model charging and discharging power states of any EV, respectively.

3. Method of scenario generation

In this section, the Monte Carlo simulation method is used to manufacturing different scenarios due to the uncertainties of primary SOC, AT, and ET for any EV. To reduce the number of scenarios, it is necessary to use the scenario reduction technique to reduce the number of scenarios. In this work, the number of scenarios produced by the Monte Carlo simulation method was initially 1000, which is reduced to 10 scenarios in GAMS software using the SCENRED tool. This tool includes two methods: the backward method and the forward method. The results obtained in the forward method are more accurate, but this method requires more computational time. SCENRED tool has the ability to select the desired number of preserved scenarios, which is named Red_num_leaves. This chapter has utilized a fast-backward reduction method according to the operating time and performance precision with the red_num_leaves factor of 10.

4. Simulation results

The stochastic programming model is provided for solving the optimal scheduling problem of EV aggregator for 24 h. Furthermore, the driving pattern of EV owners, AT and DT, battery capacity, charge and discharge rates are some factors that affect the scheduling process. In this study, it is assumed that EVs leave the house during the workday and drive directly to the workplace, and other ancillary activities of the EVs such as parking and so on are ignored. The amount of charge and discharge efficiency is considered to be 0.95 and 0.94. In this work, 10 scenarios and five types of EV have been used during 24 h period. The assumptions of the problem, which include data on the probability of occurrence of scenarios and the probability of the type of EVs, as well as other values of charge and discharge, the energy level of EVs, the FRP market price, and electricity price for 24 h are given in the following tables (Tables 10.1–10.4).

4.1 Optimal self-scheduling of EVs aggregator to participate in the energy market (case 1)

To assess the performance of the proposed approach, some numerical results are reported via implementing the proposed stochastic model on an EV aggregator by the potential of collecting 1000 vehicles. Tables 10.5 and 10.6 show the scheduled charge and discharge power of EVs in different scenarios. It is observed that the hourly charge and discharge scheduling of EVs depends on the arrival and exit time of EVs from the parking lot, as well as their initial state of charge (SOC) during arriving at parking. In general, the EVs aggregator in off-peak hours, when the price of electricity is lower (for example, between $t = 10$ and $t = 13$), prefers to buy power from the market to charge EVs. On the other hand, during peak hours, when electricity prices are higher (for example, between $t = 14$ and $t = 16$), the aggregator sells the stored energy in EVs to the market for achieving more profit, and in this status, the EVs are discharged.

TABLE 10.1 Characteristics of the EV type.

EV type	$\overline{p^c}$(kw)	$\overline{p^d}$(kw)	Percentage(%)	$\underline{p_c}$ (kw)	$\underline{p_d}$ (kw)	Capacity (kwh)
V1	17.2	17.2	21.81	3.2	3.2	100
V2	17.2	17.2	18.64	3.2	3.2	100
V3	20	20	15.00	5	5	41
V4	11.5	11.5	25.52	1.5	1.5	40
V5	12.5	12.5	19.03	2.5	2.5	25

TABLE 10.2 Characteristics of EVs in the different scenarios.

Scenario	At	ET	SOC	Occurrence probability
S1	5	20	0.58	0.09
S2	7	17	0.57	0.05
S3	6	22	0.22	0.057
S4	5	19	0.50	0.06
S5	9	20	0.20	0.077
S6	5	22	0.12	0.12
S7	6	17	0.34	0.13
S8	8	22	0.42	0.083
S9	11	18	0.69	0.183
S10	7	18	0.24	0.15

TABLE 10.3 Characteristics of the electricity price during time.

Time(h)	Electricity price (cent/kwh)	Time(h)	Electricity price (cent/kwh)	Time(h)	Electricity price (cent/kwh)	Time(h)	Electricity price (cent/kwh)
1	6.5	7	7.6	13	6.7	19	4.9
2	6.3	8	7	14	7.6	20	5.1
3	6.2	9	6.6	15	7.8	21	6.2
4	6.9	10	6.5	16	6.5	22	6.8
5	7	11	5.8	17	5.5	23	6.9
6	7.3	12	6.2	18	6.3	24	4.5

TABLE 10.4 Characteristics of the FRP market price during time.

Time(h)	The FRP market price (cent/kwh)	Time(h)	The FRP market price (cent/kwh)	Time(h)	The FRP market price (cent/kwh)	Time(h)	The FRP market price (cent/kwh)
1	0.1	7	0.2132	13	0.1	19	0.246
2	0.12	8	0.1	14	0.205	20	0.123
3	0.2009	9	0.246	15	0.2091	21	0.2501
4	0.205	10	0.2501	16	0.2132	22	0.2542
5	0.2091	11	0.1025	17	0.123	23	0.123
6	0.1	12	0.2255	18	0.2378	24	0.246

TABLE 10.5 Scheduled charge power of EVS in different scenarios.

Time	S_1	S_2	S_3	S_4	S_5	S_6	S_7	S_8	S_9	S_{10}
1	0	0	0	0	0	0	0	0	0	0
2	0	0	0	0	0	0	0	0	0	0
3	0	0	0	0	0	0	0	0	0	0
4	0	0	0	0	0	0	0	0	0	0
5	0	0	0	0	0	750	0	0	0	0
6	0	0	0	0	0	0	0	0	0	0
7	0	0	0	0	0	0	0	0	0	0
8	0	0	0	0	0	0	0	0	0	0
9	0	5221.327	0	0	0	0	8527.658	0	0	6948.800
10	6706.602	8992.500	8879.305	8879.305	6598.275	6478.034	11,117.800	0	0	11,117.800
11	15,267.800	15,267.800	15,267.800	15,267.800	15,267.800	15,267.800	15,267.800	15,267.800	13,449.642	15,267.800
12	15,267.800	15,267.800	14,766.484	14,766.484	13,696.747	14,766.484	14,766.484	15,267.800	6631.832	14,766.484
13	0	0	0	0	0	0	6948.800	0	0	6948.800
14	0	0	0	0	0	0	0	0	0	0
15	0	0	0	0	0	0	0	0	0	0
16	1292.800	15,267.800	2498.738	8895.747	1292.800	4900.009	15,267.800	0	3595.727	10,826.252
17	15,267.800	15,267.800	15,267.800	14,991.484	15,267.800	15,267.800	15,267.800	15,267.800	15,267.800	15,267.800
18	8895.747	0	6948.800	10,642.800	7445.463	6948.800	0	0	14,766.484	14,766.484
19	15,267.800	0	15,267.800	15,267.800	15,267.800	15,267.800	0	15,267.800	0	0
20	15,267.800	0	15,267.800	0	15,267.800	15,267.800	0	15,267.800	0	0
21	0	0	8895.747	0	0	8895.747	0	8370.847	0	0
22	0	0	0	0	0	0	0	0	0	0
23	0	0	0	0	0	0	0	0	0	0
24	0	0	0	0	0	0	0	0	0	0

TABLE 10.6 Scheduled discharge power of EVs in different scenarios.

Time	S_1	S_2	S_3	S_4	S_5	S_6	S_7	S_8	S_9	S_{10}
1	0	0	0	0	0	0	0	0	0	0
2	0	0	0	0	0	0	0	0	0	0
3	0	0	0	0	0	0	0	0	0	0
4	0	0	0	0	0	0	0	0	0	0
5	0	0	0	0	0	0	0	0	0	0
6	9587.648	0	1405.920	5471.349	0	0	0	0	0	0
7	15,189.230	15,267.800	11,320.552	15,015.800	0	7611.462	13,376.440	0	0	11,717.984
8	0	0	0	0	0	0	0	7909.060	0	0
9	0	0	0	0	0	0	0	0	0	0
10	0	0	0	0	0	0	0	0	0	0
11	0	0	0	0	0	0	0	0	0	0
12	0	0	0	0	0	0	0	0	0	0
13	0	0	0	0	0	0	0	0	0	0
14	14,763.800	11,819.461	14,763.800	14,540.550	14,763.800	14,763.800	11,013.923	14,763.800	14,763.800	14,697.867
15	15,267.800	15,267.800	15,267.800	15,267.800	15,267.800	15,267.800	15,267.800	15,267.800	15,267.800	15,267.800
16	0	0	0	0	0	0	0	0	0	0
17	0	0	0	0	0	0	0	0	0	0
18	4153.625	0	4153.625	0	4153.625	4153.625	0	4153.625	0	0
19	0	0	0	0	0	0	0	0	0	0
20	0	0	0	0	0	0	0	0	0	0
21	0	0	0	0	0	0	0	0	0	0
22	0	0	0	0	0	0	0	0	0	0
23	0	0	0	0	0	0	0	0	0	0
24	0	0	0	0	0	0	0	0	0	0

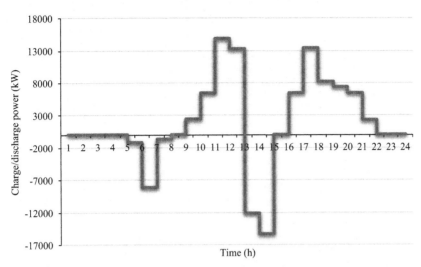

FIGURE 10.1 Hourly generation self-scheduling power of EVs.

Fig. 10.1 shows the hourly generation self-scheduling power of EVs. As shown in the figure, the operator tends to sell power to the network during peak hours, which is the discharge power (for example, between $t = 14$ and $t = 16$) and to buy power from the network during off-peak hours, which is the charging power (for example, between $t = 10$ and $t = 13$).

Table 10.7 also shows the SOC of EVs in parking lots for different scenarios. As can be observed, the SOC of EVs depends on the amount of charge and discharge of EVs in each scenario. During the EVs' arrival hours into the parking lot, the EVs' SOC is at its lowest value, but when the EVs leave the parking lots, their SOC is at its highest, and the EVs' batteries are fully charged. Additionally, for example, during hours $t = 1$ to $t = 4$ and $t = 23$ to $t = 24$, SOC of EVs in the parking lot is zero since there are not any EVs in it.

4.2 Case 1 with FRP market (case 2)

In this case, the effects of participation of EVs in the FRP market on charging and discharging power are discussed. Fig. 10.2 shows the comparison of the expected charge and discharge power of EVs in cases 1 and 2. According to this figure, charging and discharging power of EVs is different in two cases due to the participation of EVs aggregator in the FRP market. However, as can be demonstrated in Tables 10.8 and 10.9, in both cases, EVs aggregator in off-peak hours (between $t = 10$ and $t = 13$) purchases power from the market to charge EVs. On the other hand, during peak hours (between $t = 14$ and $t = 16$), the aggregator sells the stored energy in EVs to the market for obtaining more profit by applying EVs in discharging mode.

TABLE 10.7 SOC of EVs in parking lot under different scenarios.

Time	S_1	S_2	S_3	S_4	S_5	S_6	S_7	S_8	S_9	S_{10}
1	0	0	0	0	0	0	0	0	0	0
2	0	0	0	0	0	0	0	0	0	0
3	0	0	0	0	0	0	0	0	0	0
4	0	0	0	0	0	0	0	0	0	0
5	35,693.200	0	0	30,770.000	0	8097.300	0	0	0	0
6	25,493.574	0	12,043.140	24,949.416	0	8097.300	20,923.600	0	0	0
7	9334.819	18,835.460	0	8975.161	0	0	6693.345	0	0	2303.660
8	9334.819	18,835.460	0	8975.161	0	0	6693.345	17,432.906	0	2303.660
9	9334.819	23,795.720	0	8975.161	12,308.000	0	14,794.620	17,432.906	0	8905.020
10	15,706.091	32,338.595	8435.340	17,410.501	18,576.361	6154.132	25,356.530	17,432.906	0	19,466.930
11	30,210.501	46,843.005	22,939.750	31,914.911	33,080.771	20,658.542	39,860.940	31,937.316	55,239.760	33,971.340
12	44,714.911	61,347.415	36,967.910	45,943.071	46,092.681	34,686.702	53,889.100	46,441.726	61,540.000	47,999.500
13	44,714.911	61,347.415	36,967.910	45,943.071	46,092.681	34,686.702	60,490.460	46,441.726	61,540.000	54,600.860
14	29,008.740	48,773.520	21,261.740	30,474.400	30,386.511	18,980.532	48,773.520	30,735.556	45,833.830	38,964.831
15	12,766.400	32,531.180	5019.399	14,232.060	14,144.170	2738.191	32,531.180	14,493.215	29,591.489	22,722.490
16	13,994.560	47,035.590	7393.200	22,683.020	15,372.330	7393.200	47,035.590	14,493.215	33,007.430	33,007.430
17	28,498.970	61,540.000	21,897.610	36,924.930	29,876.740	21,897.610	61,540.000	28,997.625	47,511.840	47,511.840
18	32,531.180	0	24,080.220	47,035.590	32,531.180	24,080.220	0	24,578.875	61,540.000	61,540.000
19	47,035.590	0	38,584.630	61,540.000	47,035.590	38,584.630	0	39,083.285	0	0
20	61,540.000	0	53,089.040	0	61,540.000	53,089.040	0	53,587.695	0	0
21	0	0	61,540.000	0	0	61,540.000	0	61,540.000	0	0
22	0	0	61,540.000	0	0	61,540.000	0	61,540.000	0	0
23	0	0	0	0	0	0	0	0	0	0
24	0	0	0	0	0	0	0	0	0	0

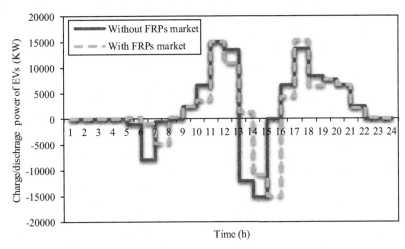

FIGURE 10.2   Charge and discharge power of EVs in case 1 and 2.

Tables 10.10 and 10.11 represent the scheduled FRUR and FRDR in the various scenarios. The FRUR and FRDR provided by aggregator affect the charging and discharging power of the EVs, their SOC, as well as the total profit of the EVs aggregator. Fig. 10.3 shows the hourly FRP provided by EVs. According to the figure, when the aggregator applies EVs in discharge mode (t = 6, t = 7, and t = 14), EVs can provide upward reserve by increasing their production. When EVs are used in charge mode (t = 9 and t = 10), they can give a downward reserve by increasing their consumption.

5. Conclusions

This chapter introduced a stochastic optimization method for the optimal scheduling of EVs aggregator. The associated uncertainties of EV owners' driving models are simulated through the scenario-based stochastic approach. The hourly charge and discharge schedule of EVs depended on the arrival and exit time of EVs from the parking lot, as well as their initial SOC, when arrived at parking. Also, the SOC of EVs depended on the amount of charge and discharge of EVs in each scenario. The effects of participation of EVs aggregator in the FRP market on charge and discharge, as well as the SOC of EVs, have been significant. Furthermore, the influence of the participation of EVs aggregator in the energy and FRP market is considered on the EVs' profit. According to the results, the total profit amount of EVs aggregator by participating in the FRP and energy market was 60,417.352 Cent, while without the FRP market, the profit amount of EVs aggregator was 52,796.851 Cent, which this indicated that in the presence of the FRP market, EVs aggregator profit has increased by about 14.433%.

TABLE 10.8   Scheduled charge power of EVs in different scenarios.

Time	S_1	S_2	S_3	S_4	S_5	S_6	S_7	S_8	S_9	S_{10}
1	0	0	0	0	0	0	0	0	0	0
2	0	0	0	0	0	0	0	0	0	0
3	0	0	0	0	0	0	0	0	0	0
4	0	0	0	0	0	0	0	0	0	0
5	1225.000	0	0	1225.000	0	2901.800	0	0	0	0
6	0	0	0	0	0	384.000	0	0	0	0
7	0	0	0	0	0	0	0	0	0	0
8	0	1292.800	1676.800	384.000	0	1676.800	1676.800	0	0	1676.800
9	2901.800	2901.800	2901.800	2901.800	2901.800	2901.800	2901.800	2901.800	0	2901.800
10	2901.800	2901.800	2901.800	2901.800	2901.800	2901.800	8557.800	2901.800	0	5983.420
11	15,267.800	15,267.800	15,267.800	15,267.800	15,196.747	15,267.800	15,267.800	15,267.800	13,449.642	15,267.800
12	12,608.933	13,020.521	12,710.989	13,541.484	9894.463	12,710.989	13,541.484	2911.117	6631.832	13,541.484
13	1676.800	1292.800	1676.800	1676.800	1676.800	1676.800	1676.800	1292.800	0	1676.800
14	0	0	0	0	0	0	0	0	0	0
15	0	0	0	0	0	0	0	0	0	0
16	1676.800	5744.117	1676.800	2151.800	1676.800	1676.800	10,297.509	1676.800	2901.800	8557.800
17	15,267.800	15,267.800	15,267.800	14,991.484	15,267.800	15,267.800	15,267.800	15,267.800	15,267.800	15,267.800
18	6017.200	0	7332.800	10,142.554	7332.800	7332.800	0	1676.800	14,766.484	14,766.484
19	15,267.800	0	15,267.800	15,267.800	15,267.800	15,267.800	0	15,267.800	0	0
20	15,267.800	0	15,267.800	0	15,267.800	15,267.800	0	15,267.800	0	0
21	0	0	8127.747	0	0	8127.747	0	6123.789	0	0
22	0	0	0	0	0	0	0	430.011	0	0
23	0	0	0	0	0	0	0	0	0	0
24	0	0	0	0	0	0	0	0	0	0

TABLE 10.9 Scheduled discharge power of EVs in different scenarios.

Time	S_1	S_2	S_3	S_4	S_5	S_6	S_7	S_8	S_9	S_{10}
1	0	0	0	0	0	0	0	0	0	0
2	0	0	0	0	0	0	0	0	0	0
3	0	0	0	0	0	0	0	0	0	0
4	0	0	0	0	0	0	0	0	0	0
5	1676.800	0	0	1676.800	0	0	0	0	0	0
6	3338.405	0	1676.800	2901.800	0	1292.800	1676.800	0	0	0
7	15,267.800	8088.600	8385.373	9228.725	0	5918.832	5160.440	0	0	5338.912
8	1676.801	384.000	0	1292.800	0	0	0	3815.770	0	0
9	0	0	0	0	0	0	0	0	0	0
10	0	0	0	0	0	0	0	0	0	0
11	0	0	0	0	0	0	0	0	0	0
12	0	0	0	0	0	0	0	0	0	0
13	0	0	0	0	0	0	0	384.000	0	0
14	14,763.800	2901.800	14,763.800	12,077.199	12,334.952	14,763.800	2901.800	14,763.800	14,144.123	6547.800
15	15,267.800	15,267.800	15,267.800	15,267.800	15,267.800	15,267.800	13,866.243	15,267.800	15,267.800	15,267.800
16	0	0	0	0	0	0	0	0	0	0
17	0	0	0	0	0	0	0	0	0	0
18	4153.625	0	4153.625	475.000	4153.625	4153.625	0	4153.625	0	0
19	0	0	0	0	0	0	0	0	0	0
20	0	0	0	0	0	0	0	0	0	0
21	0	0	0	0	0	0	0	384.000	0	0
22	0	0	0	0	0	0	0	0	0	0
23	0	0	0	0	0	0	0	0	0	0
24	0	0	0	0	0	0	0	0	0	0

TABLE 10.10 Scheduled FRUR of EVs in different scenarios.

Time	S_1	S_2	S_3	S_4	S_5	S_6	S_7	S_8	S_9	S_{10}
1	0	0	0	0	0	0	0	0	0	0
2	0	0	0	0	0	0	0	0	0	0
3	0	0	0	0	0	0	0	0	0	0
4	0	0	0	0	0	0	0	0	0	0
5	8216.000	0	0	8216.000	0	0	0	0	0	0
6	11,929.395	0	7500.289	12,366.000	0	4700.841	8216.000	0	0	0
7	0	7179.200	2551.179	5656.000	0	2834.361	8216.000	0	0	6425.226
8	7237.757	2560.000	0	5656.000	0	0	0	10,518.500	0	0
9	0	0	0	0	0	0	0	0	0	0
10	0	0	0	0	0	0	0	0	0	0
11	0	0	0	0	0	0	0	0	0	0
12	0	0	0	0	0	0	0	0	0	0
13	0	0	0	0	0	0	0	2560.000	0	0
14	504.000	12,366.000	504.000	3190.601	2932.848	504.000	12,366.000	504.000	1123.677	8720.000
15	0	0	0	0	0	0	1401.557	0	0	0
16	0	0	0	0	0	0	0	0	0	0
17	0	0	0	0	0	0	0	0	0	0
18	687.500	0	687.500	1900.000	687.500	687.500	0	687.500	0	0
19	0	0	0	0	0	0	0	0	0	0
20	0	0	0	0	0	0	0	0	0	0
21	0	0	0	0	0	0	0	2560.000	0	0
22	0	0	0	0	0	0	0	0	0	0
23	0	0	0	0	0	0	0	0	0	0
24	0	0	0	0	0	0	0	0	0	0

TABLE 10.11 Scheduled FRDR of EVs in different scenarios.

Time	S_1	S_2	S_3	S_4	S_5	S_6	S_7	S_8	S_9	S_{10}
1	0	0	0	0	0	0	0	0	0	0
2	0	0	0	0	0	0	0	0	0	0
3	0	0	0	0	0	0	0	0	0	0
4	0	0	0	0	0	0	0	0	0	0
5	3414.250	0	0	4150.000	0	12,366.000	0	0	0	0
6	0	0	0	0	0	2560.000	0	0	0	0
7	0	0	0	0	0	0	0	0	0	0
8	0	5656.000	8216.000	2560.000	0	8216.000	8216.000	0	0	8216.000
9	12,366.000	12,366.000	12,366.000	12,366.000	12,366.000	12,366.000	12,366.000	12,366.000	0	12,366.000
10	12,366.000	12,366.000	12,366.000	12,366.000	12,366.000	12,366.000	6710.000	12,366.000	0	9284.380
11	0	0	0	0	71.053	0	0	0	0	0
12	932.551	0	830.495	0	1423.337	830.495	0	8103.660	0	0
13	8216.000	631.959	8216.000	7672.136	8216.000	8216.000	6178.621	5656.000	0	8073.974
14	0	0	0	0	0	0	0	0	0	0
15	0	0	0	0	0	0	0	0	0	0
16	8216.000	9523.683	8216.000	10,116.000	8216.000	8216.000	4970.291	8216.000	12,366.000	6710.000
17	0	0	0	276.316	0	0	0	0	0	0
18	3875.600	0	2560.000	2750.246	2560.000	2560.000	0	8216.000	0	0
19	0	0	0	0	0	0	0	0	0	0
20	0	0	0	0	0	0	0	0	0	0
21	0	0	0	0	0	0	0	2560.000	0	0
22	0	0	0	0	0	0	0	0	0	0
23	0	0	0	0	0	0	0	0	0	0
24	0	0	0	0	0	0	0	0	0	0

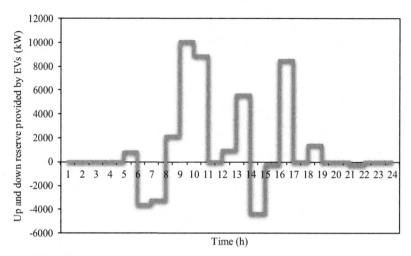

FIGURE 10.3   Hourly upward and downward reserve provided by EVs.

References

[1] L. Xu, D. Tretheway, Flexible Ramping Products: Draft Final Proposal, Calif. ISO, 2012, pp. 1–51.

[2] Midcontinent ISO (MISO). [Online]. Available: availble:%0A. https://www.misoenergy.org/pages/home.aspx.

[3] Q. Wang, B.-M. Hodge, Enhancing power system operational flexibility with flexible ramping products: a review, IEEE Trans. Ind. Informatics 13 (4) (2016) 1652–1664.

[4] W. Liu, H. Guo, J. Li, Y. Chen, Q. Chen, Optimal bidding strategy for energy storage systems in energy and flexible ramping products markets, in: 2016 IEEE Innovative Smart Grid Technologies-Asia (ISGT-Asia), 2016, pp. 776–780.

[5] M.A. Mirzaei, M. Nazari-Heris, B. Mohammadi-Ivatloo, M. Marzband, Consideration of hourly flexible ramping products in stochastic day-ahead scheduling of integrated wind and storage systems, in: 2018 Smart Grid Conference (SGC), 2018, pp. 1–6.

[6] C. Quinn, D. Zimmerle, T.H. Bradley, The effect of communication architecture on the availability, reliability, and economics of plug-in hybrid electric vehicle-to-grid ancillary services, J. Power Sources 195 (5) (2010) 1500–1509.

[7] S.I. Vagropoulos, A.G. Bakirtzis, Optimal bidding strategy for electric vehicle aggregators in electricity markets, IEEE Trans. Power Syst. 28 (4) (2013) 4031–4041.

[8] M. Alipour, B. Mohammadi-Ivatloo, M. Moradi-Dalvand, K. Zare, Stochastic scheduling of aggregators of plug-in electric vehicles for participation in energy and ancillary service markets, Energy 118 (2017) 1168–1179.

[9] B. Škugor, J. Deur, Dynamic programming-based optimisation of charging an electric vehicle fleet system represented by an aggregate battery model, Energy 92 (2015) 456–465.

[10] J. Zhao, C. Wan, Z. Xu, J. Wang, Risk-based day-ahead scheduling of electric vehicle aggregator using information gap decision theory, IEEE Trans. Smart Grid 8 (4) (2015) 1609–1618.

[11] M.G. Vayá, G. Andersson, Optimal bidding of plug-in electric vehicle aggregator in day-ahead and regulation markets, Int. J. Electr. Hybrid Veh. (IJEHV) 7 (3) (2015) 209–232.

[12] N. Korolko, Z. Sahinoglu, Robust optimization of EV charging schedules in unregulated electricity markets, IEEE Trans. Smart Grid 8 (1) (2015) 149–157.

[13] Y. Cao, L. Huang, Y. Li, K. Jermsittiparsert, H. Ahmadi-Nezamabad, S. Nojavan, Optimal scheduling of electric vehicles aggregator under market price uncertainty using robust optimization technique, Int. J. Electr. Power Energy Syst. 117 (2020) 105628.

[14] A.H. Einaddin, A.S. Yazdankhah, A novel approach for multi-objective optimal scheduling of large-scale EV fleets in a smart distribution grid considering realistic and stochastic modeling framework, Int. J. Electr. Power Energy Syst. 117 (2020) 105617.

[15] Z. Wang, P. Jochem, W. Fichtner, A scenario-based stochastic optimization model for charging scheduling of electric vehicles under uncertainties of vehicle availability and charging demand, J. Clean. Prod. 254 (2020) 119886.

[16] P. Aliasghari, B. Mohammadi-Ivatloo, M. Abapour, Risk-based scheduling strategy for electric vehicle aggregator using hybrid Stochastic/IGDT approach, J. Clean. Prod. 248 (2020) 119270.

[17] Z. Xu, Z. Hu, Y. Song, J. Wang, Risk-averse optimal bidding strategy for demand-side resource aggregators in day-ahead electricity markets under uncertainty, IEEE Trans. Smart Grid 8 (1) (2015) 96–105.

[18] A.T. Al-Awami, E. Sortomme, Coordinating vehicle-to-grid services with energy trading, IEEE Trans. Smart Grid 3 (1) (2011) 453–462.

[19] S. Shojaabadi, S. Abapour, M. Abapour, A. Nahavandi, Simultaneous planning of plug-in hybrid electric vehicle charging stations and wind power generation in distribution networks considering uncertainties, Renew. Energy 99 (2016) 237–252.

Electric vehicles as means of energy storage: participation in ancillary services markets

Mohammad Taghi Ameli[1], *Ali Ameli*[2]

[1]Department of Electrical Engineering, Shahid Beheshti University, Tehran, Iran;
[2]Entrepreneurship Incubator, Shahid Beheshti University, Tehran, Iran

1. Introduction

The entrance of EVs as an additional load in the distribution grid could cause new challenges for the grid operators. Some of these challenges could be likely line congestion and as a result, additional operating costs arise from the expansion of the network and the reinforcement of new grid components with higher capacity [1]. A promising solution for overcoming this unavoidable situation seems to be the management of the EVs charging process [2]. The batteries of the EVs could be considered as a virtual energy source in the power grid, the available potentials and possible advantages of which should be examined [3].

There are two noticeable characteristics for EVs: (a) EVs can charge their batteries, using the emission-free energy of renewable energy sources at times where there is power surplus in the grid; and (b) they have batteries that are capable of producing or storing electricity [10].

A study of the behavior of EV owners done demonstrates that the vehicles are parking 90% of a day [8]. In another study, there can be seen that the average time of driving includes just 4% of the day, which means that the parking duration of the applied EVs in the city will be about 23 h in a day [9]. Since the batteries of the EVs are almost full due to their short usage time, EVs can be considered as a potential source of energy exchange in the grid (supply and storage) [7]. The ability to inject power back to the grid is known as vehicle to grid (V2G).

The V2G concept is currently a fundamental part of the smart grid definition in most references. With the expansion of the smart grids, the speed of the influence of EVs on the grid is also increased [10]. As expected, a great part of the generated power in industrialized countries is going to be delivered by renewable energy resources (mainly wind and solar plants). The EVs batteries could play an essential role for catching the variabilities of renewable energy resources to ensure power stability in the whole grid [4]. Ancillary services will be applied to keep the power network frequency at its nominal value [5]. Along with this, it is of great importance that the driving plans of the EVs should absolutely not be interrupted.

Another feature of the battery besides its storage capacity is the state of charge (SoC) that presents the ratio of stored energy in a battery to its useable capacity, which is mostly expressed in percent [7,11]. The energy storage capacity of a battery decreases as a result of permanent switching of the charging flow. To avoid this storage capacity reduction, the battery could stop charging at a SoC of 85%–90%, which optimizes the performance of the battery and increases the lifetime of the battery [7,11].

In this chapter, the storage potentials of V2G and different types of ancillary services will be introduced, and after that, the available storage technologies are compared with the EVs for providing these ancillary services from the technical and economic point of view. Finally, a case study for the participation of EVs in ancillary market and providing different kind of services is performed.

2. EVs as energy storage systems

2.1 Storage potentials of V2G

Because of the EVs' expected and remarkable role in the future, the topic V2G is becoming of great importance, due to the following reasons and assumptions:

- The large application of renewable energy sources, which have a fluctuating nature for covering most of the system demand in developed countries.
- The EVs are parked most of the time (95% of the day) [22].
- According to [18], the major group of EVs will charge at home with a single-phase connection (3.7 kW for a 230 V distribution system), rather than fast charging.

Based on the assumption earlier, considering the use of the EVs' full battery capacity, a significant potential would be available for energy storage.

Thus, due to the V2G concept, possible services that could be provided by EVs include providing base load, peak load, and power reserves. In the following, the mentioned services are described.

In this context, important definitions should be explained for the EVs:

- Unidirectional charging: One-way power injection from the grid to the EVs.
- Bidirectional charging: Two-way power transmission between the grid and the EVs (V2G).
- Positive power reserve: Providing additional power reserve by the EVs in case of lack of power in the system.
- Negative power reserve: Imposing additional load to the system by the EVs in case of excess power in the system.

Regarding providing base load, there has been a study done in [12] on the price of providing electricity by three different groups of EVs (full electric, hybrid, and fuel cell) in California. The results of this study show that the price of power generation by the EVs is more than the baseload power. Thus, participating in the base load market would not be economically feasible. Therefore, it can be concluded that V2G technology is not a suitable solution to meet the requirements of the baseload market. These results have been confirmed in [13]. Additionally, the amount of energy that has to be traded in these energy market is enormous and availability should be guaranteed for a long time period. Providing this service by the EVs (V2G) is actually not possible because of the unreliable predicted data. In this regard, a large number of EVs with specified travel plans are needed for a two-way energy exchange. However, the additional load that will be imposed by the EVs to the power grid will increase the baseload and consequently could lead to a price reduction in the total energy prices. Regarding this, vehicles can be plugged into the grid at times with lower electricity prices (mainly at nights) where the system demand is low and cheap energy could be bought from the grid for battery charging purposes. This action flattens the load curve while increasing the total system efficiency through increasing the baseload, which is attractive for the power system operators and producers. On the other hand, the large amount of fossil fuel can be used up in power plants to generate power for charging EVs with high efficiency instead of being consumed in internal combustion engine vehicles, with low efficiency [13].

Gas turbines or diesel generators are commonly used for covering the peak load, which is more expensive than the one for baseload. The application of power plants that could provide peak power for a long time (a few hours) is not economic. The use of EVs for supplying power at peak hours could be a feasible solution. For the first time, in 1997, the economic study of providing peak power by different kinds of EVs was presented in reference [10]. In this study, peak power provision by EVs and the electric grid are compared, and it shows that peak power prices in the electricity market are more than double the needed monthly investment cost for applying the storage capacity of the EVs batteries. The peak power supply through EVs has economic benefits for the EV owners and the power grid operators according to the results of the calculations in the mentioned study. Different factors determine the economic operation of EVs for supplying peak power, including battery type, capacity, and degradation cost [14]. Due to reference [13], some factors such as battery and electricity price reduction next to various incentives for EV owners make EVs economic and qualified sources for supplying peak power. EVs participation in the peak load market decreases the need to set up common power plants [16]. There should be enough incentives for the car owners to plug in their EVs and participate in supplying peak power anytime needed. It was shown in study [15] that EV owners could not be motivated to plug in their cars to the power grid without a practical incentive program. Therefore, under current condition, this market could not be considered for EV participation.

The provision of control reserve seems to be adequate for the EVs, since lower amount of energy needs to be provided or consumed. Moreover, additional income could be earned by the car owners due to the procurement fee, which is paid to the participants in this kind of markets. This will be further discussed in the next sections.

2.2 Ancillary services

Power grid operators and regulators are in charge of power supply with high reliability while maintaining the balance between supply and demand at any time. This will be achieved by guaranteeing long-term availability of power generation capacity any time needed and fast response time of the system to demand deviation and outages. Thus, system imbalances could be avoided by ensuring the capability of increasing/decreasing the supply or demand of the system any time required. For this purpose, the communication with generator operators and other market participants who offer different kind of grid services is necessary for the grid operators. These grid services are mostly offered in energy, capacity, and ancillary service markets. Maintaining the balance between power supply and demand in real time is a worldwide technical issue.

Regarding this, ancillary service's primary purpose is to provide acceptable power quality to maintain the grid stability and support the reliability of the power system, which could be provided either through loads or power generation units.

There are different types of ancillary services that could be offered to the power grid. The main services are frequency regulation, operating reserves, reactive power regulation (voltage control), and restoration.

To manage the fluctuation of the frequency and voltage, which resulted from the imbalances between supply and demand, the ancillary services are essential to be available in the system within seconds or minutes.

2.2.1 Frequency regulation

The reserve capacity of a power system is applied to maintain the grid frequency at its standard range. Unforeseen events such as unexpected load increase, the outage of generators, transmission lines, and other equipment are resulting in an imbalance between power supply and demand of the power system. These events could lead to unwanted load shedding when there is not sufficient reserve capacity in the system.

The responsible generation unit for providing spinning reserve should react instantly after the incident and be able to supply its maximum capacity in less than 10 min. The capacity of the largest generation unit in addition to a part of the peak load capacity is generally considered as the reserve capacity, which is the most valuable part of the power generation system.

The cost for the provided spinning reserve is calculated based on the time on which the units are ready to provide power and the amount of energy that they can deliver to the grid [12]. Because the EVs are parked most time of the day, the participation in the related market is highly eligible for EV owners, and hence they can receive the reserve fee for almost a whole day. From the technical point of view, since the EVs have actually no start-up time, they can easily connect to the grid and get online in less than 10 min [12]. Therefore, the EVs could participate either in short-term electricity trading market or control reserve market.

In follow, a summary of the basics of power system balance will be presented.

Electrical systems are designed to be run at a nominal frequency value (e.g., 50 Hz in most of the countries). For example, in the United States of America the overall system frequency value is 60 Hz (cycles per second) whereas in the European countries and China, the frequency value is tried to be kept at 50 Hz. Deviation from the desired frequency value hurts the lifetime of the grid-connected equipment and also has a negative impact on their operation quality. Therefore, it is essential for gird operators to keep the frequency of their grid at its designated frequency. For this purpose, the balance between power supply and demand in the power system has to be maintained continuously. If the generated power is more than the system load, the frequency value rises and vice versa, the frequency drops if the demand is higher than the supply.

The balance between the power supply and demand of the system could be maintained as follows:

1. The excess of the generated power can be stored anytime the supply is too high or the power supply rate can be decreased by possibility. Moreover, both approaches could also be combined.
2. Applying power reserves to meet the excess of the power demand or increase the power supply by possibility relative to the additional load.

As it can be seen in Fig. 11.1, the frequency control and power balancing are categorized according to the European Network of Transmission System Operators for Electricity (ENTSO-E) as follows: primary control, secondary control, and tertiary control reserve.

Different types of control reserves are listed as follows:

- The primary control power is being prepared trough the power plants within seconds after the observation of a frequency deviation in the grid. For example, in Germany, this type of service starts to compensate the power loss right after the failure with an initialization time of 30 s and could remain active from 0 to 15 min per individual case.
- The secondary control power will be automatically activated after the failure remains in the system and could not be covered by the primary control power. In the German power system, this reserve will be ready in 5 min and could remain active from 30 s to 15 min per individual accident.
- The tertiary control power is a type of power reserve that will be activated manually just in case that the secondary control power could not compensate the occurred incident. The initialization time of this service depends on the TSO's (transmission system operator) decision, which is usually within 15 min. Its activation time is below 15 min up to as long as required.

It's noticeable that usually the secondary and tertiary control power could bring back the frequency to its nominal value where the primary control power is not able to do.

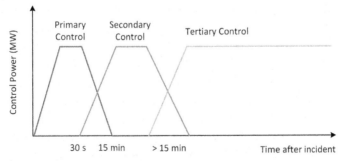

FIGURE 11.1   Different types of control reserves due to ENTSO-E. *Based on Amprion GmbH, Procurement of Control Power and Energy in Germany [online]. Available: http://www.amprion.net/en/control-energy.*

2.3 EVs as power grid controller

Due to the following reasons, EVs are well suited to provide different types of ancillary services [23]:

- EVs will have higher value in ancillary services markets regarding their fast response time.
- Generally participating power plants in the ancillary services market will lose part of their profits, since they have to reserve some generating capacity. Unlike this, EVs are able to offer their services to the market at no additional costs.
- The maintenance costs caused by repeated charging and discharging of the EVs batteries could be neglected, if the cycles are a small percentage of the total capacity of the batteries.

Generally, the primary control is provided by the power plants at a low price and because of the rare demand of tertiary control reserve [17], these markets are not appropriate for EVs' participation under current condition. Thus, the secondary control reserve market seems to be the most promising market for the EVs.

2.4 Power reserve storage systems

Different kinds of storage technologies that are currently applied for power reserve including pump storage, compressed-air energy storage (CAES), lead—acid, NaS and Li-ion batteries, with low self-discharge rate [17]. In the last decades, the pumped hydroelectric storage system (PHS) was the most adjusted storage system with over 127 GW installed capacity worldwide (more than 90% of the total world's storage capacity) [20]. The PHS system would stay the most applicable storage technology under current conditions. This is due to its high efficiency (more than 80%) [21], capability of high cycle (thus resulting in meager specific energy costs), and very low self-discharge rate. However, the construction of PHS plants is constrained due to its dependence on geographical and environmental condition. Furthermore, due to the expected wide integration of renewable energy resources, which have an uncertain and unpredictable nature, the volatility of the system would be increased, and therefore the need for other storage units next to the PHS technology becomes inevitable for maintaining the power system security and reliability.

2.4.1 Economic comparison

As discussed in the previous parts, the PHS power plants are the leading storage technology installed around the world. An economic comparison between V2G and the PHS technology could prepare a very useful viewpoint in this issue. Due to reference [17], the specific generation cost of the PHS power plants for providing reserve energy lays between 0.03 and 0.04 € per kWh. Thus, the related costs for offering same services by the EVs in the secondary control market should be logically less or at least equal to the similar costs for PHS. One important factor in the final generation costs of the EVs is the so-called gradation cost of the battery, which is described in detail in the next part.

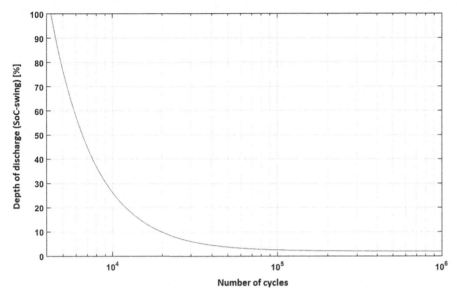

FIGURE 11.2 Relation between Li-ion battery cycle and the DoD. *Derived from D. Dallinger, J. Link, M. Büttner, Smart grid agent: plug-in electric vehicle. IEEE Trans. Sust. Energy, 5(3) (2014) 710–717.*

2.4.2 Battery degradation cost

Regarding the nonlinear specification of the EVs batteries, the depth of discharge (DoD) of the battery is very significant in the cost of the generated power through the EVs. Moreover, the DoD also affects the life time of the battery. For example, in Fig. 11.2, the relation between the life cycle and DoD of a Li-ion battery for a case scenario in 2030 is demonstrated. As it can be concluded from this figure, the life cycle of the battery is very low at high DoD and vice versa.

The number of cycles N^{cycle} is related to the battery's DoD and could be expressed with Eq. (11.1) according to [11]:

$$N^{cycle} = a \times DoD^b \tag{11.1}$$

Based on reference [11] a case study for 2030 is considered, assuming a = 4000 and b = −1.632 in (1) for a Li-ion battery.

The degradation cost of the battery in a discharging mode depends on the status of the DoD at the beginning (DoD^{start}) and at the end (DoD^{end}). By dividing one single cycle to the number of all cycles, the total battery cost could be calculated [11].

In Eq. (11.2), the discharge cost of the battery at a specific DoD is calculated due to reference [11]:

$$c^{dis}(0, DoD) = \frac{C^{bat}}{N^{cycle}(DoD)}, \tag{11.2}$$

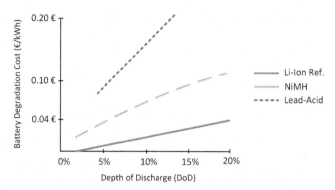

FIGURE 11.3   Battery degradation costs for different battery technologies considering the DoD [17].

where the cost of the battery is shown by C^{bat}.

Using (3), the degradation costs per kWh, which are shown in Fig. 11.3, can be derived as follows:

$$c^{\text{dis,energy}}(0, DoD) = \frac{C^{\text{bat}} \times DoD \times E^{\text{bat}}}{N^{\text{cycle}}(DoD)}, \tag{11.3}$$

where E^{bat} is defined as the usable energy of the battery.

The general degradation costs of the battery's discharging procedure can be concluded by (4):

$$c^{\text{dis,energy}}(DoD^{\text{start}}, DoD^{\text{end}}) = c^{\text{dis}}(0, DoD^{\text{end}}) - c^{\text{dis}}(0, DoD^{\text{start}}), \forall DoD^{\text{end}} > DoD^{\text{start}} \tag{11.4}$$

As presented in Fig. 11.3, the Li-ion battery technology has the gentlest slope compared to the other battery technologies (NiMH and lead acid). The investment costs for this storage technologies are considered as 250 €/kWh for the Li-Ion Ref., 500 €/kWh for the NiMH technology, and 100 €/kWh for the lead–acid batteries regarding [17]. Therefore, Li-ion battery technologies are potential candidates for the EVs in order to be economically feasible compared to PHS power plant. In 2030, the EVs are regulated to serve the secondary control reserve market under the DoD's range of 0%–20%. In case of the negative control market, EV's DoD would not be considered, as there is no need to supply power to the grid. The implementation costs of the specialized charging infrastructure and the degradation cost of the battery should be assumed in the negative control reserve.

With the invention of the Li-ion battery, the second issue would not be a significant problem anymore.

The Li-ion batteries have the lightest tilt in comparison to the other battery types, as shown in Fig. 11.3. As a result, Li-ion batteries seem to be economically competitive with the PHS power plants and hence the most attractive battery technology to be applied in the EVs. Therefore, under the DoD's range of 0%–20%, it is economically feasible for the EVs to participate in the secondary control reserve market and offer positive control power. The EVs DoD is neglected in case of offering negative control reserves, as power injection from the EVs is not needed. Furthermore, due to the low memory effect of the Li-ion technology, the battery degradation costs could also be ignored, as the maximum battery charge remains the same over time due to its low self-discharge.

Therefore, the participation of the EVs with Li-ion batteries or other similar technologies in the negative secondary control reserve market will be profitable and increase the benefits of the system.

3. Review on power regulation methods

A review of the latest studies on providing ancillary services to the electricity grid by the EVs is gathered in this section.

As discussed at the beginning of this chapter, the uncontrolled integration of EVs into the power grid will lead to many problems and challenges, such as increased peak load, frequency and voltage fluctuation, and even a system shutdown. On the other hand, with proper management and control of the charging and discharging EVs, they can be applied for optimal operation of the grid while being charged and also help in maintaining power grid stability. Therefore, papers that have recently been published in the field of the EV mainly examine different services that could be provided by them such as frequency regulation, peak shaving, load balancing, and demand response, voltage profile improvement, network loss reduction, active and reactive power management, etc.

The participation of EVs in network frequency regulation to support the operation of the power system is known as one of the most essential ancillary services. EV dispatching strategies determine the feasibility and efficiency of their participation in network frequency regulation [19].

In reference [19], an optimal strategy for EVs' participation in secondary frequency regulation is presented, and at the same time, the demand for charging EVs is met. EVs' charging demand is tracked by minimizing their energy requests along with online frequency control so that the settings are optimally adjusted according to the expected charge of the EVs. Therefore, after frequency control, the expected charge of the EV can be met optimally.

For the fair allocation of dispatching from the control center between EVs according to their charge demand, two optimal real-time strategies based on area control error (ACE) and area regulation requirement (ARR) have been proposed and compared as regulatory signals [19].

In most studies, ACE and ARR have been used as control signals for EVs to participate in frequency regulation. The ARR signal is obtained when the ACE signal passes through a PI controller.

The two optimal strategies are compared precisely in terms of improving the quality of the frequency control and ensuring the expected charge. But participating in the power reserve control market is not being investigated.

A hierarchical charging plan and control structure is presented in [24] for the efficient coordination of EVs charging and discharging procedure for providing ancillary services while meeting the travel needs of vehicles. The proposed control structure includes the coordination layer and the vehicle layers. At the vehicle layer, each set of electric vehicle supply equipment (EVSE) is equipped with a controller that estimates the charging energy and energy flexibility of the cars based on vehicle characteristics, SoC, and future travel information. The controller also controls the car's charging capacity in real time. The coordination layer applies the charge flexibility model received from each EVSE controller and updates the power system information, and the central coordinator allocates optimal power based on required ancillary service delivery and improvement of the distribution system operation or service delivery. The proposed method involves two types of control agents: local EVSE controllers and a central coordinator. Each local controller creates a simple charge flexibility model, and the central coordinator considers the flexibility models and formulates and resolves the problem of optimal charge allocation. The proposed structure is based on a hierarchical control structure that has clear advantages over a fully centralized or decentralized control for EV charging in the field of computing and communications [24]. This study considers coordinated charging for load shifting and, by moving the load from peak hours to off-peak hours, ideally provides a relatively flat load curve. This strategy helps in reducing charging cost and the annual peak load, which avoids additional expansion cost of the distribution system. Case studies using a practical testing system and actual travel patterns showed that the proposed method could meet the travel needs of the vehicle to accommodate the charge for load management.

In [25], a three-zone power system is considered, and to coordinate plug-in EVs to support primary frequency control, a dual-level consensus-based frequency control method (public satisfaction) is provided. High-level control distributes commands in different control areas intending to minimize frequency deviation in all control areas. After receiving the command from the high-level controller, the low-level controller reacts to coordinate the charging and discharging activities of the individual EVs in the area to agree on the frequency adjustment cost and minimize the cost of battery degradation. Simultaneously, PEV charging and discharging limits are also considered. The incremental cost functions are developed based on the goals of the dual control levels, in which case no objective function needs to be solved, and instead, by achieving an incremental cost of consensus between adjacent sections, optimal results can be obtained. Graph theory has been used to create communication networks in adjacent control areas in high-level control and among adjacent PEVs in low-level control. The consensus algorithm is used to specify updates and information exchange between adjacent sections. Besides, dynamic models of multizone power systems and PEVs have been developed to implement and control the controller. In the proposed control method, the consensus algorithm is used to reach an agreement on the specified incremental cost by exchanging information from adjacent sections. The design of random matrices in the consensus algorithm is done to ensure convergence of incremental cost. The proposed method in [25] can be used to simulate real-time simulations, and optimal charge instructions can be obtained and updated at any time step based on current deviation conditions and electricity prices. As a result, plugged-in EVs can be effectively coordinated, through a dual-level controller, to provide primary frequency control services.

In [26] the potential of the EVs as an alternative way for helping in a fast restoration process is investigated. An optimization approach for a two-stage restoration strategy is being considered. The optimization model is applied to maximize the needed power for the restoration and minimize the deviations of the needed power for black start. The goal of this study is to provide reliable restoration power for black start generators by coordinating the EVs and wind turbines, as well as catching the frequency fluctuation by the electric vehicle.

Applying smart grid capabilities could provide real-time information about the SoC and the amount of available power capacity from all plugged-in vehicles to help the grid operators to control their charging procedures and available regulation power capacity [27]. Therefore, a fast smart charging and regulation (SCRM) management method, which offers an optimized charging plan for the available EVs in the distribution grid considering grid operational constraints and also provides regulation power anytime needed, is presented in [28]. This proposed solution helps grid operators to have a more reliable power network and also the EV owners to meet their driving wishes with no concerns.

4. Case study

In order to show the performance of the EVs as a grid regulation service and the resulting economic values, a wind power outage of about 4.8 MW at 10:00 p.m. in the distribution network is assumed as a study case. Considering a number of 500 EVs assuming 20% of the DoD as explained in Section 2, only 0.4 MW of the 4.8 MW power failure could be compensated by the plugged-in EVs in this distribution network.

This means that in this case, not all EVs can contribute in mitigation of this error and consequently the activation of other regulation services is needed to cover the occurred power lack. After maintaining the network balance, the grid frequency will be restored to its nominal value (in this case 50 Hz).

To overcome this amount of failure without additional regulation units, a fleet of 6000 electric vehicles would be required.

4.1 Economic calculation

The appropriate market for the EVs could be defined through calculating the costs and obtained profits for providing ancillary services in different regulation markets. Some of these costs include bidirectional plug-in devices, infrastructure equipment, and battery degradation costs [17], (as discussed in section 2).

According to Fig. 11.3, for a specific investment cost of 250 €/kWh for the batteries in 2030 as well as 20% DoD (discussed in section 2), a battery degradation cost of about 0.04 €/kWh is extracted. Furthermore, throughout the depreciation time of vehicle investment, an approximate €500 infrastructure cost could be assumed (i.e., equal to 6 years in Germany (85 €/a) [17], which can be neglected due to the longer lifetime of battery compared to the vehicle.

Due to the small share of tertiary controls in the reserve market, the participation of primary and secondary control is considered in the market.

- Primary regulation:

An average energy price of 2500 €/MW/week [6] is considered for positive and negative primary regulation services. A regulation service of 0.4 MW is assumed for both positive and negative reserve in a typical work day. The cost and revenues are calculated as follows (It should be mentioned that for negative services there is no extra infrastructure and battery degradation costs.).

- Positive reserve for a typical work day:

$$\text{Cost} = (85 \text{ €} \times 500 \text{ EVs}) + \left(0.4 \text{ MWh} \times 40 \frac{\text{€}}{\text{MWh}} \right) = 42,516 \text{ €/a}$$

$$\text{Revenue} = \left(0.4 \text{ MW} \times 2500 \frac{\text{€}}{\text{MW}} \times \frac{365}{7} \right) = 52,000 \text{ €/a}$$

- Negative reserve for a typical work day:

$$\text{Cost} = 0$$

$$\text{Revenue} = \left(0.4 \text{ MW} \times 2500 \frac{\text{€}}{\text{MW}} \times \frac{365}{7} \right) = 52,000 \text{ €/a}$$

- Secondary regulation:

In order to calculate the cost and revenues for the secondary regulation market, following assumptions have been made [6]: (i) average procurement fee of 1000 €/MW/Week, (ii) average energy price of 130 €/MWh, and (iii) average procurement fee for negative regulation service of 560 €/MW.

- Positive reserve for a typical work day:

$$\text{Cost} = (85 \text{ €} \times 500 \text{ EVs}) + \left(0.4 \text{ MWh} \times 40 \frac{\text{€}}{\text{MWh}} \right) = 42,516 \text{ €/a}$$

$$\text{Revenue} = \left(0.4 \text{ MWh} \times 130 \frac{\text{€}}{\text{MWh}} \right) + \left(0.4 \text{ MW} \times 1000 \frac{\text{€}}{\text{MW}} \times \frac{365}{7} \right) = 20,852 \text{ €/a}$$

- Negative reserve for a typical work day:

$$\text{Cost} = 0$$

$$\text{Revenue} = \left(0.4 \text{ MWh} \times 130\frac{\text{€}}{\text{MWh}}\right) + \left(0.4 \text{ MW} \times 560\frac{\text{€}}{\text{MW}} \times \frac{365}{7}\right) = 11,700 \text{ €/a}$$

The results indicate the profit of offering both negative primary and secondary control reserves. The positive primary control reserve is profitable compared to the negative primary reserve. Although participating in the positive secondary control market would not be reasonable due to higher costs compared to the revenues. In general, neglecting the low DoD of 20% would result in profits through all the regulation services.

It is also valuable to mention that the negative markets become attractive by neglecting the infrastructure investment costs in the future due to lower battery prices as well as fast battery technology development [17].

5. Conclusion

The potential of applying EVs as energy storage systems has been investigated in this chapter. Based on the V2G concept and the characteristic of the EVs batteries, a significant potential would be available for offering various grid services including load management, demand response, and grid regulation services such as frequency control. The provision of control reserve is adequate for the EVs, since lower amount of energy needs to be provided or consumed. Moreover, additional income could be earned by the car owners due to the procurement fee, which is paid to the participants in the secondary control markets. As discussed, the PHS power plants are the leading storage technology installed around the world. The economic comparison between V2G and the PHS technology shows that the EVs' specific generation cost for providing reserve energy must lay between 0.03 and 0.04 € per kWh to be competitive for offering same services in the secondary control market. Considering this assumptions, the calculation results show that the primary control reserve market is profitable for the EVs for offering both negative and positive services. The negative secondary market is also interesting for the EVs' participation. Although, the positive secondary control is not a suitable option, under current assumptions for 2030 due to the high cost compared to the revenues.

However, all the regulation services would be profitable after neglecting the low DoD of 20% and infrastructure investment costs due to the fast-developing battery technology and lower battery prices in the future.

Additionally, increased reserve demand in the regulation market, due to the raise of unpredictable electricity generation by renewables in the future around the world, will lead to higher market prices and profits for the offered regulation services by the EVs.

References

[1] A.S. Masoum, S. Deilami, P.S. Moses, M. Masoum, Smart load management of plug-in electric vehicles in distribution and residential networks with charging stations for peak shaving and loss minimisation considering voltage regulation, IET Gener., Transm. Distrib. 5 (8) (2011) 877–888.

[2] J. Kim, J. Jeon, S. Kim, C. Cho, K. Nam, Cooperative control strategy of energy storage system and microsources for stabilizing the microgrid during islanded Operation, IEEE Trans. Power Electron. 25 (12) (2010) 3037–3048.

[3] C. Wu, H. Mohsenian-Rad, J. Huang, PEV- based combined frequency and voltage regulation for smart grid, Innovative Smart Grid Technologies (ISGT), IEEE Power Energy Soc. (2012) 1–6.

[4] A. Ameli, H. Ameli, S. Krauter, R. Hanitsch, An optimized load frequency control of decentralized energy system using plug-in electric vehicles, in: International Conference of Innovating Energy Access for Remote Areas: Discovering Untapped Resources, 10-12th of April, University of California, Berkeley, 2014, pp. 14–18.

[5] S. Han, S. Han, S. Kaoru, Estimation of achievable power capacity from plug-in electric vehicles for V2G frequency regulation: case studies for market participation, IEEE Transactions on Smart Grid 2 (4) (2011) 632–641.

[6] Germany's internet platform for providing regulation power services [online]. Available: https://www.regelleistung.net.

[7] T. Yiyun, L. Can, C. Lin, L. Lin, Research on vehicle-to-grid technology, in: International Conference on Computer Distributed Control and Intelligent Environmental Monitoring (CDCIEM), 2011, pp. 1013–1016.

[8] A.J. Markel, K. Bennion, W. Kramer, J. Bryan, J. Giedd, Field testing plug-in hybrid electric vehicles with charge control technology in the xcel energy territory, NREL Technical Report/ TP-550-46345 (2009), https://doi.org/10.2172/963562.

[9] P.S. Hu, J.R. Young, Summary of Travel Trends: 1995 Nationwide Personal Transportation Survey, 1999.

[10] W.J. Smith, Plug-in hybrid electric vehicles- A low-carbon solution for Ireland? Energy Pol. 38 (2010) 1485–1499.

[11] D. Dallinger, J. Link, M. Büttner, Smart grid agent: plug-in electric vehicle, IEEE Trans. Sust. Energy 5 (3) (2014) 710–717.

[12] R.J. Bessa, M.A. Matos, Economic and technical management of an aggregation agent for electric vehicles: a literature survey, Eur. Trans. Electr. Power 22 (2012) 334–350.

[13] W. Kempton, T. Kubo, Electric-drive vehicles for peak power in Japan, Energy Pol. 28 (2000) 9–18.

[14] L. Wang, Potential impacts of plug-in hybrid electric vehicles on locational marginal prices, in: IEEE Energy 2030 Conference, Atlanta, GA, 2008, pp. 1–7.

[15] S.B. Peterson, J. Whitacre, J. Apt, The economics of using plug-in hybrid electric vehicle battery packs for grid storage, J. Power Sources 195 (2010) 2377–2384.

[16] B.V. Mathiesen, H. Lund, K. Karlsson, 100% Renewable energy systems, climate mitigation and economic growth, Appl. Energy 88 (2011) 488–501.

[17] P. Jochem, T. Kaschub, W. Fichtner, How to integrate electric vehicles in the future energy system?, in: Working Paper Series in Production and Energy Karlsruhe Institute of Technology (KIT), 2013.

[18] M. Schreiber, P. Hochloff, Capacity-dependent tariffs and residential energy management for photovoltaic storage systems, in: IEEE Power & Energy Society General Meeting, Vancouver, BC, 2013, pp. 1–5.

[19] H. Liu, K. Huang, N. Wang, J. Qi, Q. Wu, S. Ma, C. Li, Optimal dispatch for participation of electric vehicles in frequency regulation based on area control error and area regulation requirement, Appl. Energy 240 (2019) 424–439.

[20] Electric Power Research Institute, Electricity Energy Storage Technology Options, White Paper Primer on Applications, Cost and Benefits, Stanford University, 2010.

[21] Energy Storage Association, Pumped Hydroelectric Storage [online]. Available: http://energystorage.org/energy-storage/technologies/pumped-hydroelectric-storage.

[22] D. Palmer, C. Ferris, Parking Measures and Policies Research Review, Project report for British Department for Transport. Transport Research Laboratory, 2010.

[23] M.A. Fasugba, P.T. Krein, Cost Benefits and Vehicle-To-Grid Regulation Services of Unidirectional Charging of Electric Vehicles, IEEE Energy Conversion Congress and Exposition, Phoenix, AZ, 2011, pp. 827–834.

[24] D. Wu, N. Radhakrishnan, S. Huang, A hierarchical charging control of plug-in electric vehicles with simple flexibility model, Appl. Energy 253 (2019) 113490.

[25] L. Wang, B. Chen, Dual-level consensus-based frequency regulation using vehicle-to-grid service, Elec. Power Syst. Res. 167 (2019) 261–276.

[26] C. Zhang, H. Zhang, S. Liu, Z. Lin, W. Fushuan, Two-stage restoration strategies for powersystems considering coordinated dispatch between plug-in electric vehicles and wind power units, IET Smart Grid 3 (2) (2020) 123–132.

[27] A. Ameli, S. Krauter, M.T. Ameli, Smart charging management system of plugged-in EVs based on user driving patterns in micro-grids, Int. J. Eng. Res. Innovat. 10 (1) (2018) 12–17.

[28] A. Ameli, S. Krauter, M.T. Ameli, H. Ameli, Proposing a smart charging decision mechanism for EVs applying smart grid capabilities, in: NEIS Conference, Conference on Sustainable Energy Supply and Energy Storage Systems, Hamburg, 2017.

Evaluating the advantages of electric vehicle parking lots in day-ahead scheduling of wind-based power systems

Mohammad Hemmati[1], Amin Mansour-Saatloo[1], Masoumeh Ahrabi[2], Mohammad Amin Mirzaei[1], Behnam Mohammadi-Ivatloo[1,3], Kazem Zare[1]

[1]Faculty of Electrical and Computer Engineering, University of Tabriz, Tabriz, East Azerbaijan, Iran; [2]Department of Electrical Engineering, Amirkabir University of Technology, Tehran, Iran; [3]Department of Energy Technology, Aalborg University, Aalborg, Denmark

Nomenclature

Index

b Bus index
i Thermal units index (NU)
j Load demand (NJ)
L Transmission line index (NL)
n Electrical vehicle index (NEV)
s Scenario index (NS)
t Time intervals (NT)
u Index of minimum on/off time limits from 1 to max$\{MUT_i, MDT_i\}$
w Wind turbine index (NW)

Constant

γ^{dis}/γ^{ch} Discharge/charge rate of the parking lot

S_{pl}^{max} Max allowable apparent power traded between parking and network

Energy Storage in Energy Markets
https://doi.org/10.1016/B978-0-12-820095-7.00013-3

N_{pl}^{max} Max EV capacity in parking

η^{ch}/η^{dis} Charging/discharging efficiency of EVs

Z_L The impedance of transmission line

S_L^{max} Maximum allowable apparent power flow in the transmission line

$SOC_{pl}^{min}/SOC_{pl}^{max}$ Min/max state of charge of the parking lot

V_b^{min}/V_b^{max} Min/max voltage magnitude

MDT_i/MUT_i Min down/uptime for thermal plant

P_i^{min}/P_i^{max} Min/max active power generated by the thermal plant

Q_i^{min}/Q_i^{max} Min/max reactive power generated by the thermal plant

R_i^{up}/R_i^{dn} Ramp up/down of the thermal plant

Variable

OF_1 Objective function

$SUC_{i,t}/SDC_{i,t}$ Cost of start-up/shutdown for thermal plant

π_s Scenario probability

$F_i^c(P_{i,t,s})$ Cost function of the thermal plant

$P_{i,t,s}$ Active power generated by the thermal plant

$Q_{i,t,s}$ Reactive power generated by the thermal plant

$P_{PL,t,s}^{PL2G}$ Active power flow in PL2G mode

$Q_{pl,t,s}^{PL2G}$ Reactive power flow in PL2G mode

$P_{PL,t,s}^{G2PL}$ Power injected into the parking from the grid

$D_{j,t,s}^r$ Hourly active value of load demand

$Q_{j,t,s}^r$ Hourly reactive value of load demand

$P_{w,t,s}$ Active power generated by wind unit

$Q_{w,t,s}$ Reactive power of wind unit

$TU_{i,u}/TD_{i,u}$ Number of consecutive hours that thermal unit is ON/OFF

$I_{i,t}$ Binary variable for the thermal unit state

$N_{pl,t,s}^{arv}$ Number of arrival EVs in parking

$N_{pl,t,s}^{dep}$ Number of EVs departed from the parking

$t_{n,s}^{arv}/t_{n,s}^{dep}$ Number of arrival and departure times

$Cap_{pl,t,s}^{dep}$ The capacity of departed EVs from the parking

$Cap_{t_{n,s}^{arv},t_{n,s}^{dep}}^{PEV}$ Total capacity of parking considering arrival and departed EVs

$Cap_{pl,t,s}^{arv}$ The capacity of arrived EVs into the parking

$Cap_{pl,t,s}$ The total battery capacity of parking

$U_{pl,t,s}^{PL2G}/U_{pl,t,s}^{G2PL}$ Binary variable represents the state of parking operation

$SOE_{pl,t,s}^{dep}$ Energy value subtracted from the parking when EVs departed

$SOE_{pl,t,s}^{arv}$ Energy value added to parking when EVs arrived

$PF_{L,t,s}$ Transmission line active power flow

$QF_{L,t,s}$ Transmission line reactive power flow

$V_{b,t,s}$ Voltage of bus b

$\delta_{b,t,s}$ Angle magnitude of bus b

1. Introduction

Extreme motorization of transportation as the largest source of carbon dioxide and other greenhouse gasses plays the most significant role in the climate and weather pattern changes. One of the opportunities to reduce emissions from transportation fleets is turning to electrification transportation. The advent of electric vehicles (EVs) in order to electrify transportation causes the current power system to face new challenges. Increment development of technology brings many concepts related to EVs [1,2]. Power flow from an EV to the grid, i.e., vehicle-to-grid (V2G), and vice versa, i.e., grid-to-vehicle (G2V) are two capabilities that enable EVs to have bidirectional relation with the power grid [3,4]. In Refs. [5,6] data mining approaches and in Ref. [7] information gap decision theory (IGDT) along with stochastic approaches have been proposed to tackle some of these challenges.

However, because of electricity market rules, EVs cannot participate in the electricity market individually [8]. Hence, EVs parking lots (EVPLs) are established to manage the hundreds of EVs charging/discharging energy schemes with the advantage of EVs owners' satisfaction and minimum disturbance of the power grid. Uncertainties associated with the arrival time, departure time, initial state-of-charge (SoC), and final SoC are some of those challenges that the power system is facing. Furthermore, technical issues such as transformers overloading, power losses, voltage deviation, etc. On the other hand, EVPLs can be very helpful for managing renewable energy sources (RESs) integration with the power system [9,10]. The integration of RESs with the power system due to their stochastic natures is one of the important challenges that must be managed [11,12]. Since future transport must be electrified, EVPLs unlock the flexibility to use RESs, and as a consequence, the future power system can benefit properly from maximum use of variable RESs if the intermittency of RESs is taken into account [13,14].

EVPLs are a profit-based entity, which participates in the market to minimize its cost or maximize its profit [15]. So far, a lot of papers have studied the optimal scheduling of EVPLs using scenario-based approaches. Integration of EVPLs with multienergy systems under the energy hub concept to participate in both energy and reserve markets with considering the EVs owners' behavior using stochastic programming was introduced in Ref. [16]. A two-stage scenario-based market-clearing model to the scheduling of integration wind power with EVPLs for delivering both energy and reserve services was proposed in Ref. [17]. An energy management model of EVPLs for peak load shaving was presented in Ref. [18], where the model was introduced as a stochastic problem because of the uncertainty of EVs owners' behavior and SoC of EVs at the time when arriving in the parking lots. The environmental aspect of integration EVPLs with the power grid was studied in Ref. [19] through a multiobjective optimization model. In Ref. [20], the integration of solar energy and EVPLs was analyzed with the benefit of reducing power consumption and losses in the distribution system. A stochastic control mechanism for V2G and G2V features of EV fleets in the parking lots integrated with hydrogen storage was proposed in Ref. [21]. In addition to stochastic programming, other optimization models, such as a robust optimization method, can be applied to consider electricity price uncertainty [22], and in a similar context, IGDT is proposed in Ref. [23] to deal with electricity price uncertainty. The planning and operation of EVPLs were studied in Ref. [24], which EVPLs were proposed to allocate in distribution system feeder to minimize the power loss and increasing reliability. A two-level model for scheduling of EVPLs integrated with RESs was proposed in Ref. [25], where the first level deals with the EVs' characteristics, including EVs owners' uncertainty, and the second level ensures that distribution-level constraints are met. In Refs. [26,27], a control strategy using optimization techniques to handle the high penetration of RESs in the microgrid infrastructure was proposed. In Ref. [28], a battery-swapping station of EVs in the optimal microgrid scheduling as the bilevel optimization approach was developed.

Taking into account the reviewed scholars and to the best of the authors' knowledge, there is no accurate model of cooperating RESs and EVPLs considering different uncertainties of EVs' behavior and RESs generation under an AC unit commitment. So, this work aims to model the integration of EVPLs and wind power. EVs' arrival/departure time, initial and final SoCs level of EV fleets along with forecasted wind energy are the uncertain parameters that alleviated through a stochastic network-constrained UC considering AC power flow (AC-NCUC). This study introduces an AC-NCUC problem for the power system integrated with EVPL and high penetration of wind power. It should be noted that the active and reactive power injection by the EVPL is considered in the model.

Apart from the current section, Section 2 provides the mathematical model of the proposed scheduling problem, including objective function and all constraints associated with unit commitment, EVPLs, and electric network. Section 3 provides the simulation and numerical results of the proposed model. Finally, Section 4 concludes the paper.

2. Problem formulation

2.1 Objective function

The objective function of the two-stage transmission networked unit commitment model integrated with EVPL and wind energy is formulated as (1). There are four terms in the objective function. The first and second terms of (1) represent the start-up and shutdown costs. The generation cost related to the thermal unit is expressed by the third term of (1). The last term of (1) shows the operation cost of the EVPL in discharging mode.

$$OF_b = \min \sum_{t=1}^{NT} \left[SUC_{i,t} + SDC_{i,t} + \sum_{s=1}^{NS} \pi_s \left[\sum_{i=1}^{NU} F_i^c(P_{i,t,s}) + \sum_{PL=1}^{NPL} \lambda_{PL,t}^{b,dis} P_{PL,t,s}^{PL2G} \right] \right] \qquad (12.1)$$

2.2 Problem constraints

The objective function (1) is restricted to multiple limitations associated with the thermal unit operation, electrical parking lot constraints, and network constraints based on the AC-OPF, which are provided as follows.

2.2.1 *Thermal unit operation limits*

The optimal operation of thermal units is limited by multiple constraints as represented by (2)–(13). Constraints (2) and (3) limit the value of active and reactive power produced by the thermal power plant based on the upper and lower allowable values. Constraints (4) and (5) define the up and down ramp limits, respectively. The minimum up/downtime limits for the thermal unit are respectively expressed by constraints (6)–(7) and (8)–(9). The start-up and shutdown limitations for the thermal power plant are represented by (10)–(11) and (12)–(13), respectively.

$$P_i^{\min} I_{i,t} \le P_{i,t,s} \le P_i^{\max} I_{i,t} \tag{12.2}$$

$$Q_i^{\min} I_{i,t} \le Q_{i,t,s} \le Q_i^{\max} I_{i,t} \tag{12.3}$$

$$P_{i,t,s} - P_{i,t-1,s} \le [1 - I_{i,t}(1 - I_{i,t-1})]R_i^{up} + I_{i,t}(1 - I_{i,t-1})P_i^{\min} \tag{12.4}$$

$$P_{i,t-1,s} - P_{i,t,s} \le [1 - I_{i,t-1}(1 - I_{i,t})]R_i^{dn} + I_{i,t-1}(1 - I_{i,t})P_i^{\min} \tag{12.5}$$

$$I_{i,t} - I_{i,t-1} \le I_{i,t+TU_{i,u}} \tag{12.6}$$

$$TU_{i,u} = \begin{cases} u & u \le MUT_i \\ 0 & u > MUT_i \end{cases} \tag{12.7}$$

$$I_{i,t-1} - I_{i,t} \le 1 - I_{i,t+TD_{i,u}} \tag{12.8}$$

$$TD_{i,u} = \begin{cases} u & u \le MDT_i \\ 0 & u > MDT_i \end{cases} \tag{12.9}$$

$$SU_{i,t} \ge SUC_i(I_{i,t} - I_{i,t-1}) \tag{12.10}$$

$$SU_{i,t} \ge 0 \tag{12.11}$$

$$SD_{i,t} \ge SDC_i(I_{i,t-1} - I_{i,t}) \tag{12.12}$$

$$SD_{i,t} \ge 0 \tag{12.13}$$

2.2.2 EVs parking lot constraints

Thanks to both G2PL and PL2G capability, EVs can participate in the energy market with active and reactive power exchanged. The SOC of each EV without energy exchanging in the parking is established by (14). According to the arrival and departure times, the total number of EVs into the parking lot is calculated as a constraint (15). The numbers of EVs arrived or departed from the parking lot are respectively calculated by (16) and (17). Constraint (18) determines the number of EVs at each time. The maximum capacity of the parking lot, which restricts the number of available EVs, is established by (19). The total battery capacity in the parking, according to the number of departed and arrived EVs, is calculated in (20)−(22). The value of power exchanged between a parking lot and grid in charging and discharging modes is defined by (23)−(27). Constraints (23) and (24) limit the current charged and discharged power between the parking and grid, respectively. The logical relationship that separates the charging and discharging modes is represented by (25). The maximum kVA power exchanged between the parking and grid based on grid-to-parking and parking-to-grid capabilities is defined based on (26). The current capacity of parking for arrived and departed EVs is respectively calculated by (27) and (28). The total energy capacity level of the parking lot is a function of previous capacity level, arrived or departed EVs, as well as energy exchange between parking and grid is defined by (29). Finally, the energy capacity of the parking lot is bounded as (30).

$$
soc^{EV}_{n,t_n^{arv},t_n^{dep},s} = \begin{cases} soc^{EV}_{n,s} & t_{n,s}^{arv} \le t \le t_{n,s}^{dep} \\ 0 & \text{else} \end{cases} \tag{12.14}
$$

$$
N^{EV}_{t_n^{arv},t_n^{dep},s} = \sum_{n=1}^{NEV} EV_{n,t_n^{arv},t_n^{dep},s} \quad t_{n,s}^{arv} \le t \le t_{n,s}^{dep} \tag{12.15}
$$

$$
N^{arv}_{pl,t,s} = \sum_{t \in t_n^{dep}} N^{EV}_{t_{n,s}^{arv},t_{n,s}^{dep}} \tag{12.16}
$$

$$
N^{dep}_{pl,t,s} = \sum_{t \in t_n^{arv}} N^{EV}_{t_{n,s}^{arv},t_{n,s}^{dep}} \tag{12.17}
$$

$$
N_{pl,t,s} = N_{pl,t-1,s} + N^{arv}_{pl,t,s} - N^{dep}_{pl,t,s} \tag{12.18}
$$

$$
N_{pl,t,s} \le N^{max}_{pl} \tag{12.19}
$$

$$
Cap^{arv}_{pl,t,s} = \sum_{n=1}^{NEV} \sum_{t \in t_n^{dep}} Cap^{PEV}_{t_{n,s}^{arv},t_{n,s}^{dep}} \tag{12.20}
$$

$$Cap_{pl,t,s}^{dep} = \sum_{n=1}^{NEV} \sum_{t \in t_n^{arv}} Cap_{t_{n,s}^{arv}, t_{n,s}^{dep}}^{PEV} \tag{12.21}$$

$$Cap_{pl,t,s} = Cap_{pl,t-1,s} + Cap_{pl,t,s}^{arv} - Cap_{pl,t,s}^{dep} \tag{12.22}$$

$$P_{pl,t,s}^{PL2G} \le \gamma^{dis} N_{pl,t,s} U_{pl,t,s}^{PL2G} \tag{12.23}$$

$$P_{pl,t,s}^{G2PL} \le \gamma^{ch} N_{pl,t,s} U_{pl,t,s}^{G2PL} \tag{12.24}$$

$$U_{pl,t,s}^{PL2G} + U_{pl,t,s}^{PL2G} \le 1 \tag{12.25}$$

$$\left(P_{pl,t,s}^{PL2G} \right)^2 + \left(Q_{pl,t,s}^{PL2G} \right)^2 \le \left(S_{pl}^{max} \right)^2 \tag{12.26}$$

$$SOE_{pl,t,s}^{arv} = \sum_{n=1}^{NEV} \sum_{t \in t_n^{dep}} Cap_{t_{n,s}^{arv}, t_{n,s}^{dep}}^{EV} SOC_{n,t_{n,s}^{arv}, t_{n,s}^{dep}}^{EV} \tag{12.27}$$

$$SOE_{pl,t,s}^{dep} = \sum_{n=1}^{NEV} \sum_{t \in t_n^{arv}} Cap_{t_{n,s}^{arv}, t_{n,s}^{dep}}^{EV} SOC_{n,t_{n,s}^{arv}, t_{n,s}^{dep}}^{EV} \tag{12.28}$$

$$SOE_{pl,t,s} = SOE_{pl,t-1,s} + SOE_{pl,t,s}^{arv} - SOE_{pl,t,s}^{dep} + \eta_{ch} P_{pl,t,s}^{G2PL} - \frac{P_{pl,t,s}^{PL2G}}{\eta_{dis}} \tag{12.29}$$

$$SOC_{pl}^{min} Cap_{pl,t,s} \le SOE_{pl,t,s} \le SOC_{pl}^{max} Cap_{pl,t,s} \tag{12.30}$$

2.2.3 AC-OPF and network constraints

The power balance constraints for active and reactive power at each electrical bus are calculated based on (31) and (32), respectively. The voltage magnitude of each bus is restricted by the lower and upper amounts in p.u. as a constraint (33). The thermal capacity of each transmission line is limited by constraint (34). The AC-OPF model for power flow through the transmission line is established by (35) and (36).

$$\sum_{i=1}^{NU_b} P_{i,t,s} + \sum_{w=1}^{NW_b} P_{w,t,s} + \sum_{PL=1}^{NPL_b} P_{PL,t,s}^{PL2G} - \sum_{PL=1}^{NPL_b} P_{PL,t,s}^{G2PL} - \sum_{j=1}^{NJ_b} D_{j,t,s}^r = \sum_{L=1}^{NL_b} PF_{L,t,s} \tag{12.31}$$

$$\sum_{i=1}^{NU_b} Q_{i,t,s} + \sum_{w=1}^{NW_b} Q_{w,t,s} - \sum_{PL=1}^{NPL_b} Q_{PL,t,s} - \sum_{j=1}^{NJ_b} Q^r_{j,t,s} = \sum_{l=1}^{NL_b} QF_{L,t,s} \tag{12.32}$$

$$V_b^{\min} \leq V_{b,t,s} \leq V_b^{\max} \tag{12.33}$$

$$PF^2_{L,t,s} + QF^2_{L,t,s} \leq \left(S_L^{\max}\right)^2 \tag{12.34}$$

$$PF_{L,t,s} = \frac{V^2_{b,t,s}}{Z_L}\cos(\theta_L) - \frac{V_{b,t,s}V_{b',t,s}}{Z_L}\cos(\delta_{b,t,s} - \delta_{b',t,s} + \theta_L) \tag{12.35}$$

$$QF_{L,t,s} = \frac{V^2_{b,t,s}}{Z_L}\sin(\theta_L) - \frac{V_{b,t,s}V_{b',t,s}}{Z_L}\sin(\delta_{b,t,s} - \delta_{b',t,s} + \theta_L) - \frac{b_L V^2_{b,t,s}}{2} \tag{12.36}$$

3. Simulation and numerical results

3.1 Case study

The proposed EVPL scheduling is examined on the modified six-bus IEEE system. All the characteristics of bus, line, and load data of the proposed test system can be found in Refs. [26,27]. The wind power plant and EPVL are respectively located at bus number 5. All required data and characteristics about the thermal power plant, parking, forecasted load, and wind profile could be found in Refs. [29,30]. The proposed model is formulated as the MINLP model and all the codes are carried out in GAMS software that is solved by DICOPT solver. As previously discussed, wind power variability is handled under a stochastic framework. To this end, 1000 scenarios are produced by the Monte Carlo simulation based on the Weibull distribution function. The characteristics of the Weibull distribution can be found in Ref. [31]. The generated scenarios are reduced to the desired scenarios via the SCENRED tool in the GAMS.

3.2 Numerical results

An EV parking lot is embedded at bus number 5. To reveal the effectiveness of EVPL on total operational cost reduction, the EVPL is considered as a passive load. In other words, there is no reactive power exchange between EVPL and grid, and it only operates in G2PL mode. Using the proposed strategy, the expected operational cost equals $ 75,895.36. The power dispatch values of three thermal units are depicted in Fig. 12.1. According to Fig. 12.1, since G1 is cheaper generation unit, it is operated on the whole time horizon while G_2 as the higher-cost unit is turned on only in 8 h.

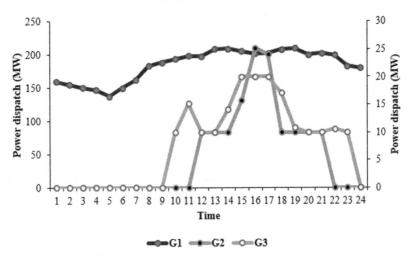

FIGURE 12.1 Hourly power dispatch of thermal units.

Let's consider the active power injection capability for the parking lot, which enables EVPL to operate in both G2PL and PL2G modes. Fig. 12.2 shows the optimal parking lot scheme. According to Fig. 12.2, in off-peak hours, EVPL operates in the G2PL mode while it works in PL2G mode over peak hours, consequent sells electricity to the grid. This interaction leads to $149.30 cost reduction compared with the previous case when there is no EVPL is in the grid (the expected operation cost equals $75,469.39).

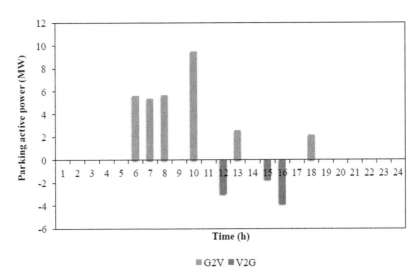

FIGURE 12.2 Optimal hourly scheme of EVPL.

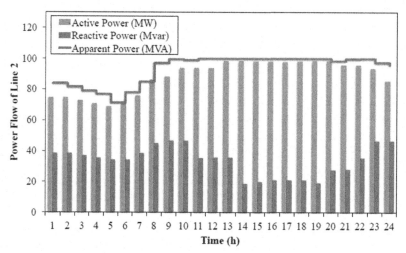

FIGURE 12.3   Power flow rate of transmission line 2.

The power flow (active and reactive) in line number 2 is given in Fig. 12.3. It is obvious from Fig. 12.3 that since line 2 is connected to the lower-cost unit (G_1), it operates with the maximum thermal capacity (100 MVA) during the time horizon.

The effect of the reactive power injection capability for the EVPL on the active and reactive power flow through the line number 2 is respectively shown in Figs. 12.4 and 12.5. Considering the reactive power injection capability in the EVPL operation results in more operational cost reduction. While the reactive power injection capability is considered for the EVPL, due to the power factor improvement, more capacity of the transmission line is

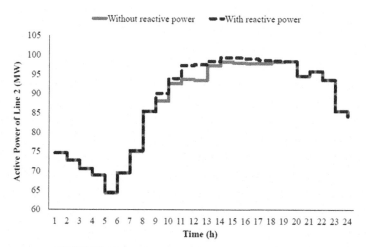

FIGURE 12.4   Active power flow of line number 2.

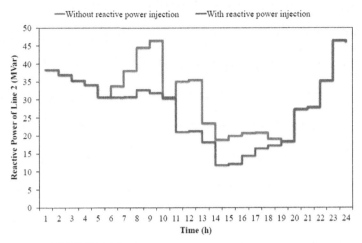

FIGURE 12.5 Reactive power flow of line number 2.

TABLE 12.1 Comparison of total operational cost and energy dispatch for two cases: with and without reactive power injection.

	Operation cost ($)	G_1(MWh)	G_2(MWh)	G_3(MWh)
EVPL without Q injection	75,469.39	4441.01	166.25	188.25
EVPL with Q injection	74,707.56	4473.79	96.16	175.48

available. Therefore, the lower-cost generation unit is committed more than before and provides more benefits. This action leads to the expected operational cost equal to $ 74,707.56, which shows the $ 911.13 reduction compared with the previous case.

The total system operation cost, as well as power generation by thermal units, is given in Table 12.1. According to Table 12.1, as the value of reactive power injection is considered for the EVPL, the total operation cost is reduced. Also, the total power dispatch by G_2 and G_3 is reduced, while the total power dispatch by G_1 increases, due to the extra line power capacity available.

4. Conclusion

EVs as flexible emerging sources with significant features play as unspeakable parts of the power system that imposed the scheduling and planning of it. However, these resources can be considered as mobile load or generation units, which facilitate the integration of renewable power generation, especially wind power. In other words, without expansion of the transmission lines, EVs can transfer the wind power out to different places in the power network, hence the wind energy curtailment is reduced. Therefore, this chapter evaluated the impact of EVPLs and power systems integrated with wind energy, aiming to operational cost minimization considering wind power uncertainty under a stochastic framework. To achieve a more realistic model, drivers' behavior in both operation modes (including PL2G, and G2PL) was modeled separately. Furthermore, for enhancing the operation of the whole system, the active and reactive power exchanged between the network and the parking was considered. The proposed framework was formulated as an MINLP model in the AC-transmission networked unit commitment problem. The proposed scheduling was tested on the six-bus IEEE system. Numerical results verified the effectiveness of the proposed model in the operational cost reduction by 3%.

References

[1] U.C. Chukwu, S.M. Mahajan, V2G electric power capacity estimation and ancillary service market evaluation, in: 2011 IEEE Power and Energy Society General Meeting, 2011, pp. 1–8.

[2] L. Drude, L.C.P. Junior, R. Rüther, Photovoltaics (PV) and electric vehicle-to-grid (V2G) strategies for peak demand reduction in urban regions in Brazil in a smart grid environment, Renew. Energy 68 (2014) 443–451.

[3] S. Stüdli, E. Crisostomi, R. Middleton, R. Shorten, Optimal real-time distributed V2G and G2V management of electric vehicles, Int. J. Contr. 87 (6) (2014) 1153–1162.

[4] S.S. Hosseini, A. Badri, M. Parvania, The plug-in electric vehicles for power system applications: the vehicle to grid (V2G) concept, in: 2012 IEEE International Energy Conference and Exhibition (ENERGYCON), 2012, pp. 1101–1106.

[5] A. Mansour-Saatloo, A. Moradzadeh, B. Mohammadi-Ivatloo, A. Ahmadian, A. Elkamel, Machine learning based PEVs load extraction and analysis, Electronics 9 (7) (2020) 1150.

[6] H. Jahangir, S.S. Gougheri, B. Vatandoust, M.A. Golkar, A. Ahmadian, A. Hajizadeh, Plug-in electric vehicle behavior modeling in energy market: a novel deep learning-based approach with clustering technique, IEEE Trans. Smart Grid 11 (6) (2020) 4738–4748.

[7] P. Aliasghari, B. Mohammadi-Ivatloo, M. Abapour, Risk-based scheduling strategy for electric vehicle aggregator using hybrid Stochastic/IGDT approach, J. Clean. Prod. 248 (2020) 119270.

[8] C. Goebel, H.-A. Jacobsen, Aggregator-controlled EV charging in pay-as-bid reserve markets with strict delivery constraints, IEEE Trans. Power Syst. 31 (6) (2016) 4447–4461.

[9] W. Hu, C. Su, Z. Chen, B. Bak-Jensen, Optimal operation of plug-in electric vehicles in power systems with high wind power penetrations, IEEE Trans. Sustain. Energy 4 (3) (2013) 577–585.

[10] N. Masuch, J. Keiser, M. Lützenberger, S. Albayrak, Wind power-aware vehicle-to-grid algorithms for sustainable EV energy management systems, in: 2012 IEEE International Electric Vehicle Conference, 2012, pp. 1–7.

[11] M. Hemmati, B. Mohammadi-Ivatloo, S. Ghasemzadeh, E. Reihani, Risk-based optimal scheduling of reconfigurable smart renewable energy based microgrids, Int. J. Electr. Power Energy Syst. 101 (2018) 415–428.

[12] M. Hemmati, B. Mohammadi-ivatloo, M. Abapour, Day-ahead profit-based reconfigurable microgrid scheduling considering uncertain renewable generation and load demand in the presence of energy storage, J. Energy Storage 28 (2020) 101161.

[13] M.A. Mirzaei, M. Nazari-Heris, B. Mohammadi-Ivatloo, M. Marzband, Consideration of hourly flexible ramping products in stochastic day-ahead scheduling of integrated wind and storage systems, in: 2018 Smart Grid Conference (SGC), 2018, pp. 1–6.

[14] M.A. Mirzaei, A.S. Yazdankhah, B. Mohammadi-Ivatloo, M. Marzband, M. Shafie-khah, J.P.S. Catalão, Stochastic network-constrained co-optimization of energy and reserve products in renewable energy integrated power and gas networks with energy storage system, J. Clean. Prod. 223 (2019) 747–758.

[15] R.J. Bessa, M.A. Matos, Economic and technical management of an aggregation agent for electric vehicles: a literature survey, Eur. Trans. Electr. Power 22 (3) (2012) 334–350.

[16] M. Yazdani-Damavandi, M.P. Moghaddam, M.-R. Haghifam, M. Shafie-khah, J.P.S. Catalão, Modeling operational behavior of plug-in electric vehicles' parking lot in multienergy systems, IEEE Trans. Smart Grid 7 (1) (2015) 124–135.

[17] E. Heydarian-Forushani, M.E.H. Golshan, M. Shafie-khah, Flexible interaction of plug-in electric vehicle parking lots for efficient wind integration, Appl. Energy 179 (2016) 338–349.

[18] İ. Sengör, O. Erdinç, B. Yener, A. Tascıkaraoğlu, J.P.S. Catalão, Optimal energy management of EV parking lots under peak load reduction based DR programs considering uncertainty, IEEE Trans. Sustain. Energy 10 (3) (2018) 1034–1043.

[19] J. Jannati, D. Nazarpour, Optimal performance of electric vehicles parking lot considering environmental issue, J. Clean. Prod. 206 (2019) 1073–1088.

[20] M.T. Turan, Y. Ates, O. Erdinc, E. Gokalp, J.P.S. Catalão, Effect of electric vehicle parking lots equipped with roof mounted photovoltaic panels on the distribution network, Int. J. Electr. Power Energy Syst. 109 (2019) 283–289.

[21] R. Razipour, S.-M. Moghaddas-Tafreshi, P. Farhadi, Optimal management of electric vehicles in an intelligent parking lot in the presence of hydrogen storage system, J. Energy Storage 22 (2019) 144–152.

[22] A. Mansour-Saatloo, M. Agabalaye-Rahvar, M.A. Mirzaei, B. Mohammadi-Ivatloo, K. Zare, Robust scheduling of hydrogen based smart micro energy hub with integrated demand response, J. Clean. Prod. (2020) 122041.

[23] J. Liu, C. Chen, Z. Liu, K. Jermsittiparsert, N. Ghadimi, An IGDT-based risk-involved optimal bidding strategy for hydrogen storage-based intelligent parking lot of electric vehicles, J. Energy Storage 27 (2020) 101057.

[24] M. Rahmani-Andebili, H. Shen, M. Fotuhi-Firuzabad, Planning and operation of parking lots considering system, traffic, and drivers behavioral model, IEEE Trans. Syst. Man, Cybern. Syst. 49 (9) (2018) 1879–1892.

[25] M. Shafie-Khah, P. Siano, D.Z. Fitiwi, N. Mahmoudi, J.P.S. Catalao, An innovative two-level model for electric vehicle parking lots in distribution systems with renewable energy, IEEE Trans. Smart Grid 9 (2) (2017) 1506–1520.

[26] T. Wu, Q. Yang, Z. Bao, W. Yan, Coordinated energy dispatching in microgrid with wind power generation and plug-in electric vehicles, IEEE Trans. Smart Grid 4 (3) (2013) 1453–1463.

[27] H.N.T. Nguyen, C. Zhang, M.A. Mahmud, Optimal coordination of G2V and V2G to support power grids with high penetration of renewable energy, IEEE Trans. Transp. Electrif. 1 (2) (2015) 188–195.

[28] M. Hemmati, M. Abapour, B. Mohammadi-ivatloo, Chapter 1 Optimal Scheduling of Smart Microgrid in Presence of Battery Swapping Station of Electrical Vehicles, 2020, pp. 1–14.

[29] M. Amin, A. Sadeghi, Stochastic Network-Constrained Co-optimization of Energy and Reserve Products in Renewable Energy Integrated Power and Gas Networks with Energy Storage System, vol. 223, 2019, pp. 747–758.

[30] E. Heydarian-forushani, M.E.H. Golshan, P. Siano, Evaluating the benefits of coordinated emerging flexible resources in electricity markets, Appl. Energy 199 (2017) 142–154.

[31] M. Hemmati, B. Mohammadi-ivatloo, A. Soroudi, Uncertainty Management in Power System Operation Decision Making, 2020.

13

Large-scale energy storages in joint energy and ancillary multimarkets

Jaber Fallah Ardashir, Hadi Vatankhah Ghadim

Department of Electrical Engineering, Tabriz Branch, Islamic Azad University, Tabriz, East Azarbaijan, Iran

1. Introduction

Any energy grid has its actors, which are categorized as generators, transmission and distribution line operators, and consumers. These sections are usually the same in different regions; some might have minor differences. With developments in grid technologies, some sections have improved their capabilities and also different components have appeared in them. For example, some of the consumption parties have been equipped with renewable energy power generation and also storage units. Distribution grids have faced challenges of different small-scale distributed generation resources and new demand-side management issues, forcing them to give more attention to ancillary services and communication infrastructures. Generation section is also facilitated with improved fossil fuel power plants and emerging technologies for large-scale renewable energy power units and energy storage systems (ESSs) (Fig. 13.1).

The structure of power systems was not in the form of what we know nowadays. Previously, state-owned generation units had the responsibility of maintaining the need for energy. This energy was transferred through the transmission lines of which the operator of it was the national dispatching office, a branch from the department of energy. Eventually, after regulating the voltage level of transmitted power in transformer stations, this energy was delivered to consumers of any type, industrial or commercial or residential.

There was not any opportunity for customers to participate in the generation of energy as there was not any suitable technology, regulation, or feed-in-tariffs (FIT) introduced. The billing system was directly connecting the consumers to the government and costs of generation, transmission, and distribution of energy would be paid to the government. Fig. 13.2 shows the explained one-way approach of power systems.

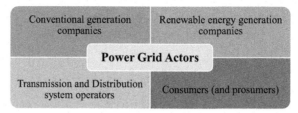

FIGURE 13.1　Actors of an electrical power grid.

After changes in energy policies that resulted from changes in environmental and economic concerns of communities all around the world, different and somehow innovative approaches were introduced. Due to horizontal unbundling of generation and distribution market of power systems, private companies found the chance to participate in these markets. Distributed generation stations, renewable energy power plants, retail market business companies, and appearance of prosumers—consumers that have the capability of generating energy with small-scale renewable energy power plants—are some of the results of applying horizontal unbundling in the power system.

Besides, due to the increasing attention of system operators to the stability and reliability of the grid, storage systems are widely considered in different points of the grid, which they can participate in any markets such as energy, reserve, and regulation markets in the grid. On

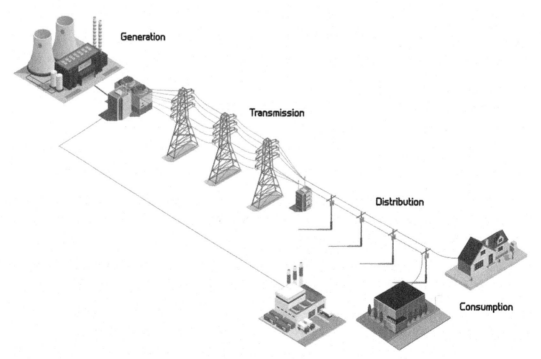

FIGURE 13.2　Traditional power system isometric view.

the other hand, vertical unbundling made the privatization of the grid much easier. The generation and retail energy market are separated from state-owned parts of the grid. As a result, only transmission infrastructures remained under the control of the state in order to maintain a stable connection between different regions of the grid. Fig. 13.3 shows how the energy system is changed after deregulating it via unbundling.

Reformed power systems have different and new challenges in comparison with traditional ones. The presence of third-party companies in the energy market has created new opportunities to solve these problems. Maintaining energy balance, acquiring reactive power for the system, regulation of voltage and frequency, etc. are part of the services that companies provide in order to supply the stability of the power system.

As mentioned earlier, new components have entered the power system. Meanwhile, storage systems have been one of the most important issues in restructured systems, because, in the presence of renewable energies such as wind or solar, which have uncertainties and cannot be used in energy plannings effectively, storages can perform as a solution to solve these kinds of problems and propound the use of green energy. Storage technologies of any kind are capable of providing ancillary services for the power system in the time of need. In the following sections, the role of large-scale storage systems in joint energy and ancillary markets will be discussed.

2. Large-scale storage systems

When we are talking about large-scale storage systems, there are only a few choices of storage systems that actually fit in this definition. Pumped hydro, compressed air, flywheel, and

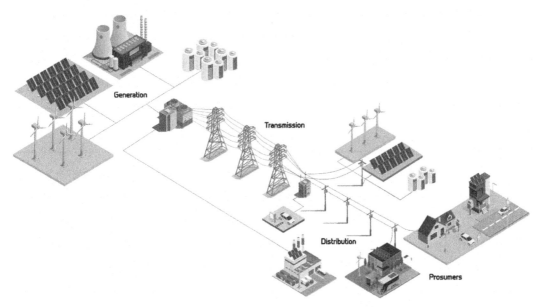

FIGURE 13.3   Deregulated power system isometric view.

battery energy storage systems (BESSs) are the ones usually accounted for as the suitable utility-sized storage technologies. Others may be considered as potential choices but because of financial constraints, they are inapplicable. Due to the incremental trend of renewable energy usage in different energy grids around the world and because of the numerous factors responsible for making these power stations unpredictable, the implementation of large-scale storage facilities seemed necessary.

Additionally, ESS can provide an economic justification for renewable energy power units that have high initial costs, reduce their construction and operation costs, and thus generate higher revenue from power plants. This makes renewable energy power plants more attractive to economic dispatch planning of power grid operators. Table 13.1 lists some of the benefits that ESS can bring to any renewable energy power unit.

Even for conventional fossil fuel power stations, ESS is beneficial. Creating the opportunity of operating at higher capacities in the energy market while providing chances of participation in ancillary service markets are some of the advantages of ESS for conventional power plants. In Table 13.2, there are the benefits of utility-sized energy storage facilities for fossil fuel–based energy power stations.

In further sections, the participation of large-scale ESS in different markets will be discussed and their responses will be explained.

2.1 Market participation

Large-scale ESSs are considered as one of the necessary elements in offering different energy and ancillary services in the energy grid. There is a categorization of ESSs operation forms, which is detailed in Fig. 13.4.

As we can see in Fig. 13.4, storage systems are usually involved in different groups depending on where they are located and what services they are assigned to perform. In the first and second categories, independent system operator (ISO) and merchant storage operator (MSO) can use ESSs as a tool to offer ancillary services to the grid, such as energy time shifting. These categories usually consist of grid operators, power station owners,

TABLE 13.1 ESS beneficial features for renewable energy power units.

Features and services	Explanations
Firming capacity	The intermittent output of renewable energy power units can be controlled by a storage unit, smoothing the output of units and limiting the output rate in order to prevent instability of the power grid.
Renewable energy time shift	Due to uncertainties of climate condition and dependency of renewable energy power units to it, storage units can store exceeded energy during the generation peaks, releasing it to the grid when renewable power units are unable to operate.
Intermittent integration	Due to challenges in intermittent integration of renewable energy units to grid for conventional power plants, storage units can offer the opportunity to mitigate the need for fossil fuel power units.

TABLE 13.2 ESS beneficial features for conventional power plants.

Features and services	Explanations
Black start	In the case of blackouts (and also brownouts), storage units can be used in order to initiate the restart of large-scale power plants
Voltage support	In order to provide voltage stability in the grid and to maintain the proportion between supply and demand of active and reactive power, storage units can be implemented into the grid.
Spinning and supplemental reserves	Storage units can be used as rapid response assets for providing both spinning and supplemental reserves in the case of contingency events at different levels.
Frequency regulations	Storage units will aim to stabilize the frequency of output power by providing a synthetic inertial response to ensure the exact adaptation of generation and consumption
Energy arbitrages	Charging the storage units in low-demand (low-price) time intervals of the energy system and injecting the stored energy to grid in peak hours with higher price will expand the profit.
Ramping products	Storage units (especially supercapacitors) can respond to immediate large-scale contingency events in grid in a short timescale (minutes). This feature can prevent the generator from slipping out of the stable mode and it can eliminate the chance of blackouts.

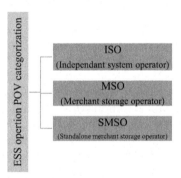

FIGURE 13.4 Energy storage system's operation POV categorization.

electric vehicle charging stations, which all are called EV aggregators, etc., which have the ability to act in higher frequency and capacity when it is needed.

In the third category, we have the merchant operators of stand-alone ESSs, which are not considered as the main players of ancillary service markets. These operators are capable of being active in both energy and ancillary services because of their independent operation from the generation section. They can offer energy arbitrage services while they are able to perform voltage or spinning reserve. From this point, we can understand that any kind of storage system can be active in joint ancillary and energy service markets if they are being operated independently.

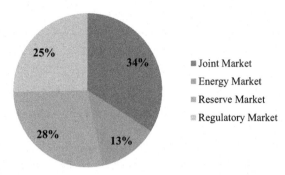

FIGURE 13.5   Market revenues for battery energy storage systems [1].

In [1], a BESS with a capacity of 30 MWh is simulated in order to evaluate the best operation of this system in both ancillary and energy service markets. In this study, different aspects of a battery storage system construction and installation are discussed and new regulatory policies have been described. The main pricing scheme, which is used in this study, is the pay-for-performance regulation due to the FERC order no.755. The simulation of this study considers different optimal scheduling scenarios in which a BESS can operate. The results show that while the operation of BESS in the reserve market is more beneficial compared to the energy and regulatory markets, the operation of this system has more revenue if it is implemented in a joint market. Fig. 13.5 shows the comparison of each market revenues for BESS.

Due to the fast response capability and higher capacity opportunities of BESS, most of the ESS systems that are currently being used around the world are either BESS or hybrid systems including batteries. In Ref. [2], an SNS (smarter network storage) system and its duties were discussed. In this system, a 10 MWh Li-ion BESS was implemented in order to provide peak shaving, frequency regulation, operational reserve, and tolling services. This ESS is fully assigned to offer ancillary services while in Ref. [3], a hybrid flywheel—BEES, which is located in Ireland and is operated by EirGrid management system, offers both ancillary and energy services. Fast frequency responses, synchronous inertia responses, and operational reserve capabilities of this storage system can offer higher revenue because of participation in joint markets.

In another case [4], a storage system consisting of ultracapacitors and Pb—acid batteries—commercially named UltraBattery—is used in both energy and regulatory markets. This storage system, which is located in eastern Pennsylvania, has the capability to offer services in both markets up to 3 MW margin with different time levels. Because of a faster response to regulatory signals from the grid—approximately 0.75 MW/s—this system can have a higher income. The state of charge for this system differs from 45% to 55% with an average of 50 in different time intervals. Also, the usage of BESS in this storage system makes the operation more accurate than other technologies such as hydro or other approaches. Besides, the implementation of this system in demand-side management services makes it a profitable choice to use. Detailed results are available in Ref. [3].

As it is obvious, the difference in participation in separate markets and joint markets is significant. In most cases, the operation of storage systems in two or more markets can guarantee their economic justification while, in some areas, hybrid operation of ESS with renewable energy sources, for example, wind or solar farms, could not offer much profit. Since many studies were done in this field, the profit that can be achieved by the operation of storage systems in joint markets can be up to 1.5 times more than that of the operation of the same system in energy or ancillary markets separately.

2.2 Future concepts

As we are moving toward a reliable and resilient flow of energy in our grids, the need for enhancing tools such as ESS is becoming more and more inevitable. Operation of different technologies has been tested and their specifications are available in order to choose the best option for any kind of services system operators to seek to be provided.

3. Economic justifications

Construction and implementation of an ESS, whether it is utility-sized or not, require economic analysis previously. Different costs of an ESS, from design to operation, are necessary to have a realistic viewpoint of how much we are investing in technology and how could its rate of return be. In this particular case, if we want to choose the best technologies in order to store energy with it in a way that the energy excess can be used in both energy and ancillary markets, we need to consider the ability of each technology in offering services related to these markets.

Electrochemical and electromechanical approaches of ESSs have this ability to offer services in both energy and ancillary markets. In Figs. 13.6 and 13.7, we can see the global share of these two methods of saving by the main use case provided by the International Renewable Energy Agency (IRENA).

The other two methods, which are pumped hydro and thermal storage technologies, have been discarded from this study due to the fact that more than 70% of their use was in energy or reserve markets separately. Electrochemical and electromechanical storage systems are the common technologies in ESS. The most prevalent examples for both of the approaches are battery and flywheel ESSs. In the following paragraphs, costs and performances of both approaches will be explained in detail.

3.1 Flywheel energy storage system (FESS)

Flywheel storages are kinetic energy storages in which electrical energy is absorbed via a reversible machine connected to the grid. There are two types of FESSs named as low-speed and high-speed. Due to their rotational operation, there are friction losses in which magnetic or ball bearings are usually used to minimize the mentioned losses. Superconduction magnetic bearings are considered as the most preferable options to use in FESS structure because of their lowest degree of friction losses. Besides, there should be an effective cooling system to create an appropriate environment of operation for FESS but the installation of

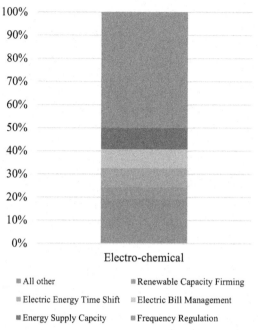

FIGURE 13.6   Global share of electrochemical storage technologies by main use case, 2017 [5].

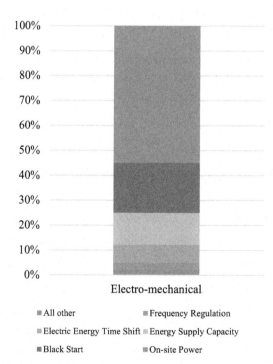

FIGURE 13.7   Global share of electromechanical storage technologies by main use case, 2017 [5].

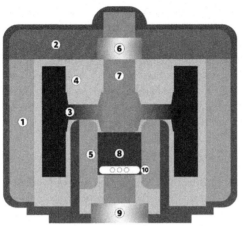

1. Steel Shield
2. Upper Vacuum Space
3. Flywheel Rotor
4. Lower Vacuum Space
5. Stator

6. Upper Electromagnet
7. Rotor Axis
8. Motor/Generator
9. Lower Electromagnet
10. Electrical Output

FIGURE 13.8 NASA G2 flywheel energy storage unit [6].

high-temperature superconduction magnetic bearings can reduce the construction costs, and also it will improve the lifespan of the storage facility. A flywheel storage unit of G2, which belongs to NASA, is available in Fig. 13.8.

Higher energy density, fast charge specifications, and outstanding lifetime limit, which is about 1 million cycles, make this technology a suitable option for short-term case uses. This storage technology can be used in both energy markets (arbitrage) and ancillary markets (frequency regulation). Even it can be used in buffering the saved energy to public electrical transportation systems such as trams in the case of need. A list of pros and cons for FESS is available in Table 13.3.

TABLE 13.3 Benefits and drawbacks of FESS [5].

Benefits	Drawbacks
Fast charging	Lower power density in comparison with BESS
Long lifetime with zero degradation of capacity	Great idle losses
High power density	Bearing related maintenances or energy consumptions
Low maintenance requirements	Vulnerable in front of unexpected dynamic loads and external shocks
Easier determination of the state of charge (SoC)	
Wide operational experience	

Installation costs for an FESS are approximately about 1500–6000 $/KWh, which with idle losses of 15% per hour will not make it a suitable option for mid- or long-term usages. After some developments and improvements in the technology of flywheels, it is expected that the installation costs will decrease by 35% on average, reaching 1000–3900 $/KWh cost interval. Also, improvements of other specifications such as lifetime and materials used to build and components such as bearings and machines are expected to increase the efficiency of these storage facilities up to 85.75% and expand their calendar life by 50%, up to 30 years on average. A brief review of the numbers is available in Fig. 13.9.

3.2 Battery energy storage system (BESS)[1]

BESS is one of the most common yet immature technologies suitable for utility-sized energy storage. These storage systems are based on electrochemical reactions to save energy, and they consist of numerous materials that specify their operating conditions. Sodium–Sulfur, Li-Ion, Lead–Acid, Sodium–Metal Halide, Zinc–Hybrid Cathode, and Redox Flow

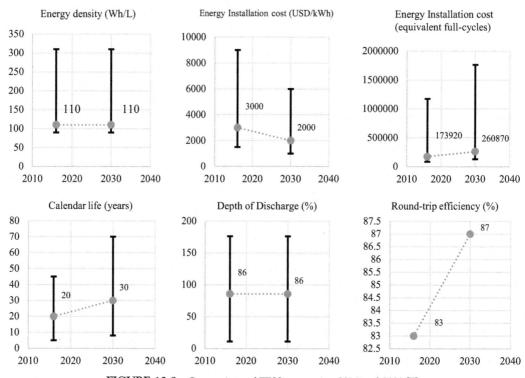

FIGURE 13.9 Comparison of FESS properties, 2016 and 2030 [5].

[1] There are a lot of implemented projects of BESS in the world. Toshiba is one of the companies that is active in this field. For more info of a project of this company in USA, visit https://www.toshiba.co.jp/about/press/2015_04/pr2101.htm

batteries are the technologies that are common in the construction of ESS, and they will be analyzed in this section.

First, we need to define an equation to calculate the total cost of installation and operation for ESSs. In order to have the most accurate calculation for building, operation, and maintenance (BOM) costs of these facilities, the equation is written as follows:

$$[(Capitol\ Cost + C\&C + Var.\ O\&M) \times E/P\ rate] + (PCS + BOP) + (Fix.\ O\&M \times 1\ yr.)$$
$$= Total\ BOM\ cost \tag{13.1}$$

C&C: Construction & Commissioning cost ($/kWh)
Var. O&M: Variable Operation and Maintenance (cents/kWh)
E/P ratio: energy to power ratio (h)
PCS: Power conversion system ($/kW)
BOP: Balance of plant ($/kW)
Fix. O&M: Fixed Operation and Maintenance ($/kw.yr)
BOM: Building, Operation, and Maintenance costs ($/kW)

In this study, E/P ratio is considered 4 h for BESS. The resulting costs will determine the average total costs of BOM of a BESS. As it is mentioned in Energy Storage Technology and Cost Characterization Report of the U.S. Department of Energy, the capital cost is a category in BESS technologies economic analysis, which indicates the expenses of obtaining a battery storage unit and its components such as electrodes, electrolytes, and separators and does not include PCS, BOP, or C&C costs.

C&C costs, which are also named as EPC costs, include site design, equipment acquirement and transportation, and the labor costs necessary for installation of storage units. Variable O&M expenses consist of costs that are accounted as necessary for operation of storage systems. Costs for the inverter system and its control facility and also the packaging of it are included in PCS costs. These costs can be reduced if the system voltage increases, though less current will be for the same power output of the storage system.

BOP expenses are related to wirings, interconnections of the storage system and transformers, and also the ancillary service equipment, which might be installed on the system. Fixed O&M costs indicate the costs that are as necessary as variable O&M to keep the system operational, but the difference of fixed and variable costs is that fixed one has no fluctuation in amount.

According to the calculations extracted via Eq. (13.1) about the BOM costs of storage systems, in the following sections, there will be a comparison between the costs of technology improvements between the years 2018 and 2025 separately.

(a) *Sodium—Sulfur Battery*

About this battery technology, a brief description could be that this battery is made of sodium and sulfur substances, and it is operational in high temperatures with less than 4 h' application duration capability. During the upcoming years, some developments are being expected, which will reduce the overall BOM cost of sodium—sulfur BESS. In Fig. 13.10, the cost intervals and their average values for both of the mentioned years are available.

BOM expenses and detailed costs of an ESS equipped with this technology in both of the 2018 and 2025 years are available in Table 13.4.

As it is obvious from the chart, the BOM cost of a Sodium–Sulfur BESS will have a downward trend until 2025 by about 27%, which will increase the chance of implementation due to suitable economic circumstances. E/P ratio of this BESS is considered 4 h because of its maximum available time of action.

(b) *Lithium-Ion Battery*

Lithium-Ion BESS is mainly based on the chemical reaction between cathode and anode of battery in which a metal coated with lithium oxide reacts with the graphite component in it. With a wide range of applications from quadcopters and mobile phones to utility-sized energy storage facilities, this technology has the potential to perform optimized and hand out desired results. A comparison between the BOM cost components of this technology in the years 2018 and 2025 is available in Fig. 13.11.

BOM expenses and detailed costs of an ESS equipped with this technology in both of the 2018 and 2025 years are available in Table 13.5.

As it is obvious from the chart, the BOM cost of a Li-Ion BESS will have a downward trend until 2025 by about 23%, which will increase the chance of implementation due to suitable economic circumstances. E/P ratio of BESS from now on will be considered 4 h in order to have reliable calculation outcomes.

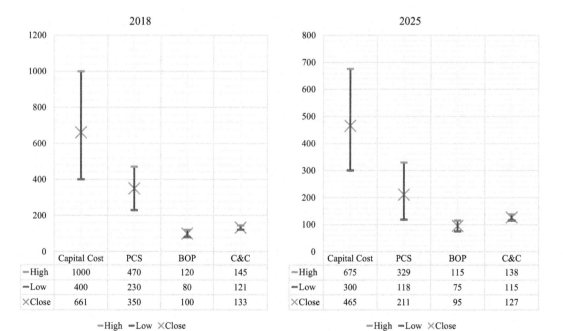

FIGURE 13.10   Sodium–sulfur BESS BOM cost components changing trend between 2018 and 2025 [7].

TABLE 13.4 Sodium−sulfur BESS BOM cost in 2018 and 2025 [7].

Parameters	Sodium sulfur battery	
	2018	**2025**
Capital cost ($/kWh)	661	465
PCS ($/kW)	350	211
BOP ($/kW)	100	95
C&C ($/kWh)	133	127
Fix. O&M ($/kw.yr)	10	8
Var. O&M (cents/kWh)	0.39	0.39
E/P (h)	4 h	4 h
BOM ($/kW)	**3636.016**	**2682.016**

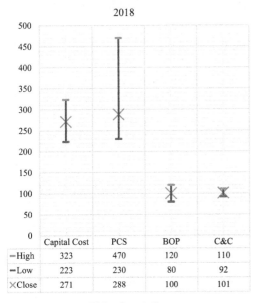

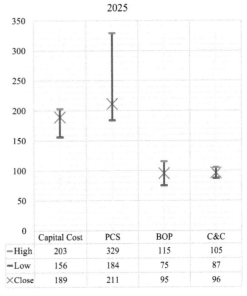

2018	Capital Cost	PCS	BOP	C&C
High	323	470	120	110
Low	223	230	80	92
Close	271	288	100	101

High Low Close

2025	Capital Cost	PCS	BOP	C&C
High	203	329	115	105
Low	156	184	75	87
Close	189	211	95	96

High Low Close

FIGURE 13.11 Lithium-ion BESS BOM cost components changing trend between 2018 and 2025 [7].

TABLE 13.5 Li-ion BESS BOM cost in 2018 and 2025 [7].

Parameters	Li-ion battery	
	2018	**2025**
Capital cost ($/kWh)	271	189
PCS ($/kW)	288	211
BOP ($/kW)	100	95
C&C ($/kWh)	101	96
Fix. O&M ($/kw.yr)	10	8
Var. O&M (cents/kWh)	0.445	0.445
E/P (h)	4	4
BOM ($/kW)	**1886.018**	**1454.018**

(c) *Lead—Acid Battery*[2]

The lead—acid battery is one of the technologies used in grids in order to provide services as BESS. This technology can be used in two variations, which are named VLA (vented lead—acid) and VRLA (valve-regulated lead—acid). The operation range of this technology is up to a few megawatts, and it can provide as high as 10 MWh energy for the grid. A comparison between the BOM cost components of this technology in the years 2018 and 2025 is available in Fig. 13.12.

BOM expenses and detailed costs of an ESS equipped with this technology in both of the 2018 and 2025 years are available in Table 13.6.

As it is obvious from the chart, the BOM cost of a lead—acid BESS will have a downward trend until 2025 by about 16%, which will increase the chance of implementation due to suitable economic circumstances.

(d) *Sodium—Metal Halide Battery*

Sodium—Metal Halide Battery, which is also known as Zebra Battery previously, was implemented on electric vehicles. The sizing of battery is smaller than other variations of chemical-based battery technologies, and it also has a higher level of flexibility and scalability, which makes it suitable for numerous applications from transportation to residential and commercial usages. Compared to the sodium—sulfur battery, Zebra Battery requires less temperature to operate; howsoever, it still needs independent heaters in order to provide ideal environmental operation constraints. With respect to the facts mentioned earlier, this technology is usually accounted for as a suitable and appropriate option for an ESS construction. A

[2] Hitachi delivered a sample of commercial lead—acid battery "LL1500-WS" for renewable hybrid energy storage systems to County Offaly, Ireland. For more information, visit http://schwungrad-energie.com/vs/2017/06/Schwungrad-Energie_EirGrid-Demonstration-Project-Report.pdf, page 7.

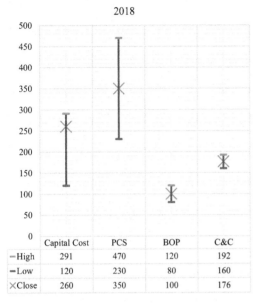

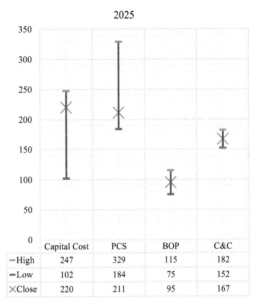

	Capital Cost	PCS	BOP	C&C
─High	291	470	120	192
─Low	120	230	80	160
✕Close	260	350	100	176

─High ─Low ✕Close

	Capital Cost	PCS	BOP	C&C
─High	247	329	115	182
─Low	102	184	75	152
✕Close	220	211	95	167

─High ─Low ✕Close

FIGURE 13.12 Lead─acid BESS BOM cost components changing trend between 2018 and 2025 [7].

TABLE 13.6 Lead─acid BESS BOM cost in 2018 and 2025 [7].

	Lead-acid battery	
Parameters	**2018**	**2025**
Capital cost ($/kWh)	260	220
PCS ($/kW)	350	211
BOP ($/kW)	100	95
C&C ($/kWh)	176	167
Fix. O&M ($/kw.yr)	10	8
Var. O&M (cents/kWh)	0.375	0.375
E/P (h)	4	4
BOM ($/kW)	**2204.015**	**1862.015**

comparison between the BOM cost components of this technology in the years 2018 and 2025 is available in Fig. 13.13.

BOM expenses and detailed costs of an ESS equipped with this technology in both of the 2018 and 2025 years are available in Table 13.7.

As it is obvious from the chart, the BOM cost of a sodium─metal halide BESS will have a downward trend until 2025 by about 28%, which will increase the chance of implementation due to suitable economic circumstances.

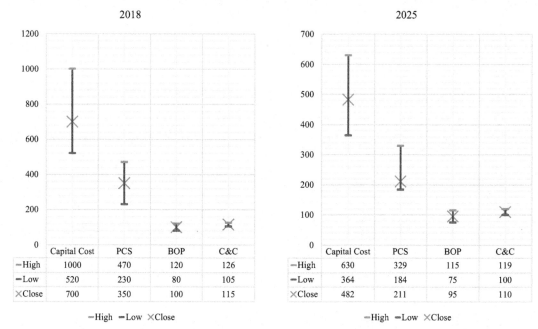

FIGURE 13.13 Sodium−metal halide BESS BOM cost components changing trend between 2018 and 2025 [7].

TABLE 13.7 Sodium−metal halide BESS BOM cost in 2018 and 2025 [7].

Parameters	Sodium−metal halide battery	
	2018	2025
Capital cost ($/kWh)	700	482
PCS ($/kW)	350	211
BOP ($/kW)	100	95
C&C ($/kWh)	115	110
Fix. O&M ($/kw.yr)	10	8
Var. O&M (cents/kWh)	0.43	0.43
E/P (h)	4	4
BOM ($/kW)	**3720.017**	**2682.017**

(e) *Zinc−Hybrid Cathode Battery*

This technology, which is also famous for its abbreviation, "Znyth", is a developing and commercialized battery technology primarily introduced by Eos energy storage company. With the low-cost production of this battery, it captured the attention of operators and investors of the grid rapidly. The nonabsorbent capability of battery in front of CO_2 pollution

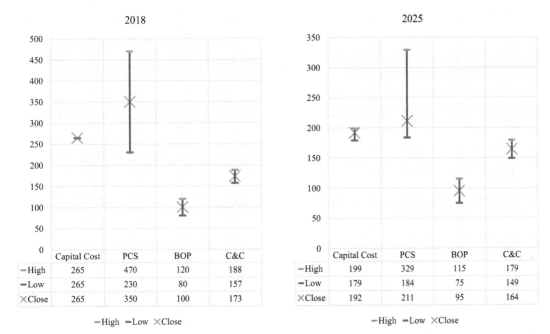

2018	Capital Cost	PCS	BOP	C&C
—High	265	470	120	188
—Low	265	230	80	157
✕Close	265	350	100	173

—High —Low ✕Close

2025	Capital Cost	PCS	BOP	C&C
—High	199	329	115	179
—Low	179	184	75	149
✕Close	192	211	95	164

—High —Low ✕Close

FIGURE 13.14   Zinc—hybrid cathode BESS BOM cost components changing trend between 2018 and 2025 [7].

prevents clogging issues to come up. Although his technology sounds so suitable for future planning of ESS, it is immature and more developments are needed. Yet, it is one of the potential options of energy supply in storage systems. A comparison between the BOM cost components of this technology in the years 2018 and 2025 is available in Fig. 13.14.

BOM expenses and detailed costs of an ESS equipped with this technology in both of the 2018 and 2025 years are available in Table 13.8.

As it is obvious from the chart, the BOM cost of a sodium—metal halide BESS will have a downward trend until 2025 by about 22%, which will increase the chance of implementation due to suitable economic circumstances.

(f) *Redox Flow Battery*[3]

Redox flow battery is an innovative yet immature technology of storing energy that offers competitive specialties in contrast to other chemical-based technologies that are mature and more commercialized such as Li-ion and lead—acid. Higher life cycle duration, low operational temperature requirements, and easy scalabilities are just some of the advantages of this technology. A comparison between the BOM cost components of this technology in the years 2018 and 2025 is available in Fig. 13.15.

---

[3] San Diego Gas & Electric (SDGE) has unveiled a vanadium redox battery storage pilot project in coordination with Sumitomo Electric, stemming from a partnership between Japan's New Energy and Industrial Technology Development Organization (NEDO) and the California Governor's Office of Business and Economic Development (GO-Biz). For more information, visit https://sdgenews.com/article/innovative-battery-storage-technology-connected-california-grid

TABLE 13.8   Zinc—hybrid cathode BESS BOM cost in 2018 and 2025 [7].

Parameters	Zinc—hybrid cathode battery	
	2018	**2025**
Capital cost ($/kWh)	265	192
PCS ($/kW)	350	211
BOP ($/kW)	100	95
C&C ($/kWh)	173	164
Fix. O&M ($/kw.yr)	10	8
Var. O&M (cents/kWh)	0.375	0.375
E/P (h)	4	4
BOM ($/kW)	**2212.015**	**1738.015**

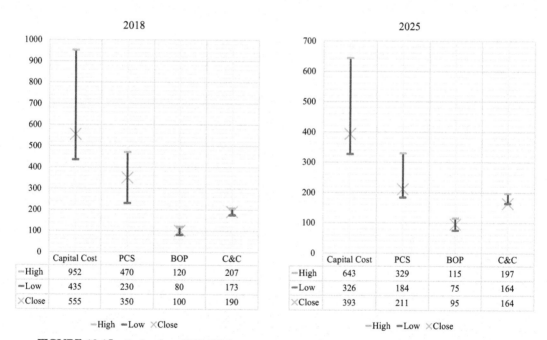

FIGURE 13.15   Redox flow BESS BOM cost components changing trend between 2018 and 2025 [7].

BOM expenses and detailed costs of an ESS equipped with this technology in both of the 2018 and 2025 years are available in Table 13.9.

As it is obvious from the chart, the BOM cost of a redox flow BESS will have a downward trend until 2025 by about 25%, which will increase the chance of implementation due to suitable economic circumstances.

TABLE 13.9 Redox flow BESS BOM cost in 2018 and 2025 [7].

Parameters	Redox flow battery	
	2018	2025
Capital cost ($/kWh)	555	393
PCS ($/kW)	350	211
BOP ($/kW)	100	95
C&C ($/kWh)	190	164
Fix. O&M ($/kw.yr)	10	8
Var. O&M (cents/kWh)	0.3525	0.365
E/P (h)	4	4
BOM ($/kW)	3440.014	2606.015

What we have seen until now were the overall BOM costs of utility-sized FESS and BESS. To understand which of these approaches are suitable for that specific grid that we are tending to install ESS on it, there is a necessity to identify the revenue streams that are common for these storage systems. There are three different revenue markets—Energy, Reserve, and Ancillary markets—in which each ESS can operate in them. To predict the potential revenue of an energy storage facility, a process of four actions need to be done.

At first, a mathematical model is needed for any ESS we are willing to analyze. Then a complete review of the historical data for the target market is needed to be done. After this step, an ideal and perfect prediction of the future is needed to be assumed. Besides, there should be some test strategies that are not relying on the ideal thought, which shaped our prediction in the third step. After these main steps, so many procedures such as optimizations and joint market applications can be applied in order to ensure the maximum economic outcome.

Insomuch we are focusing on joint energy and ancillary multimarket applications of utility-sized energy storage facilities. There should be a comparison between an ESS and a stand-alone power plant for each application. There should be a reminder that a hydropower plant is a suitable option for both energy arbitrage and frequency or voltage regulation wherever there is a suitable position to set this kind of power plant. But as long as there is not that much appropriate location in every country in order to build a hydro plant and also in the near future, the price of at least three technologies of BESS will drop down to a level in which hydro plants will not be competitive with them, considering ESS as the ultimate solution for both energy and ancillary needs of the grid is rational.

Except for wind and solar PV power plants, which are known as renewables with uncertainties are not eligible to be used as reliable solutions for the special demands of the grid, other renewable power plants such as biomass and geothermal usually cost average BOM expenses of 5729 up to 8735 $/kW (Table 13.10). The calculations are available further. This number is much higher than what an ESS system can cost, which makes the storage systems a suitable option for operation in both energy and ancillary markets. The integration of ESS

TABLE 13.10 Financial information of biomass and geothermal power units.

Technology	Total overnight cost (TOC) $/KW	Fuel cost $/KW.yr	Fixed O&M cost $/KW.yr	Variable O&M cost $/MWh	Lifespan (yr)	Total BOM cost $/KW
Geothermal	2680	0	113.29	1.16	30	6358.492
Biomass (BFB/CFB/Stoker)	2800	45.007	126	4.2	25	7919.375
Biomass (Gasifiers)	3250	43.415	146.25	3.7	25	8735.325
Biomass (Anaerobic digestion)	1975	125.47	79	4.2	25	7930.95
Landfill gas	1557	69.433	20.02	6.17	25	5728.794

with renewables in different spots around the world caused more reduction in expenses of a utility-sized ESS, making it an attractive option for grid operators and investors.

$$TOC(\$/KW) + \left[Fuel\ (\$/KW \cdot yr) + \left(\frac{Var \cdot O\&M(\$/KWh) \times 24(h) \times 335^*(day)}{1000} \right) \right.$$
$$\left. + Fix. O\&M\ (\$/KW \cdot yr) \right] \times Lifespan(yr) = Total\ BOM\ cost\ (\$/KW)$$

TOC: Total Overnight Cost
O&M: Operation and Maintenance
BOM: Building, Operation, and Maintenance
* Because of sparing 30 days (1 month) for the overhaul process in power plants, the period of a year is considered 335 days.

4. Conclusion

There is an increasing trend of need for ESS in energy systems, specially deregulated ones. Despite technical issues such as voltage support or frequency regulation, etc., which are usual in any analysis related to the implementation of ESS, there should be an independent study about its economic justification and the revenue that it can create in triple markets of energy systems (energy, reserve, and ancillary markets). By rational futurology of ESS in energy systems, there can be complete planning of renewable energy integrated grids that can offer numerous services in joint energy and ancillary multimarkets.

Beneficial financial studies besides technical feasibility analysis are important for the installation of ESS in developed countries while it is usually accounted for as the most effective motive in initiating such projects in the least developed and developing countries. The important fact that is needed to be cared is that before any kind of action in energy grids, there should be a vivid picture of which markets are needing this technology and what services are needed in this project. For joint ancillary and energy multimarkets, there are just a few technologies suitable, but it is expected to see significant improvements in other variations of ESSs to make them both economically and technically compatible with energy grids.

References

[1] M. Kazemi, H. Zareipour, N. Amjady, W.D. Rosehart, M. Ehsan, Operation scheduling of battery storage systems in joint energy and ancillary services markets, IEEE Trans. Sustain. Energy 8 (2017) 1726–1735.
[2] S. Bradbury, J. Hayling, P. Papadopoulos, N. Heyward, Smarter Network Storage Electricity Storage in GB: SNS 4.7 Recommendations for Regulatory and Legal Framework, 2015 (SDRC 9.5).
[3] L. Meng, J. Zafar, S.K. Khadem, A. Collinson, K.C. Murchie, F. Coffele, et al., Fast frequency response from energy storage systems—a review of grid standards, projects and technical issues, IEEE Trans. Smart Grid 11 (2) (2020) 1566–1581, https://doi.org/10.1109/TSG.2019.2940173.
[4] J. Seasholtz, Grid-Scale Energy Storage Demonstration of Ancillary Services Using the UltraBattery Technology, East Penn Mfg. Co., Inc., Lyons, PA (United States), 2015.
[5] International Renewable Energy Agency, Electricity Storage and Renewables: Costs and Markets to 2030, 2017. Report Abu Dhabi.
[6] A.S. Nagorny, R.H. Jansen, M.D. Kankam, Experimental Performance Evaluation of a High Speed Permanent Magnet Synchronous Motor and Drive for a Flywheel Application at Different Frequencies, 2007.
[7] K. Mongird, V.V. Viswanathan, P.J. Balducci, M.J.E. Alam, V. Fotedar, V.S. Koritarov, et al., Energy Storage Technology and Cost Characterization Report, Pacific Northwest National Lab.(PNNL), Richland, WA (United States), 2019.

Further reading

[1] IRENA, Innovation Landscape Brief: Utility-Scale Batteries, International Renewable Energy Agency, Abu Dhabi, 2019.
[2] IRENA, Electricity Storage Valuation Framework: Assessing System Value and Ensuring Project Viability, International Renewable Energy Agency, Abu Dhabi, 2020.
[3] Q. Huang, Y. Xu, C. Courcoubetis, Financial incentives for joint storage planning and operation in energy and regulation markets, IEEE Trans. Power Syst. 34 (2019) 3326–3339.
[4] A. Eller, D. Gauntlett, Energy Storage Trends and Opportunities in Emerging Markets, Navigant Consulting Inc, Boulder, CO, USA, 2017.
[5] T. Soares, H. Morais, P. Faria, Z. Vale, Smart grid market using joint energy and ancillary services bids, in: 2013 IEEE Grenoble Conference: IEEE, 2013, pp. 1–6.
[6] A. Berrada, K. Loudiyi, I. Zorkani, Valuation of energy storage in energy and regulation markets, Energy 115 (2016) 1109–1118.
[7] J. Cho, A.N. Kleit, Energy storage systems in energy and ancillary markets: a backwards induction approach, Appl. Energy 147 (2015) 176–183.
[8] U.S. Energy Information Administration, U.S., Battery Storage Market Trends, May 21, 2018.

Electric energy storage systems integration in energy markets and balancing services

Vahid Vahidinasab[1], Mahdi Habibi[2]

[1]School of Engineering, Newcastle University, Newcastle upon Tyne, United Kingdom;
[2]Faculty of Electrical Engineering, Shahid Beheshti University, Tehran, Iran

1. Introduction

There is a global endeavor for decarbonization where the energy sector has a critical role toward this goal. Decarbonization strategy in energy systems asks for a rapid increase in uptaking the low carbon renewable resources that will cause some operational issues as these renewable generations are less predictable and flexible than conventional fossil-fueled generations and therefore cause an increasing challenge for the security of supply. In such an environment, energy storage systems (ESSs) are identified as a priority to cope with this intermittency and create a more efficient electricity system. Conventional regulatory regimes behave with ESSs like a generator. This issue is the main barrier that affected investment in ESSs.

To remove the mentioned investment barrier, a market structure is required to value the flexibility offered by ESSs by considering it as a complementing service rather than a competing element with the current network and generation assets.

This chapter is an effort to analyze the current progress in the integration of electric energy storage systems (EESSs) into the existing energy and balancing mechanisms. We start by a systematic review of the publications in this field and then by providing a detailed formulation of the problem, highlight the status quo, challenges, and outlook of the works in this area.

1.1 Technologies of the electric energy storage systems

In this section, a number of technologies of EESSs are listed (see Fig. 14.1) and briefly explained in the following.

1.1.1 Pumped hydroelectricity storage

Pumped hydroelectricity storage (PHS) is a technology that is based on pumping water to an upstream reservoir during off-peak or the times that there is redundant electricity produced by renewable energy sources (RESs), and when electricity is needed, it is released through the hydro turbines. The PHS technology is a suitable option for large-scale applications to cope with intermittency of RESs. Currently, PHS has the highest share of grid-scale electricity storage around the world. For instance, the United Kingdom has about 2.75 Gigawatts of installed PHS capacity. While the PHS is still used globally, it has some restrictions such as high investment cost and geological limitations that make it noncompetitive for private investors.

1.1.2 Compressed air energy storage

Compressed air energy storage (CAES) is based on storing high-pressure air (often underground) that can be used to generate electricity via a turbine in time of needs. It is one of the suitable EESSs for large-scale applications but has some geological restrictions.

1.1.3 Pumped heat electricity storage

In the pumped heat electricity storage (PHES), the electricity is applied to drive an engine (that indeed can work in two modes of heat pump and heat engine) linked to two thermal store tanks. So, for storing the electricity, the engine works in the mode of the heat pump

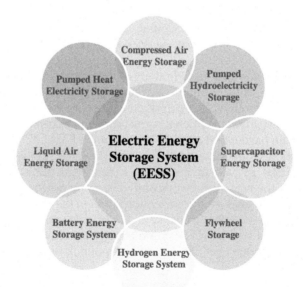

FIGURE 14.1   Technologies of the electric energy storage systems.

and pumps heat from the cold tank to the hot one, and for releasing the stored energy, the heat engine mode will be activated that drives a generator for producing the electricity.

1.1.4 Liquid air energy storage

The liquid air energy storage (LAES) cools the air until it turns into a liquid. Then the liquid air will be stored in a tank. In the reverse process, by exposing to ambient air or the waste heat from a process, the liquid air will be brought back to the gaseous state that is able to turn a turbine for power generation.

1.1.5 Battery energy storage system

Battery energy storage systems (BESSs) include lead—acid, lithium-ion (Li-ion), and sodium—sulfur designs. Sodium—sulfur batteries are suited to grid-scale applications, while Li-ion batteries are used in electric vehicles (EVs).

1.1.6 Hydrogen energy storage system

Hydrogen energy storage system (HESS) is a storage device in which it can be charged by injecting the hydrogen produced by redundant electricity generations. This energy can then be discharged by using the gas as fuel in a combustion engine or a fuel cell. HESS can be used in the different scales of households, vehicles, or at large scale.

1.1.7 Flywheel and supercapacitor

Flywheel and supercapacitor are among the other types of EESSs, which are able to deliver power rapidly but with a limited capacity. They are also potential options for energy harvesting from braking in cars and trains. Moreover, these storage systems can provide short-term fast-response backup electricity for large energy consumers in the time of interruptions.

1.2 Literature survey

In this section, a systematic review of the works in this area is presented. A comprehensive analysis is done on the works in this field, and a list of selected high-impact publications is proposed and precisely reviewed.

1.2.1 Surveying methodology

We started this literature survey by gathering scholarly publications including technical reports, journals and conference papers, and books from different areas. To this end, we considered a wide range of publications that have the keywords of *"storage"* <and> *"energy"* <and> *"market"* in their title since we plan to bring a broader overview of the works presented in this field and categorize and analyze them in a systematic way. We used the Web of Science database and only peer-reviewed documents in English were analyzed where this exploration led to a total number of 306 documents over the last two decades (i.e., 2000 − 20).

1.2.2 A statistical overview

These publications are from 44 countries and 366 institutions. As shown in Table 14.1, among them, the United States, China, and Canada have the most published works with 80, 41, and 29 publications, respectively. Also, citations per publication of these countries are 17, 6, and 17, respectively.

Additionally, based on Table 14.2 *Sandia National Laboratories, RWTH Aachen University, and University of California, Riverside* with respectively 8, 7, and 7 publications are the institutions with the highest number of works in this field with a citations per publication of 3, 20, and 25, respectively. Also, the journals with the maximum number of published works in the area are listed in Table 14.3.

In Fig. 14.2, a quantitative representation of the works presented in this field over the last two decades is demonstrated along with the number of citations to the published of each year.[1]

After the aforementioned quantitative analysis, all the journal publications are selected and analyzed in a qualitative assessment process. To this end, every publication was

TABLE 14.1   The countries with the maximum number of publication in this field.

Countries	Number of publications	Average citations
United States	80	17
China	41	6
Canada	29	17

TABLE 14.2   The institutions with the maximum number of publication in this field.

Name of institution	Number of publications	Average citations
Sandia National Laboratories	8	3
RWTH Aachen University	7	20
University of California, Riverside	7	25

TABLE 14.3   Journals with the maximum number of works in this field.

Name of the journal	Number of publications	Average citations
Applied Energy	18	26
Energy	16	34
IEEE Transactions on Power Systems	15	18

[1]The reported information of the year 2020 is not for the whole year. It is based on the information received on October 22, 2020 from the *Web of Science* website: https://www.webofknowledge.com/.

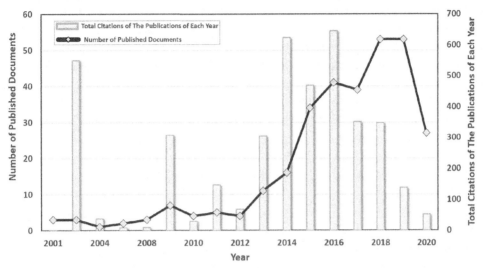

FIGURE 14.2   A quantitative comparison of the number of works published in each year along with the number of citations to the publication of that specific year.

associated with a technical category and then analyzed to assess its relevance to the topic. A label is assigned to each paper to highlight the papers from the quality and relevance point of view. At this stage, a total of 53 documents remained. Among them, there are 20 scholarly articles with more than 50 citations that make the core references of the works presented in this area. For comparison of the technical contents of the selected works, they are tabulated in Table 14.4.

The most highlighted indices regarding the applications and issues of scheduling the EESSs in the operational horizon are collected in Table 14.4. The services provided by EESSs can be used for either strength operational resilience or uncertainty handling. The most important features of the models include the ancillary services consisting of the optimality of deploying, the delivery check of services, and features of balancing services. Also, the technical features of EESSs including the check for feasibility of reservoir for compensation, the nonanticipativity constraints (NAC), and reflection of degradation costs are categorized and checked in Table 14.4.

1.2.3 Overview of the topics

The conventional generators are not a desirable source for power supply based on economic and environmental aspects [54]. Also, renewable generations are not permanently available for power supply, and the intermittency of their output cannot be avoided [55]. In this way, the application of EESSs can benefit the power system operation in the market perspectives. Here, we review some of the arising applications and issues of EESS scheduling in different aspects including the uncertainty covering, reserve provision, operational resilience, spot market perspectives, and balancing services.

TABLE 14.4 Taxonomy of publications on EESSs participation in energy and balancing services.

References	Model	Uncertainty Modeling	EESS Technology	EESS for									
				Energy[a]	Resilience	Flexibility	Reserve	ROC[b]	RDC[c]	Balancing	ERFC[d]	NAC[e]	DCM[f]
[1]	MILP[g]	Deterministic	BESS-Li-ion	✓	–	–	✓	✓	–	–	–	–	✓
[2]	MINLP[h]	Stochastic	GS[i]	–	–	–	✓	✓	✓	–	–	–	–
[3]	MILP	Stochastic	GS	–	–	✓	✓	✓	✓	–	✓	✓	–
[4]	MINLP	Stochastic	CAES	–	–	✓	✓	✓	–	✓	–	–	–
[5]	NLP[j]	Max-Min	GS	–	–	–	✓	✓	✓	–	–	✓	–
[6]	MILP	Robust	GS	–	–	✓	✓	✓	✓	–	✓	✓	–
[7]	MILP	Stochastic	SGB[k]	–	–	✓	✓	✓	✓	–	–	–	–
[8]	MILP	Robust	GS	–	–	–	✓	✓	–	–	–	✓	–
[9]	NLP	Robust	BESS-EV	✓	–	–	–	–	–	–	–	–	–
[10]	MILP	Deterministic	BESS	✓	–	✓	✓	✓	✓	–	–	✓	–
[11]	MILP	Deterministic	CAES	✓	–	✓	✓	✓	✓	–	–	✓	–
[12]	MILP	Stochastic	BESS-PEV	–	–	✓	✓	–	✓	–	–	–	–
[13]	MILP	Stochastic	BEST	–	–	✓	–	–	✓	–	–	–	–
[14]	MINLP	Stochastic	GS	–	–	✓	✓	✓	✓	–	✓	✓	–
[15]	MINLP	Stochastic	GS	–	–	✓	✓	✓	✓	–	–	✓	–
[16]	MILP	Stochastic	GS	–	–	✓	✓	✓	✓	–	–	✓	–
[17]	MILP	Stochastic	GS	–	–	✓	✓	✓	✓	–	–	–	–
[18]	MILP	Stochastic	CAES	✓	–	✓	✓	✓	✓	✓	✓	✓	–
[19]	MILP	Stochastic	GS	–	✓	✓	✓	✓	✓	–	–	–	–
[20]	MILP	Stochastic	GS	–	–	✓	✓	✓	✓	–	✓	✓	–
[21]	MILP	Stochastic	GS	–	–	✓	✓	✓	✓	–	–	✓	–
[22]	MILP	Stochastic	GS	✓	–	–	✓	✓	✓	–	–	–	–

Ref.	Model	Approach	Storage type											
[23]	MILP	Hybrid	BESS-EV	✓	—	—	—	—	—	—	✓	—	—	✓
[24]	MILP	Deterministic	GS	—	✓	✓	✓	✓	—	✓	—	—	—	—
[25]	MILP	Stochastic	MES	—	✓	—	✓	—	✓	—	—	—	—	—
[26]	NLP	Deterministic	GS	—	✓	—	✓	—	✓	—	—	—	—	—
[27]	MILP	Stochastic	GS	—	✓	✓	✓	✓	✓	—	—	—	—	—
[28]	MILP	Stochastic	BESS-Li-ion	✓	✓	✓	✓	✓	—	—	—	—	—	✓
[29]	MILP	Robust	GS	—	✓	—	✓	✓	—	—	—	—	—	—
[30]	MILP	Stochastic	GS	—	✓	✓	✓	—	—	—	—	—	—	—
[31]	NLP	Deterministic	PEV	✓	✓	—	✓	✓	—	—	—	—	—	—
[32]	NLP	Deterministic	GS	✓	✓	—	✓	—	—	—	—	—	—	—
[33]	LP[1]	Deterministic	Multiple types	✓	✓	—	—	✓	—	—	—	—	—	—
[34]	MILP	Deterministic	CAES	✓	—	✓	✓	✓	—	—	✓	—	—	—
[35]	MILP	Deterministic	CAES	✓	—	✓	✓	✓	—	—	✓	—	—	—
[36]	MILP	Deterministic	EV	—	—	—	✓	✓	✓	—	✓	✓	—	—
[37]	MILP	Robust	GS	✓	—	✓	✓	✓	—	—	✓	✓	—	✓
[38]	MINLP	Deterministic	Multiple types	✓	—	✓	✓	✓	—	—	—	✓	—	—
[39]	MILP	Stochastic	PHES	—	—	✓	✓	—	—	—	✓	—	—	—
[40]	MILP	Deterministic	CAES	✓	—	—	✓	✓	—	—	✓	✓	—	—
[41]	MILP	Deterministic	CAES	✓	—	✓	✓	—	—	—	✓	✓	—	—
[42]	MILP	Real-time control	BESS	✓	—	✓	✓	—	—	—	—	—	—	—
[43]	NLP	Deterministic	GS	—	—	✓	—	—	—	—	—	—	—	—
[44]	NLP	Deterministic	CAES	✓	—	—	—	—	—	—	—	—	—	—
[45]	MILP	Hybrid	BEST	—	—	—	—	—	✓	—	✓	—	—	—
[46]	MILP	Stochastic	GS	—	—	—	—	—	✓	—	—	—	—	—

(Continued)

TABLE 14.4 Taxonomy of publications on EESSs participation in energy and balancing services.—cont'd

References	Model	Uncertainty Modeling	EESS Technology	EESS for									
				Energy[a]	Resilience	Flexibility	Reserve	ROC[b]	RDC[c]	Balancing	ERFC[d]	NAC[e]	DCM[f]
[47]	NLP	Deterministic	BESS	✓	—	—	—	—	—	—	—	—	—
[48]	MILP	Stochastic	GS	—	—	✓	—	—	—	—	—	—	—
[49]	NLP	Hierarchical	GS	✓	—	—	—	—	—	—	—	—	—
[50]	MILP	Stochastic	PHS	—	—	✓	✓	✓	✓	✓	—	—	—
[51]	NLP	Deterministic	PHS and CAES	✓	—	—	—	—	—	—	—	—	—
[52]	MILP	Robust	CRES	—	—	✓	✓	✓	✓	—	—	—	—
[53]	MILP	Stochastic	PHS	—	—	✓	✓	✓	✓	—	—	✓	—
			Frequency (%)	42	9	55	70	51	51	28	9	21	8

[a]Energy Backup.
[b]Reserve optimality check.
[c]Reserve deliverability check.
[d]EESS reservoir feasibility check for uncertainty covering.
[e]Nonanticipativity constraints check.
[f]Degradation cost model.
[g]Mixed-integer linear programming.
[h]Mixed-integer nonlinear programming.
[i]Generic Storage.
[j]Nonlinear programming.
[k]Second generation EV Battery.
[l]Linear programming.

1.2.3.1 EESS application for uncertainties covering

The application of ESSs for compensation of uncertainties is desirable for system operators [56,57]. In this regard, the NACs are very important in stochastic models, and they mean EESSs cannot be operated in both charging and discharging modes at the same time. Also, the scheduling result must contain feasible values of decision variables at least for one scenario, which is mainly defined as the scenario of base case [18].

Reference [8] explicitly considers transmission reserves in addition to generation reserves for the accommodation of uncertainties, while the compensation of EESSs is used to reduce the prices of energy and reserves. In that study, uncertainty marginal prices are introduced to find the reserve price for covering uncertainties, and the energy price is calculated using the locational marginal prices [58]. Reference [9] employs a robust model for active and reactive power management of EVs for distribution systems, which it considers the revenue of reactive power exchange with the network for the worst-case scenario as the problem. A time–space robust–stochastic (hybrid) model for describing the railway transport network of the BEST system is proposed in Ref. [45], which is developed for the day-ahead market clearing.

A coordinated multi-timescale scheduling of EESSs in the presence of a large share of renewable generations is evaluated in Ref. [29], while NACs are ignored in the proposed model. The provision of ancillary services from utility-scale EESSs for covering transmission-line contingencies is proposed in Ref. [17], in which the optimality of fast regulation services and the NACs are not evaluated in the decomposed stochastic model. The uncertainty of RESs is addressed by proper management of charging/discharging of mobile plug-in electric vehicle (PEV) in Ref. [12]. In that model, PEVs provide ancillary services and release the congestion of the transmission system by transferring energy between locations, and the NACs are properly addressed by the stochastic model. The battery-based energy storage transportation (BEST) system is used for compensation of variable output of wind generation in Ref. [13]. That study considers the random interruptions forced by component outages of both power and railway systems, and NACs are considered in the stochastic model.

1.2.3.2 EESS application for reserve provision

The exploitation of reserve services from EESSs can bring them additional revenue, and some studies introduce models for reserve provision through charging, discharging, and idling modes of EESSs. In this regard, six modes for reserve provision of EESSs can be defined to optimize the allocation of reserve capacity in addition to energy in a joint market [2,3]. The assessment of CAES conducted in Ref. [34] reveals that the conventional type could earn the additional revenue 23 ± 10/kw-yr, and the adiabatic CAES 28 ± 13/kw-yr by providing the operational reserves. The cryogenic energy storage (CRES), which stores energy as liquid gas and vaporizes it to generate energy by driving a turbine, is used for load shifting, arbitraging, and providing operational reserves in Ref. [52]. Reference [16] studies a stochastic security-constrained unit commitment (UC) that includes the probabilities of generation and transmission contingencies for optimal allocation of reserves. The participation of EESSs in providing reserve is interesting, but there is a concern about the unforeseen sequence of scenarios in real time, which can make the EESS scheduling infeasible. In this regard, Refs. [20,28] raise the potential issue of successive reserve provision from EESSs. A multi-timescale stochastic model for coordinated participation of EESSs and flexible loads as a network service provider is presented in Ref. [20], in which flexible loads back up the energy consumption of EESSs for compensation of renewable resources. In Ref. [28] a stochastic model for limited look-ahead scheduling is proposed to address the issue of successive reserve provision of EESSs against the unforeseen sequence of scenarios. The adequate regional reserves needed to address a set of contingencies as well as load-following reserves for ramping between time steps are determined in Ref. [15], in which the delivery of services is guaranteed by checking the transmission capacity. That model considers a secure range of operation for EESSs' reservoir based on limited look-ahead scheduling.

1.2.3.3 EESS application for operational resilience

The concept of operational resilience can be defined as the ability of the power systems to accommodate severe situations because of large imbalances, ramp events, or network congestion that may happen during the operation period. The ramp events are defined as the significant variation in generation or consumption within a short period. The frequency of the occurrence of rare events is low, but the corresponding impact can dramatically disrupt the balance between generation and consumption. This situation can lead to locational or whole system blackouts, and such cases are reported in the United Kingdom, Australia, and the United States [59]. The fast dynamic of ramp events requires appropriate flexibility services to ensure the stability of operation [60]. The extra reserves provided by conventional generators increase the total cost and greenhouse gas emissions [61,62]. So, these services can be supplied by EESSs, which can improve the robustness of the daily operation alongside the environmental aspects.

In [25], the network reconfiguration and truck-mounted mobile energy storage (MESs) are used for critical load restoration posthurricane. Reference [10] applies the BEST of the railway system for enhancing the resilience of the power system in disaster areas with a congested network. The short look-ahead forecast of ramp events followed by compensation of EESSs including the upward/downward regulations is applied in Ref. [24].

The ancillary services including operational reserves and ramping products offered by different producers can be used to alleviate the potential risks of ramp events and to prevent price spikes in real-time operation [23]. Reference [19] employs services provided by EESSs for compensation of wind ramp events, while enough spinning reserves are supplied for covering uncertainties during the normal situation. The coordinated scheduling of EESSs based on ahead adjustments against the wind ramp events is proposed in Refs. [26], and [27] strengthens a continuous stochastic UC using distributed EESSs for potential intrahour regulation of wind ramp events.

1.2.3.4 EESS application for balancing services

The services with fast dynamics are important for the balancing of the power system in near real-time markets. In this way, a series of ancillary services can be assigned to the EESSs for selling their redundant available capacity in balancing markets. This can bring them additional revenue, while the more accurate forecasts of uncertain parameters are available to the market operator.

The concepts of active network management (ANM) and virtual power plants are employed for designing principles of microgrid considering the operational model in Ref. [49]. That model considers both island and grid-connection modes and hierarchical control, which are applied for balancing of EESSs near delivery time. A real-time scheme for controlling BESSs for managing the energy sold in the balancing market is presented in Ref. [42].

The optimal dispatch of EESSs considering the regulation in short-term compensation for imbalances of wind farms is evaluated in Ref. [32]. Reference [33] compares the optimal capacities for power and energy in various balancing markets evaluating different technologies of EESSs.

The look-ahead scheduling of EESSs of [41] considers the physical characteristics of CAESs such as compressor, expander, and dynamics of performance. The balancing look-ahead model could enable additional revenues by using arbitrage opportunities. The ancillary services provided by private large-scale EESSs for relieving transmission congestion in near real-time optimal dispatch are studied in Ref. [40], which produce revenues by identifying arbitrage opportunities in electricity day-ahead prices.

The scheduling of EVs with consideration of the energy aggregators in both day-ahead and balancing markets is studied in Ref. [36]. That study defines a protocol for communication among the market players and shows how to integrate the scheduling of EESSs in day ahead and real time. Two robust models for offering spinning reserve and regulation services alongside the day-ahead energy are studied in Refs. [37,41].

1.2.3.5 Spot markets from the perspective of private investors of EESSs

Reference [63] reviews the various types of mechanisms for rewards of PHSs in electrical energy and reserve, which can be used by investors and policymakers. A day-ahead market for energy and reserve offering from independently operated EESSs with a private investor is developed in Ref. [30], in which the fluctuations of wind and market prices are investigated. A similar coordinated operation of distributed private EESSs is investigated in Ref. [46]. A competitive market considering the optimal charge/discharge scheduling community energy storage (CES) is proposed for day ahead and real time in Ref. [31].

Reference [14] proposes a leader—follower multiobjective problem for expansion planning of active distribution network (ADN) with EESSs and high penetration of the RESs. A bilevel stochastic model for determination of the optimal size of EESSs including charging/discharging rate and the reservoir capacity from the perspective of strategic investor is investigated in Ref. [48]. The value of arbitrage for PHSs and CAESs in European markets and strategies for selection and sizing are evaluated in Ref. [51]. Reference [38] executes an economic comparison among different energy storage technologies considering competitive energy and ancillary service markets. That model evaluates the internal rate of return (IRR) index to determine the required financial supports for different technologies that can be used by investors. Reference [43] operates and sizes the EESSs for wind power plants, in which the model uses variations in the spot price to maximize the value of revenues.

2. Potential markets and services

EESSs are vital tools for balancing the grid at different scales, and among them, due to the widespread uptake of BESSs, they are most popular now. The main function of a BESS is to provide the network operators with a fast-response reserve and frequency-response services. BESSs normally provide a larger capacity over a short period of time rather than providing a lower capacity over a large period, and the majority of large-scale BESSs are be able to provide power for around 30—90 min now. A comparison of the EESS technologies from the

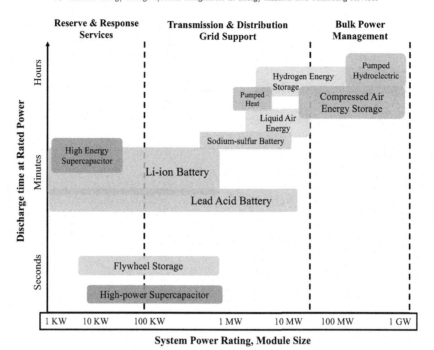

FIGURE 14.3   A comparison of the EESS technologies from the discharge time and power rating point of view [64,65].

discharge time and power rating point of view is made (data is mostly adopted from Ref. [64,65] and demonstrated in Fig. 14.3.

There are a number of ways that an EESS can participate in the energy market to support the operator for the grid balancing including the balancing and ancillary services provision and maintaining the grid frequency, which in parallel reduces the cost of the systems as well, minimizing foot room costs, and black start capability [66].

According to Ref. [65], EESS technologies are able to participate in different parts of the electricity system including generation (e.g., by providing balancing services and reserve), transmission (e.g., by providing frequency control and investment deferral), distribution (e.g., by providing voltage control and capacity support), and end users (e.g., by providing peak shaving and cost reduction).

Currently, the balancing market is a promising large-scale market for EESS technologies. Although the EESSs could participate in other ancillary services markets for fast reserve [18,21] and frequency services [67]. There are a number of markets in different electricity systems around the world that an EESS could provide their services. For instance, the experience of the United Kingdom regarding the services that an EESS could provide to the electricity market is introduced in Fig. 14.4[2] [65].

[2]By RoCoF, we mean the Rate of Change of Frequency.

	Response Time	Timescale	Role of storage
RoCoF Control	< 0.5 sec	< 15 min	Charging or discharging in response to frequency deviations
Frequency Containment	< 10 sec	< 10-30 sec	(Dis)charging in response to loss of generation or load that affects the system frequency
Frequency Replacement	< 2 min	< 30 min	(Dis)charging in response to loss of generation or load, with the aim of red that affects the system frequency
Voltage Support	< 1 sec	< 1-60 sec	(Dis)charging reactive power to stabilize voltage in the transmission and/or distribution system
Operating Reserve	240 min	< 2-24 h	Discharging at times of high demand in preference to flexible generation
Black Start	NA	NA	Contributing to restarting an electricity system following a total failure

FIGURE 14.4 The experience of the United Kingdom regarding the services that an EESS could provide to the electricity market [65].

3. Integration of electric storage in energy and ancillary services markets: model and formulation

The integration of EESSs to the market of energy and ancillary services presented as a UC problem, which is common for scheduling in the operational horizon. The uncertainties related to the variation of renewables' output power and load forecasting error are two regular sources of imbalances in stochastic UC models, and they are considered as the uncertain parameters in the formulation of this section. The base scenario is defined as the first-stage decision variable to meet the nonanticipativity condition ($s = 0$), and the second-stage variables indicate the state variables for different scenarios of uncertainties. The EESSs are employed for providing reserves in different working statuses and for dispatching in charging/discharging modes in the base scenario.

The objective function (Eq. 14.1) consists of two terms: (i) costs related to conventional generators for energy production and deploying reserves and (ii) total costs of EESSs for dispatching and providing reserves. The equations of the stochastic UC includes Eqs. (14.2)−(14.15), together with (14.16)−(14.29) for EESSs model that will be described in the following.

$$\min_{\substack{st,sd,i \\ j,p}} \sum_t \left(Cost_t^{CG} + Cost_t^{SE} \right) \tag{14.1}$$

3.1 Model of conventional generators and renewable power plants

The model of regular generators and renewable units in a stochastic UC-based market is presented here, and it will be integrated into the model of EESSs in the following section. The objective of conventional generators (Eq. 14.2) includes costs of startup/shutdown and the energy cost, which is a function of $P_{g,t}^s$ and $i_{g,t}$, and costs of reserves in up/down directions. It should be noted that the production cost of renewable generations is considered to be zero, but it can be modified based on policies of target markets. Also, the demand response is not considered, but the uncertainty of renewable generation can be managed by curtailing the extra production. The weighted sum of generation cost is calculated for $s \geq 1$, and it should be noted that the sum of probabilities of scenarios is equal to 1 ($\sum_{s \geq 1} \pi_s = 1$). Also, the coefficients of reserve cost in up/down directions are positive values.

$$Cost_t^{CG} = \sum_g \left(\underbrace{C_g^{ST} st_{g,t} + C_g^{SD} sd_{g,t} + \sum_{s \geq 1} \pi_s F\left(p_{g,t}^s, i_{g,t}\right)}_{\text{Production Costs of Generators}} + \underbrace{C_g^{\uparrow} r_{g,t}^{\uparrow} + C_g^{\downarrow} r_{g,t}^{\downarrow}}_{\text{Generators' Reserve Costs}} \right) \tag{14.2}$$

The operational constraints for conventional generators and renewable units are presented by (3)–(14). Constraints (14.3) and (14.4) are the relationship between binary variables of startup/shutdown and online/offline status of generators. The generation of upper- and lower-bound limits is checked by (Eq. 14.5), and consumable values of renewable generations are constrained by available energy per scenarios as reflected by (Eq. 14.6). The ramp rates including the startup/shutdown ramp limits are evaluated by Eqs. (14.7)–(14.10) limit the minimum online and offline duration of generators, respectively.

$$st_{g,t} - sd_{g,t} = i_{g,t} - i_{g,(t-1)} \tag{14.3}$$

$$st_{g,t} - sd_{g,t} \leq 1 \tag{14.4}$$

$$\underline{P}_g i_{g,t} \leq p_{g,t}^s \leq \overline{P}_g i_{g,t} \tag{14.5}$$

$$0 \leq p_{n,t}^{Ren,s} \leq \overline{P}_{n,t}^{Ren,s} \tag{14.6}$$

$$p_{g,t} - p_{g,(t-1)} \leq RP_g^{\uparrow} i_{g,(t-1)} + RP_g^{ST} st_{g,t} \tag{14.7}$$

$$p_{g,(t-1)} - p_{g,t} \leq RP_g^{\downarrow} i_{g,t} + RP_g^{SD} sd_{g,t} \tag{14.8}$$

$$\sum_{\tau=t}^{t+UT_g-1} i_{g,\tau} \geq UT_g st_{g,t} \quad \forall t \in \{1, ..., |\tau| - UT_g + 1\} \tag{14.9}$$

$$\sum_{\tau=t}^{t+DT_g-1} (1 - i_{g,\tau}) \geq DT_g sd_{g,t} \quad \forall t \in \{1, \ldots, |\tau| - DT_g + 1\} \tag{14.10}$$

Constraints (14.11)–(14.14) calculate the required operational reserve provided by conventional generators. The deployed reserves are the upper bound of redispatches regarding the base schedule. Also, the reserves are limited by offered values by generators.

$$r_{g,t}^{\uparrow} \geq p_{g,t}^0 - p_{g,t}^s \tag{14.11}$$

$$r_{g,t}^{\downarrow} \geq p_{g,t}^s - p_{g,t}^0 \tag{14.12}$$

$$0 \leq r_{g,t}^{\uparrow} \leq RR_g^{\uparrow} \tag{14.13}$$

$$0 \leq r_{g,t}^{\downarrow} \leq RR_g^{\downarrow} \tag{14.14}$$

The nodal power injection of buses is used to check the line flow limits by using the sensitivity coefficient $SF_{l,b}$ for DC power flows as presented by Eq. (14.15). This constraint can simultaneously check both the generation/consumption power balance and power flow limits of transmission lines.

$$\underline{FL}_l \leq \sum_b SF_{l,b} \left[\sum_{g \in G} p_{g,t}^s + \sum_{e \in \xi} \left(p_{e,t}^{Dis,s} - p_{e,t}^{Ch,s} \right) + \sum_{n \in \Gamma} p_{n,t}^{Ren,s} - \sum_{d \in \phi} p_{d,t}^s \right] \leq \overline{FL}_l \tag{14.15}$$

3.2 Model of EESS for compensation of uncertainties

This section will present the model of EESSs for scheduling in base scenario and scenarios of uncertainties, the calculation of EESSs' reserve provision, and the feasibility check of reservoir in compensations mode. The objective function of EESSs (Eq. 14.16) includes the cost of discharging and the reserve costs. Since the UC model is developed from perspective of an independent market operator and EESSs are considered as private entities, the discharging cost of EESSs is reflected in the objective function and the cost of charging will be evaluated implicitly as a load that is supplied by other resources. Additionally, the reserve costs are paid for both up and down directions.

$$Cost_t^{ES} = \sum_e \left(\underbrace{\sum_{s \geq 1} \pi_s \left(C_e^{Dis} p_{e,t}^{Dis,s} \right)}_{\text{Production Cost of EESSs}} + \underbrace{C_e^{\uparrow} r_{e,t}^{\uparrow} + C_e^{\downarrow} r_{e,t}^{\downarrow}}_{\text{EESSs' Reserve Costs}} \right) \tag{14.16}$$

Constraints (14.17)−(14.22) are related to EESSs' performance within scenarios of the base case and uncertainties. They consist of charging/discharging limits, hourly energy update based on previous values and current dispatches in charging, discharging, or idle modes, daily continuous operation constraint of EESSs, and the boundary constraints of stored energy.

$$f_{e,t}^{Ch,s} + f_{e,t}^{Dis,s} \leq 1 \tag{14.17}$$

$$\underline{P}_e^{Ch} f_{e,t}^{Ch,s} \leq p_{e,t}^{Ch,s} \leq \overline{P}_e^{Ch} f_{e,t}^{Ch,s} \tag{14.18}$$

$$\underline{P}_e^{Dis} f_{e,t}^{Dis,s} \leq p_{e,t}^{Dis,s} \leq \overline{P}_e^{Dis} f_{e,t}^{Dis,s} \tag{14.19}$$

$$SE_{e,t}^{ES,s} = SE_{e,(t-1)}^{ES,s} + p_{e,t}^{Ch,s} \beta_e^{Ch} - p_{e,t}^{Dis,s}/\beta_e^{Dis} \tag{14.20}$$

$$SE_{e,0}^{ES,s} = SE_{e,NT}^{ES,s} \tag{14.21}$$

$$\underline{SE}_e \leq SE_{e,t}^{ES,s} \leq \overline{SE}_e \tag{14.22}$$

3.2.1 Reserve provision from EESSs

The following constraints calculate reserves provided by EESSs in up/down direction within different operation modes. Based on the operation modes in the base scenario and within scenarios of uncertainties, six modes can be defined for the provision of reserves from EESSs [2]. Six modes consist of increased/decreased charging in charging mode, increased/decreased discharging in discharging mode, switching to discharging or idle mode from charging mode, and switching to charging or idle mode from discharging mode.

$$r_{e,t}^{\uparrow} \geq \left(p_{e,t}^{Dis,s} - p_{e,t}^{Dis,0}\right) + \left(p_{e,t}^{Ch,0} - p_{e,t}^{Ch,s}\right) \tag{14.23}$$

$$r_{e,t}^{\downarrow} \geq \left(p_{e,t}^{Dis,0} - p_{e,t}^{Dis,s}\right) + \left(p_{e,t}^{Ch,s} - p_{e,t}^{Ch,0}\right) \tag{14.24}$$

3.2.2 Reservoir check of EESSs within compensation mode

As stated, unknown sequences of scenarios in real time can lead to deviations in reservoir of EESSs while they participate in uncertainty covering. Here we present the model offered by Ref. [3] to ensure a secure delivery of EESSs' reserves by defining a window of multiple hours ($|C|$ = number of hours) for checking the enough headroom and floor room for EESSs so that they are able to follow dispatches within the base scenario alongside reserve provision.

$$0 \leq r_{e,t}^{UB} \geq \left(p_{e,t}^{Dis,s} - p_{e,t}^{Dis,0} \right) / \beta_e^{Dis} + \left(p_{e,t}^{Ch,0} - p_{e,t}^{Ch,s} \right) \beta_e^{Ch} \tag{14.25}$$

$$0 \leq r_{e,t}^{LB} \geq \left(p_{e,t}^{Dis,0} - p_{e,t}^{Dis,s} \right) / \beta_e^{Dis} + \left(p_{e,t}^{Ch,s} - p_{e,t}^{Ch,0} \right) \beta_e^{Ch} \tag{14.26}$$

$$R_{e,t}^{UB} \geq \sum_{\tau=0}^{C} r_{e,(t+\tau)}^{UB} + \left(SE_{e,(t+c)}^{ES,0} - SE_{e,t}^{ES,0} \right) \quad \forall C = \{1, ..., |C|\} \tag{14.27}$$

$$R_{e,t}^{LB} \geq \sum_{\tau=0}^{C} r_{e,(t+\tau)}^{LB} + \left(SE_{e,t}^{ES,0} - SE_{e,(t+c)}^{ES,0} \right) \quad \forall C = \{1, ..., |C|\} \tag{14.28}$$

$$\underbrace{SE_e + R_{e,t}^{LB}}_{\text{Reservoir Lower Bound}} \leq SE_{e,T}^{ES,0} \leq \underbrace{\overline{SE}_e - R_{e,t}^{UB}}_{\text{Reservoir Upper Bound}} \tag{14.29}$$

Constraints (14.25) and (14.26) calculate the maximum deviations of the reservoir at the base scenario regarding the compensations in up/down directions, respectively. The required energy headroom and floor room to address the total maximum hourly up and down deviations of $r_{e,t}^{UB}$ and $r_{e,t}^{LB}$ for the next window are calculated by Eqs. (14.27) and (14.28), respectively. Constraint (14.29) checks the stored energy to ensure that EESSs have enough energy headroom and floor room to operate both for dispatching in the base scenario and for providing deployed reserves in the real time.

3.3 Balancing market

The balancing market uses the most updated data of the system parameters near the delivery to reach a more accurate schedule. It is assumed that we have the actual values for uncertain parameters including load and available renewable energy. So, the objective function of the model is considered for the regulation cost as presented by Eq. (14.30), in which F' and $C_e^{Dis,BM}$ are the costs in the balancing market. The cost of EESS regulation in charging mode will be implicitly considered in the redispatch cost of generators (power charging values considered as loads). Also, the redispatches of generators and EESSs will be limited based on constraints presented by Eqs. (14.31)–(14.34). The other constraints of balancing market will be rewritten from Eqs. (14.5), (14.6), (14.15), and (14.17)–(14.22).

$$\min_p \sum_g \left(\underbrace{F'\left(p_{g,t}^{BM}, \widehat{I}_{g,t} \right) - F\left(\widehat{P}_{g,t}^0, \widehat{I}_{g,t} \right)}_{\text{Re-dispatch Cost of Generators}} \right) + \sum_e \left(\underbrace{C_e^{Dis,BM} p_{e,t}^{Dis,BM} - C_e^{Dis} \widehat{P}_{e,t}^{Dis,0}}_{\text{Re-dispatch Cost of EESSs}} \right) \tag{14.30}$$

$$\widehat{P}^{0}_{g,t} - p^{BM}_{g,t} \leq \widehat{r}^{\uparrow}_{g,t} \tag{14.31}$$

$$p^{BM}_{g,t} - \widehat{P}^{0}_{g,t} \leq \widehat{r}^{\downarrow}_{g,t} \tag{14.32}$$

$$\left(p^{Dis,BM}_{e,t} - \widehat{P}^{Dis,0}_{e,t}\right) + \left(\widehat{P}^{Ch,0}_{e,t} - p^{Ch,BM}_{e,t}\right) \leq \widehat{r}^{\uparrow}_{e,t} \tag{14.33}$$

$$\left(\widehat{P}^{Dis,0}_{e,t} - p^{Dis,BM}_{e,t}\right) + \left(p^{Ch,BM}_{e,t} - \widehat{P}^{Ch,0}_{e,t}\right) \leq \widehat{r}^{\downarrow}_{e,t} \tag{14.34}$$

4. Nomenclature

The indices used in this chapter are listed further for quick reference.

4.1 Indices

t Index of time $t \in \{1, \cdots, NT\}; (0 = \text{initial state})$.

s Index of scenarios $s \in \{0, 1, \cdots, NS\}; (0 = \text{base case scenario})$.

g Index of generator power plants $g \in \{1, \cdots, NS\}$.

e Index of energy storage devices $e \in \{1, \cdots, NE\}$.

n Index of renewable power plants $e \in \{1, \cdots, NN\}$.

l Index of transmission lines $l \in \{1, \cdots, NL\}$.

4.2 Sets and symbols

G Set of generators connected to bus b.

ξ Set of storage devices connected to bus b.

Γ Set of renewable power plants connected to bus b.

ϕ Set of active loads connected to bus b.

ST/SD Symbols of startup/shutdown.

Ren Symbols renewable generations.

ES Symbols of electric energy storage devices.

CG Symbols of conventional generators.

Ch/Dis Symbols of charging/discharging modes.

BM Symbols of values in balancing market.

LB/UB Symbols of head/floor boundaries.

$\uparrow/\downarrow$ Symbols of changing in up/down directions.

$\underline{\square}/\overline{\square}$ Symbols of minimum/maximum boundaries.

4.3 Parameters

$C_g^{ST/SD}$	Cost of startup/shutdown of generators, $(\$/h)$.
$C_g^{\uparrow/\downarrow}, C_e^{\uparrow/\downarrow}$	Cost of up/downward reserves of generators/EESSs, $(\$/MW)$.
$C_e^{Ch/Dis}$	Cost of charging/discharging of EESSs, $(\$/MWh)$.
$\underline{P}_g, \overline{P}_g$	Minimum/maximum power of generators, (MW).
$\underline{P}_e^{Ch/Dis}, \overline{P}_e^{Ch/Dis}$	Minimum/maximum charging/discharging of EESSs, (MW).
$\overline{P}_{n,t}^{Ren,s}$	Maximum available renewable generation scenario s, (MW).
$p_{d,t}^s$	Demand active power in different scenarios, (MW).
UT_g, DT_g	Minimum online/offline duration of generatiors, (h).
$RR_g^{\uparrow/\downarrow}$	Reserve limits in up/down directions, (MW).
$RP_g^{\uparrow/\downarrow}$	Ramp rate limits, (MW/h).
$RP_g^{ST/SD}$	Ramp rate limits in startup/shutdown, (MW/h).
$SF_{l,b}$	Sensitivity of power flow of lines due to injection at buses.
$\underline{SE}_e, \overline{SE}_e$	Minimum/maximum stored energy of EESSs, $(p.u.)$.
π_s	Probability of scenarios, $(\%)$.
$\beta_e^{Ch/Dis}$	Storage device charging/discharging efficiency, $(\%)$.

4.4 Variables

$st_{g,t}, sd_{g,t}$	Binary startup/shutdown status of generators.
$i_{g,t}$	Binary online/offline status of generators.
$j_{e,t}^{Ch/Dis,s}$	Binary charging/discharging status of EESSs.
$Cost_t^{CG/ES}$	Operational hourly cost of generators/EESSs, $(\$/h)$.
$p_{g,t}^s$	Power output of generators, (MW).
$p_{n,t}^{Re,s}$	Power output of renewable power plants, (MW).
$p_{e,t}^{Ch/Dis,s}$	Power charging/discharging of EESSs.
$r_{g,t}^{\uparrow/\downarrow}$	Reserve deployed by generators, (MW).
$r_{e,t}^{\uparrow/\downarrow}$	Reserve deployed by EESSs, (MW).
$r_{e,t}^{UB/LB}$	Maximum up/down deviations on reservoir of EESS, (MWh).
$R_{e,t}^{UB/LB}$	Required room to prevent EESSs' energy deviations, (MWh).
$SE_{e,t}^{ES,s}$	Stored energy of EESSs, $(p.u.)$.

5. Case studies

The model of EESS for performance as a reserve provider is evaluated using a six-bus test system presented in Fig. 14.5, and the data of the test system is available in Ref. [68]. The model is implemented using the general algebraic modeling system (GAMS) and CPLEX solver, at Intel 7 core 2.4 GHz with 8 GB memory. In this section, the model will be examined in two timeframes of the day-ahead market and the balancing market.

5.1 Day-ahead market

In the day-ahead market, EESSs will be modeled as presented by Eqs. (14.16)–(14.29). Three case studies based on Table 14.5 are defined to evaluate the EESSs' performance as a source of energy and reserve. The difference between Case 3 and Case 4 is in considering reservoir checking constraints (14.25)–(14.29), and Case 4 represents the comprehensive model presented in the previous section. The performance EESSs regarding the performance in reserve provision and the feasibility of the reservoir for compensation will be examined in the following.

5.1.1 The role of EESS in reserve provision

The performance of EESSs as a reserve provider can benefit the market by increasing the number of resources. In this regard, Table 14.6 compares the costs of generation and reserves in different cases. As can be seen, the highest generation ($77546.6) and reserve ($3963.7) costs among all cases belong to Case 1, in which EESSs are not capable of uncertainty covering. In Case 2, a lower reserve cost compared to Case 1, which is equal to $2282.5, is evaluated for

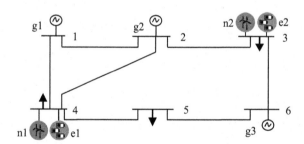

FIGURE 14.5   Six bus test system.

TABLE 14.5   Specifications of case studies.

	Case 1	Case 2	Case 3	Case 4
Reserve provision of EESSs	–	✔	✔	✔
Reserve costs for EESSs	–	–	✔	✔
Reservoir feasibility check	–	–	–	✔

TABLE 14.6 Comparison of generation and reserve costs in different cases.

	Case 1	Case 2	Case 3	Case 4
Generation cost of CG($)	77546.6	74571.2	74743.7	76560.4
Cost of EESSs' discharging ($)	518.2	948.6	821.1	536.3
Reserve cost of CGs($)	3963.7	2282.5	2479.5	3386
Reserve cost of EESs($)	—	0	1910.4	866.5
Total operation costs ($)	82028.5	77802.2	79954.6	81349.2

conventional generators. Because the large share of reserves is expected to be deployed by EESSs, where the reserve cost of EESSs is zero in this case. The comparison of costs between Case 3 and Case 4 shows that considering reservoir feasibility check in Case 4 increases the generation and reserve costs of CGs to $76560.4 and $3386, but the reserve cost is reduced to $866.5 by limiting the compensation range of EESSs.

The total dispatches of EESSs in charging, discharging, and idle at the base scenario (total hourly charge and discharge values) and also the total hourly deployed reserves by EESSs for Case 3 and Case 4 are presented in Figs. 14.6 and 14.7, respectively. Although it was observed that EESSs' discharging cost is reduced in Case 4, here we can see EESSs are charging and discharging more energy in the base scenario by applying reservoir checking constraints. The reason is that model will optimize the expected cost of scenarios, and it depends on the realization of dispatches in different scenarios. Also, higher values of reserves in up/down directions are deployed in Case 3 compared to Case 4. It should be noted that the deployments of reserves in both directions are related to different units and the scheduling for different scenarios.

A comparison between the values of deployed reserves in different cases is performed in Fig. 14.8. As expected, the lowest and highest reserves in up/down directions are deployed in Case 2 and Case 4, respectively. This result shows that the deployment of EESSs for reserve deployment is desirable based on scenarios of uncertainties and network configuration.

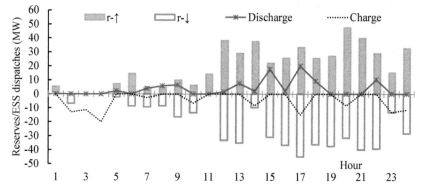

FIGURE 14.6 Total hourly charging/discharging and reserves deployed by EESSs in Case 3.

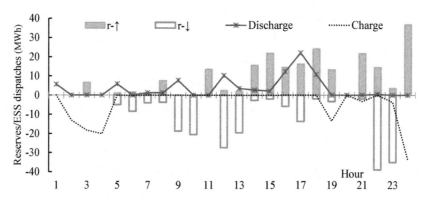

FIGURE 14.7   Total hourly charging/discharging and reserves deployed by EESSs in Case 4.

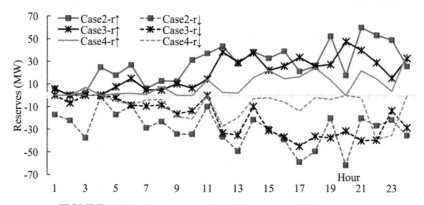

FIGURE 14.8   Comparison of EESSs' reserves in different cases.

5.1.2 The role of EESSs in uncertainty handling

The evaluation of EESS performance within scenarios at hour 22 is conducted in Table 14.7. The uncertainties of load and renewables are covered by redispatching of units in different scenarios. The deployed reserves by EESSs in upward is 8.8 and 5.9 MW, and in downward is 20 and 18.4 MW for $e1$ and $e2$, respectively. In this way, the increases of charging by 20 and 18.4 MW of $e1$ and $e2$ compared to dispatch of base scenario ($s = 0$) 0 MW are equivalent to downward reserves. On the other hand, the increases of discharging by 2.3 and 8.8 MW of $e1$ in scenarios $s = 3$ and $s = 5$ compared to dispatch of base scenario ($s = 0$) 0.5 MW are equivalent to upward reserves. It is worth to be noted that the dispatch of renewable generation $w2$ in the base scenario is 0 MW, but the weighted sum is 18.7 MW. The reason is that the model deploys this value as the operational reserve.

5.1.3 Checking energy of EESSs for feasible reserve provision

Here we analyze the energy boundaries of EESSs for reserve deployment to ensure services can be delivered in real-time operation. In this regard, Fig. 14.9 evaluates the stored energy

TABLE 14.7 Dispatches within scenarios of uncertainties in hour 22.

Scenarios	0	1	2	3	4	5
Probability	1	0.285	0.108	0.243	0.127	0.236
Load (MW)	242.4	237.4	209.8	257.7	169.2	287.2
n1 (MW)	62.5	68.9	56.7	59.2	53.1	65.7
n2 (MW)	0	24.1	7.4	16.3	14.2	22.2
g1 (MW)	179.5	144.4	184	179.4	140.4	184
Ch-e1 (MW)	0	0	−20	0	−20	0
Ch-e2 (MW)	0	0	−18.4	0	−18.4	0
Dis-e1 (MW)	0.5	0	0	2.8	0	9.3
Dis-e2 (MW)	0	0	0	0	0	5.9
Sum of dispatches (MW)	242.4	237.4	209.8	257.7	169.2	287.2

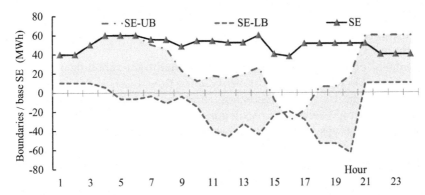

FIGURE 14.9 Evaluation of stored energy and feasible boundaries of the reservoir in case 3.

SE and the boundaries for e1 in Case 3. The boundaries of SE are not checked in Case 3, but the values are calculated after solving the model to evaluate the effect of successive compensations. As can be seen, both upper and lower boundaries are significantly dropped into negative values that show the compensation is infeasible, and the EESS will run out of energy in the real-time operation. Also, the curve of energy SE is not located between boundaries in this case.

Fig. 14.10 presents the stored energy and boundaries for Case 4. As can be seen, the boundaries of energy have not deviated in this case, and the stored energy SE is located in the feasible space between the upper and lower bounds. Additionally, the energy is mainly increased in the early hours of scheduling to store enough energy for discharging and in the last hours to address the daily continuous operation constraint.

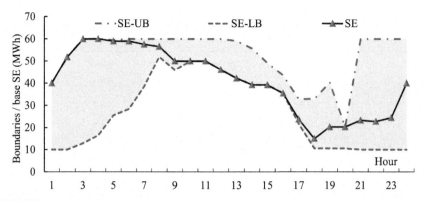

FIGURE 14.10   Evaluation of stored energy and feasible boundaries of reservoir in case 4.

5.2 The role of EESS in balancing market

The balancing market will use the schedule obtained from the day-ahead market, and here we evaluate the results based on the assumption that we have the actual values of uncertain parameters in near real time. Based on results, the total cost of regulations in the balancing market is $21764.7, which consists of $20980.8 for the share of CGs and $783.9 is the cost of EESSs regulations. It should be noted that the regulation cost of charging EESSs is calculated implicitly in CGs cost.

The performance of EESS $e1$ in balancing is compared to the schedule of the base scenario obtained from Case 4 in Fig. 14.11. As can be seen, the regulations in the balancing market do not lead to large deviations in the stored energy, and limited EESS compensation is feasible in this stage. Fig. 14.12 represents the regulations of $e1$ in the balancing market, and it is compared with the reserves in up/down directions obtained from the day-ahead market. It can be seen that regulations do not exceed the values of reserves, and we can see that adequate reserves are deployed from EESSs for the balancing market.

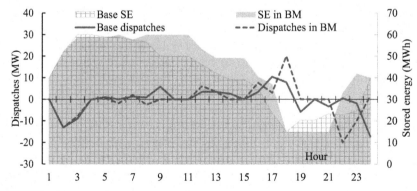

FIGURE 14.11   EESS performance in balancing market vs. the schedule of base scenario.

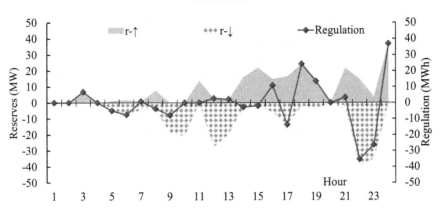

FIGURE 14.12 Evaluation of regulation services provided by EESSs based on deployed reserves.

The values of total hourly regulations of different units in the balancing market against the values of the dispatches of the base scenario of day ahead are presented in Table 14.8. As can be seen, the EESSs are deploying both upward and downward regulations in different operation modes. For example, in hours 3 and 4, EESSs are deploying 1.2 MW upward and 0.4 MW downward regulations in discharging mode, which are equivalent to decrease and increase of charging values, respectively. Also, in hours 16 and 17, EESSs work in discharging mode and deploy 4.4 MW upward and 7.4 MW downward regulations, respectively. Also, it is worth to be noted that renewable generations are also participating in deploying regulation services as it is expected based on the results of e day-ahead market.

TABLE 14.8 Regulation of units in balancing market according to the base dispatches.

	$P_t^{CG,0}$	$Reg_{gt}^{CG,\uparrow}$	$Reg_{gt}^{CG,\downarrow}$	$P_t^{ES,0}$	$Reg_{gt}^{ES,\uparrow/\downarrow}$	$P_t^{Ren,0}$	$Reg_{gt}^{Ren,\uparrow/\downarrow}$
1	100	0	0	5.7	0	69.9	+22.7
2	100	0	0	−13.1	0	76.6	+9.9
3	100	0	0	−18.3	+1.2	79.6	0
4	100	0	0	−20.1	−0.4	78.1	+4.6
5	102.7	0	−2.7	5.8	−1.8	49.1	+15.6
6	118	0	−18	−0.4	+0.8	40.5	+28.3
7	100	0	0	1.3	−3.4	63.3	0
8	126.8	0	−13.6	1.2	−5.9	50.9	+9.5
9	160	0	−11.3	7.7	0	19.2	0
10	153.5	+37.3	0	0	0	51.6	0
11	192.3	+12.4	0	0	0	35.1	0

(Continued)

TABLE 14.8   Regulation of units in balancing market according to the base dispatches.—cont'd

	$P_t^{CG,0}$	$Reg_t^{CG,\uparrow}$	$Reg_t^{CG,\downarrow}$	$P_t^{ES,0}$	$Reg_t^{ES,\uparrow/\downarrow}$	$P_t^{Ren,0}$	$Reg^{Ren,\uparrow/\downarrow}$
12	210	+15	0	10.3	+2.7	8.8	0
13	226.8	+28.7	0	3.5	0	12	0
14	225.3	+4.8	0	2.7	−2.7	25.5	0
15	241.7	+4.4	−21.7	2.2	0	21.4	0
16	213.1	+11.1	0	12.4	+4.4	27.7	0
17	223.9	0	−43.9	22.1	−7.4	5.3	+9.6
18	208.5	+15	0	10.8	+12.2	32.1	0
19	205.8	+67.1	0	−13.5	+5.9	38.8	0
20	191.5	+38	−22.5	0	0	50.2	0
21	209.7	+18.2	−22.5	−3.3	+3.3	52.9	0
22	179.5	+1	0	0.5	−20.5	62.5	0
23	130	+33.9	0	−3.7	−8.1	70.7	0
24	160	+11.1	0	−33.9	+19	84.8	0

6. Concluding remarks on status quo, challenges, and outlook

An analysis of the current progresses in the integration of EESSs into the existing energy and balancing markets was provided. Using a systematic review of the literature in this field, the challenges and opportunities are analyzed, and through a number of case studies on an example system, the role of EESSs in different situations was highlighted. This chapter was an endeavor to identify and analyze the potentials of EESSs in energy and balancing markets across the globe and to highlight future directions.

6.1 Status quo and challenges

Some analysts believe that in the long term there is uncertainty about the future need for storage outside the transport sector (which is itself a barrier to storage development) while there are several others who believe the storage technologies are the main pathway to solve energy trilemma.

While EESS technologies are recognized as the key technology in most of the countries that will be able to help in maintaining the security of supply and accelerating decarbonization of electricity, heat, and transport sectors, the cost is still the main barrier for most of these technologies. Also, current regulation and markets, which are mostly based on the conventional generation assets, could be another barrier that prevents storage systems from developing into active players in the electricity markets.

Therefore, governmental supports such as changes in legislation and regulation along with subsidies for innovation and advancement in these technologies could address this barrier and accelerate their uptaking.

6.2 Outlook

Technology analysts believe that the integration of distributed renewable generations as an intermittent source of power along with the growing volatility of demand brings technical challenges in the operation of the electrical systems that highlight the value of EESSs for participating in the provision of the ancillary services.

It means that the participation of EESSs is expected to be boosted due to the increasing need for frequency regulation, voltage support, and in general maintaining a sustainable, clean, and secure supply.

References

[1] N. Padmanabhan, M. Ahmed, K. Bhattacharya, Battery energy storage systems in energy and reserve markets, IEEE Trans. Power Syst. 35 (1) (2019) 215–226.

[2] Z. Tang, J. Liu, Y. Liu, L. Xu, Stochastic reserve scheduling of energy storage system in energy and reserve markets, Int. J. Electr. Power Energy Syst. 123 (2020) 106279.

[3] Z. Tang, Y. Liu, L. Wu, J. Liu, H. Gao, Reserve model of energy storage in day-ahead joint energy and reserve markets: a stochastic uc solution, IEEE Trans. Smart Grid 12 (1) (2020).

[4] E. Akbari, R.-A. Hooshmand, M. Gholipour, M. Parastegari, Stochastic programming-based optimal bidding of compressed air energy storage with wind and thermal generation units in energy and reserve markets, Energy 171 (2019) 535–546.

[5] S. Majumder, S.A. Khaparde, Revenue and ancillary benefit maximisation of multiple non-collocated wind power producers considering uncertainties, IET Gener. Transm. Distrib. 10 (3) (2016) 789–797.

[6] N.G. Cobos, J.M. Arroyo, N. Alguacil, J. Wang, Robust energy and reserve scheduling considering bulk energy storage units and wind uncertainty, IEEE Trans. Power Syst. 33 (5) (2018) 5206–5216.

[7] S. Zhan, P. Hou, P. Enevoldsen, G. Yang, J. Zhu, J. Eichman, M.Z. Jacobson, Co-optimized trading of hybrid wind power plant with retired ev batteries in energy and reserve markets under uncertainties, Int. J. Electr. Power Energy Syst. 117 (2020) 105631.

[8] H. Ye, Y. Ge, M. Shahidehpour, Z. Li, Uncertainty marginal price, transmission reserve, and day-ahead market clearing with robust unit commitment, IEEE Trans. Power Syst. 32 (3) (2016) 1782–1795.

[9] S. Pirouzi, J. Aghaei, V. Vahidinasab, T. Niknam, A. Khodaei, Robust linear architecture for active/reactive power scheduling of ev integrated smart distribution networks, Elec. Power Syst. Res. 155 (2018) 8–20.

[10] Y. Sun, Z. Li, M. Shahidehpour, B. Ai, Battery-based energy storage transportation for enhancing power system economics and security, IEEE Trans. Smart Grid 6 (5) (2015) 2395–2402.

[11] H. Daneshi, A. Srivastava, Security-constrained unit commitment with wind generation and compressed air energy storage, IET Gener. Transm. Distrib. 6 (2) (2012) 167–175.

[12] M.E. Khodayar, L. Wu, M. Shahidehpour, Hourly coordination of electric vehicle operation and volatile wind power generation in scuc, IEEE Trans. Smart Grid 3 (3) (2012) 1271–1279.

[13] Y. Sun, J. Zhong, Z. Li, W. Tian, M. Shahidehpour, Stochastic scheduling of battery-based energy storage transportation system with the penetration of wind power, IEEE Trans. Sustain. Energy 8 (1) (2016) 135–144.

[14] R. Li, W. Wang, M. Xia, Cooperative planning of active distribution system with renewable energy sources and energy storage systems, IEEE Access 6 (2017) 5916–5926.

[15] C.E. Murillo-Sánchez, R.D. Zimmerman, C.L. Anderson, R.J. Thomas, Secure planning and operations of systems with stochastic sources, energy storage, and active demand, IEEE Trans. Smart Grid 4 (4) (2013) 2220–2229.

[16] V. Guerrero-Mestre, Y. Dvorkin, R. Fernández-Blanco, M.A. Ortega-Vazquez, J. Contreras, "Incorporating energy storage into probabilistic security-constrained unit commitment, IET Gener. Transm. Distrib. 12 (18) (2018) 4206—4215.

[17] Y. Wen, C. Guo, H. Pandžić, D.S. Kirschen, Enhanced security-constrained unit commitment with emerging utility-scale energy storage, IEEE Trans. Power Syst. 31 (1) (2015) 652—662.

[18] M. Habibi, V. Vahidinasab, A. Pirayesh, M. Shafie-khah, J.P. Catalão, An enhanced contingency-based model for joint energy and reserve markets operation by considering wind and energy storage systems, IEEE Trans. Ind. Inf. (2020), https://doi.org/10.1109/TII.2020.3009105. Submitted for publication.

[19] M. Habibi, V. Vahidinasab, "Emergency services of energy storage systems for wind ramp events, in: 2019 Smart Grid Conference (SGC), IEEE, 2019, pp. 1—6.

[20] M. Habibi, V. Vahidinasab, A. Allahham, D. Giaouris, S. Walker, P. Taylor, Coordinated storage and flexible loads as a network service provider: a resilience-oriented paradigm, in: 2019 IEEE 28th International Symposium on Industrial Electronics (ISIE), IEEE, 2019, pp. 58—63.

[21] M. Habibi, V. Vahidinasab, A. Allahham, D. Giaouris, H. Patsios, P. Taylor, Exploitation of ancillary service from energy storage systems as operational reserve, in: UK Energy Storage Conference 2019 (UKES2019), Newcastle University, 2019.

[22] M. Habibi, A. Oshnoei, V. Vahidinasab, S. Oshnoei, Allocation and sizing of energy storage system considering wind uncertainty: an approach based on stochastic scuc, in: 2018 Smart Grid Conference (SGC), IEEE, 2018, pp. 1—6.

[23] S.A. Bozorgavari, J. Aghaei, S. Pirouzi, V. Vahidinasab, H. Farahmand, M. Korpås, Two-stage hybrid stochastic/robust optimal coordination of distributed battery storage planning and flexible energy management in smart distribution network, J. Energy Storage 26 (2019) 100970.

[24] Y. Gong, Q. Jiang, R. Baldick, Ramp event forecast based wind power ramp control with energy storage system, IEEE Trans. Power Syst. 31 (3) (2015) 1831—1844.

[25] A. Kavousi-Fard, M. Wang, W. Su, Stochastic resilient post-hurricane power system recovery based on mobile emergency resources and reconfigurable networked microgrids, IEEE Access 6 (2018) 72 311—372 326.

[26] L. Han, R. Zhang, K. Chen, A coordinated dispatch method for energy storage power system considering wind power ramp event, Appl. Soft Comput. 84 (2019) 105732.

[27] K. Hreinsson, A. Scaglione, B. Analui, Continuous time multi-stage stochastic unit commitment with storage, IEEE Trans. Power Syst. 34 (6) (2019) 4476—4489.

[28] N. Li, C. Uckun, E.M. Constantinescu, J.R. Birge, K.W. Hedman, A. Botterud, Flexible operation of batteries in power system scheduling with renewable energy, IEEE Trans. Sustain. Energy 7 (2) (2015) 685—696.

[29] Y. Tian, L. Fan, Y. Tang, K. Wang, G. Li, H. Wang, A coordinated multi-time scale robust scheduling framework for isolated power system with esu under high res penetration, IEEE Access 6 (2018) 9774—9784.

[30] H. Akhavan-Hejazi, H. Mohsenian-Rad, Optimal operation of independent storage systems in energy and reserve markets with high wind penetration, IEEE Trans. Smart Grid 5 (2) (2013) 1088—1097.

[31] R. Arghandeh, J. Woyak, A. Onen, J. Jung, R.P. Broadwater, Economic optimal operation of community energy storage systems in competitive energy markets, Appl. Energy 135 (2014) 71—80.

[32] G.N. Bathurst, G. Strbac, Value of combining energy storage and wind in short-term energy and balancing markets, Elec. Power Syst. Res. 67 (1) (2003) 1—8.

[33] K. Bradbury, L. Pratson, D. Patiño-Echeverri, Economic viability of energy storage systems based on price arbitrage potential in real-time us electricity markets, Appl. Energy 114 (2014) 512—519.

[34] E. Drury, P. Denholm, R. Sioshansi, The value of compressed air energy storage in energy and reserve markets, Energy 36 (8) (2011) 4959—4973.

[35] A. Foley, I.D. Lobera, Impacts of compressed air energy storage plant on an electricity market with a large renewable energy portfolio, Energy 57 (2013) 85—94.

[36] C. Jin, J. Tang, P. Ghosh, Optimizing electric vehicle charging with energy storage in the electricity market, IEEE Trans. Smart Grid 4 (1) (2013) 311—320.

[37] M. Kazemi, H. Zareipour, N. Amjady, W.D. Rosehart, M. Ehsan, Operation scheduling of battery storage systems in joint energy and ancillary services markets, IEEE Trans. Sustain. Energy 8 (4) (2017) 1726—1735.

[38] S.J. Kazempour, M.P. Moghaddam, M. Haghifam, G. Yousefi, Electric energy storage systems in a market-based economy: comparison of emerging and traditional technologies, Renew. Energy 34 (12) (2009) 2630—2639.

[39] H. Khaloie, A. Abdollahi, M. Shafie-Khah, A. Anvari-Moghaddam, S. Nojavan, P. Siano, J.P. Catalão, Coordinated wind-thermal-energy storage offering strategy in energy and spinning reserve markets using a multistage model, Appl. Energy 259 (2020) 114168.

[40] H. Khani, M.R.D. Zadeh, A.H. Hajimiragha, Transmission congestion relief using privately owned large-scale energy storage systems in a competitive electricity market, IEEE Trans. Power Syst. 31 (2) (2015) 1449–1458.

[41] R. Khatami, K. Oikonomou, M. Parvania, Look-ahead optimal participation of compressed air energy storage in day-ahead and real-time markets, IEEE Trans. Sustain. Energy 11 (2) (2019) 682–692.

[42] A. Khatamianfar, M. Khalid, A.V. Savkin, V.G. Agelidis, Improving wind farm dispatch in the Australian electricity market with battery energy storage using model predictive control, IEEE Trans. Sustain. Energy 4 (3) (2013) 745–755.

[43] M. Korpaas, A.T. Holen, R. Hildrum, Operation and sizing of energy storage for wind power plants in a market system, Int. J. Electr. Power Energy Syst. 25 (8) (2003) 599–606.

[44] H. Lund, G. Salgi, B. Elmegaard, A.N. Andersen, Optimal operation strategies of compressed air energy storage (caes) on electricity spot markets with fluctuating prices, Appl. Therm. Eng. 29 (5–6) (2009) 799–806.

[45] M.A. Mirzaei, M. Hemmati, K. Zare, B. Mohammadi-Ivatloo, M. Abapour, M. Marzband, A. Farzamnia, Two-stage robust-stochastic electricity market clearing considering mobile energy storage in rail transportation, IEEE Access 8 (2020) 121 780–121 794.

[46] H. Mohsenian-Rad, Coordinated price-maker operation of large energy storage units in nodal energy markets, IEEE Trans. Power Syst. 31 (1) (2015) 786–797.

[47] K. Möllersten, J. Yan, J.R. Moreira, "Potential market niches for biomass energy with CO_2 capture and storage–opportunities for energy supply with negative CO_2 emissions, Biomass Bioenergy 25 (3) (2003) 273–285.

[48] E. Nasrolahpour, S.J. Kazempour, H. Zareipour, W.D. Rosehart, Strategic sizing of energy storage facilities in electricity markets, IEEE Trans. Sustain. Energy 7 (4) (2016) 1462–1472.

[49] O. Palizban, K. Kauhaniemi, J.M. Guerrero, "Microgrids in active network management–part i: hierarchical control, energy storage, virtual power plants, and market participation, Renew. Sustain. Energy Rev. 36 (2014) 428–439.

[50] M. Parastegari, R.-A. Hooshmand, A. Khodabakhshian, A.-H. Zare, Joint operation of wind farm, photovoltaic, pump-storage and energy storage devices in energy and reserve markets, Int. J. Electr. Power Energy Syst. 64 (2015) 275–284.

[51] D. Zafirakis, K.J. Chalvatzis, G. Baiocchi, G. Daskalakis, The value of arbitrage for energy storage: evidence from European electricity markets, Appl. Energy 184 (2016) 971–986.

[52] Q. Zhang, I.E. Grossmann, C.F. Heuberger, A. Sundaramoorthy, J.M. Pinto, Air separation with cryogenic energy storage: optimal scheduling considering electric energy and reserve markets, AIChE J. 61 (5) (2015) 1547–1558.

[53] H. Alharbi, K. Bhattacharya, Participation of pumped hydro storage in energy and performance-based regulation markets, IEEE Trans. Power Syst. 35 (6) (2020).

[54] V. Vahidinasab, S. Jadid, Joint economic and emission dispatch in energy markets: a multiobjective mathematical programming approach, Energy 35 (3) (2010) 1497–1504.

[55] M. Saffari, M. Kia, V. Vahidinasab, K. Mehran, "Integrated active/reactive power scheduling of interdependent microgrid and ev fleets based on stochastic multi-objective normalised normal constraint, IET Gener. Transm. Distrib. 14 (11) (2020) 2055–2064.

[56] H. Nezamabadi, V. Vahidinasab, Market bidding strategy of the microgrids considering demand response and energy storage potential flexibilities, IET Gener. Transm. Distrib. 13 (8) (2019) 1346–1357.

[57] S. Pirouzi, J. Aghaei, T. Niknam, M. Shafie-Khah, V. Vahidinasab, J.P. Catalão, Two alternative robust optimization models for flexible power management of electric vehicles in distribution networks, Energy 141 (2017) 635–651.

[58] V. Vahidinasab, S. Jadid, Bayesian neural network model to predict day-ahead electricity prices, Eur. Trans. Electr. Power 20 (2) (2010) 231–246.

[59] S. Poudel, A. Dubey, Critical load restoration using distributed energy resources for resilient power distribution system, IEEE Trans. Power Syst. 34 (1) (2018) 52–63.

[60] C. O'Dwyer, L. Ryan, D. Flynn, Efficient large-scale energy storage dispatch: challenges in future high renewable systems, IEEE Trans. Power Syst. 32 (5) (2017) 3439–3450.

[61] R. Aghapour, M.S. Sepasian, H. Arasteh, V. Vahidinasab, J.P. Catalão, Probabilistic planning of electric vehicles charging stations in an integrated electricity-transport system, Elec. Power Syst. Res. 189 (2020) 106698.

[62] V. Vahidinasab, S. Jadid, Stochastic multiobjective self-scheduling of a power producer in joint energy and reserves markets, Elec. Power Syst. Res. 80 (7) (2010) 760–769.

[63] E. Barbour, I.G. Wilson, J. Radcliffe, Y. Ding, Y. Li, A review of pumped hydro energy storage development in significant international electricity markets, Renew. Sustain. Energy Rev. 61 (2016) 421–432.

[64] P. Taylor, R. Bolton, D. Stone, X.-P. Zhang, C. Martin, P. Upham, Pathways for Energy Storage in the UK, 2012. Report for the centre for low carbon futures, York.

[65] G.C. Gissey, P.E. Dodds, J. Radcliffe, Market and regulatory barriers to electrical energy storage innovation, Renew. Sustain. Energy Rev. 82 (2018) 781–790.

[66] UK National Grid Electricity System Operator, What is Battery Storage and How Do Batteries Help Us to Balance the Grid?, 2020 [Online]. Available from: https://www.nationalgrideso.com.

[67] B. Wooding, V. Vahidinasab, S. Soudjani, Formal controller synthesis for frequency regulation utilising electric vehicles, in: 2020 International Conference on Smart Energy Systems and Technologies (SEST), IEEE, 2020, pp. 1–6.

[68] V. Vahidinasab, M. Habibi, Data_EESS_Book, 2020 [Online]. Available from: http://vahidinasab.com/data/Data_EESS_Elsevier_04NOV2020.xlsx.

CHAPTER

15

Reliability modeling of renewable energy sources with energy storage devices

Vasundhara Mahajan, Soumya Mudgal, Atul Kumar Yadav, Vijay Prajapati

Department of Electrical Engineering, Sardar Vallabhbhai National Institute of Technology, Surat, Gujarat, India

Abbreviations

COPT Capacity Outage Probability Table
EDNS Expected Demand Not Supplied
EENS Expected Energy Not Supplied
ESD Energy Storage Devices
FOR Force Outage Rate
LOLE Loss of Load Expectation
OTS Optimal Transmission Switching
PV Photovoltaic
RES Renewable Energy Sources
RTS Reliability Test System
SOC State of Charge

Nomenclature

A Area of cylindrical transverse section (m^2)
$B_{x,y}$ Susceptance of line connected between bus 'x' and 'y'
c Scale index
C_r^q Combination of r units taken q at a time
$C_{i,j}^{out}$ Capacity outage for transition from state "i" to "j" (MW)
C_p Power coefficient of turbine
D_i Time duration for state "i"
G_N Power output for each state (MW)
I State for modeling of RES and ESD
I_x Injected current at bus "x" (A)
J Jacobian matrix

Energy Storage in Energy Markets
https://doi.org/10.1016/B978-0-12-820095-7.00019-4

M Number of RES units

N Total number of states

n_g Number of generator bus

n_p Number of load bus

$P^c_{\text{ESD}m,t}$ Charging power of ESD (MW)

$P^d_{\text{ESD}m,t}$ Discharging power of ESD (MW)

P_i Probability of state

p_{ind} Individual probability

P_{pv} Power output for solar module (MW)

P_r Rated power of wind turbine (MW)

P_w Power output for wind turbine (MW)

S Area of module (m^2)

si Solar irradiance (kW/m^2)

si_m Solar irradiance (kW/m^2)

$SOC_{\text{ESD}m,t}$ State of charge of ESD at time t

T Time period (hours)

V Wind velocity (m/s)

x Bus index for load flow

v_{ci} Cut-in velocity of wind (m/s)

v_{co} Cut-out velocity of wind (m/s)

V_x Voltage at bus "x" (volt)

v_r Rated velocity of wind (m/s)

v_m Mean wind velocity (m/s)

$Y_{x,y}$ Admittance of line between bus "x" and "y"

α Shape parameter

β Scale parameter

δ Angle at bus "i"

η Efficiency of solar module (%)

η_{ESD_c} Charging efficiency of ESS

η_{ESD_d} Discharging efficiency of ESS

μ Repair rate

ρ Density of air (kg/m^3)

$\rho_{i,j}$ Rate of transitions from state "i" to "j"

σ Standard deviation of the hourly irradiance data

λ Failure rate

1. Energy storage device and its modeling

The energy storage device (ESD) is used in a power system network to store the surplus energy during the off-peak period and utilize the stored energy during peak period. It is installed at strategic locations to improve the power system performance and meet the peak demand. Rated capacity (MWh) indicates the storage capacity, whereas rated power indicates the amount of power to be charged and discharged. The ESD can mitigate the uncertainties of RES and feed the uninterrupted power to the customer by determining the optimal charging and discharging schedules of ESD.

The charging and discharging of an ESD are an involving task. If left unmanaged, it can cause problems such as addition in cost and decreased reliability. Therefore, some constraints are considered for it. The state of charge at any given time t is given as:

$$SOC_{ESD_{m,t}} = SOC_{ESD_{m,t-1}} + \left(P^c_{ESD_{m,t}} \eta_{ESD_c} - \frac{P^d_{ESD_{m,t}}}{\eta_{ESD_d}} \right) t \tag{15.1}$$

where at time t, $SOC_{ESD_{m,t}}$ is state of charge of ESD, $P^c_{ESD_{m,t}}$ is charging power (MW) of ESD, η_{ESD_c} and η_{ESD_d} are charging and discharging efficiency of ESD respectively, $P^d_{ESD_{m,t}}$ discharging power (MW), m is ESD bus index.

2. Renewable energy sources

The drastic increase in population, fast urbanization, and rapid industrialization have substantially risen the world energy demand. Therefore, there is a constant search for newer sources of energy. Traditionally, conventional energy sources fulfilled the power demand. These included coal, oil, and gas reserves. But, the conventional sources have their disadvantages.

The conventional sources are not sustainable. That means, they're not self-replenishing and require ample time for production. Their existing reserves are depleting at alarming rates. Also, the extraction process is tedious and difficult. For example, coal is found deep under the earth's surface and has to be mined. The mining is dangerous, leads to several accidents. The land that is mined becomes barren and useless. Similarly, oil is found deep in the oceans. Its extraction is challenging.

Conventional sources are nothing but complex chains of hydrocarbon compounds. These are hazardous to the environment as their combustion releases sulfur, nitrogen, and carbon oxides (greenhouse gases) into the atmosphere. The greenhouse gases are gradually increasing the temperature of the planet resulting in global warming. This also causes air pollution, smog, and acid rain.

Due to these problems, alternate sources of energy come into the picture. Renewable energy sources (RESs) are limitless and continuously available. These have low carbon emissions, negligible as compared to the conventional sources. Hence, these are eco-friendly.

The RES has made the electricity cost stable. There is no addition of fluctuating fuel costs to the initial capital investment. Although the maintenance cost for RES is higher than that for non-RES, their production cost is substantially low.

An efficient decentralized system can be established with RES. By locally fulfilling the required power demand, the distribution losses are overcome. Thus, the power flow of the system improves. But the availability of the system is compromised. This is discussed in detail in Section 5.

In this emerging sector, new technology is advancing by the day, thereby creating plentiful opportunities for employment to many. The different types of RES are [1]:

- **Wind Energy** is an abundant source of renewable energy. Turbines convert the mechanical energy of wind into electrical energy by a generator. Several electrical powered vehicles use wind energy to save fuel.
- **Solar Energy** is generated by powering photovoltaic (pv) cells with photons from the sun rays. This has resulted in electricity supply to even the remote areas where setting up of transmission lines is difficult. Solar water heaters, pumps, lights, and calculators are a part of everyday lives now.
- **Geothermal Energy** is used by farmers to heat their greenhouses all year long. This results in improved production of fruits and vegetables. In cold countries, geothermal energy heats the walkways to prevent them from freezing.
- **Biomass** is fired in an incinerator. The heat produced generates steam that generates electricity. Also, biomass produces biofuels for transportation.

2.1 Solar energy

Energy generated from the solar radiations and heat is solar energy. An optoelectronic device called a solar cell is used for this. The following sections discuss more about generation, efficiency, and output power in the solar cell.

2.1.1 Photovoltaic cell

A pv cell is a semiconductor material that can absorb photons from the sunlight, to produce electricity [2]. In silicon atoms, the exposure to sunlight results in a change in charges across the cell. Free electrons of the atom flow through the cell, resulting in the generation of direct current (dc). An inverter then converts the dc into ac. There is the option of battery storage for usage during unavailability of the sunlight.

Several pv cells are stacked along with appropriate protection to form a solar module. These modules get arranged in series (to increase voltage) or parallel (to increase current) as per the overall requirement.

2.1.2 Composition of module

Solar modules are primarily of two types [3].

1. Monocrystalline module
   These are widely used for fabrication of solar modules as their efficiency is high (almost 24.4{%}). Single-crystal silicon is used to construct the module. Therefore, the loose electrons have more space to move freely. These have a sleek design and a blackish hue.
2. Multicrystalline/polycrystalline module
   These do not have composition of only single-crystal silicons. Multiple silicon atoms are melted into wafers and fragments and from these modules are made. As a result, these modules have lower efficiency (almost 19.8%) as well as lower cost. These have a bluish hue.

From these solar modules, the solar cells are constructed and categorized as follows.

- Hybrid cells

  These are composed of a combination of crystalline silicon as well as noncrystalline silicon atoms. Manufacturing them is a complex process, but the efficiency is improved.
- Carbon nanotube cells

  These cells have a transparent carbon nanotube (CNT) conductor material that improves the flow of current.
- Tandem cells

  These are also called multijunction solar cells due to the construction. Several individual cells of varying bandgap are stacked on top of one another to form a multijunction arrangement. This arrangement has improved efficiency as the gap energy is now increased.

2.1.3 Performance of module

The output power of the solar module depends upon irradiation and ambient temperature [4]. Insolation is the exposure to sunlight that any area experiences. The degree of insolation defines the output power of the solar cell. When the insolation becomes large, the cells get saturated, and a reverse saturation current flows. The mobility of free electrons gets reduced. The rate of generation of energy from photons is dependent upon the ambient temperature. When temperature increases substantially, there is a reduction in bandgap as a reverse saturation current flows through the cell. Therefore, temperature is a negative factor that influences the performance of the solar cell. Approximately room temperature (almost 25°C) is preferred as the ambient temperature.

2.1.4 Solar generation modeling using beta probability density function

This section discusses the method to calculate the power output of a solar panel, using beta probability density function (pdf). The output power mainly depends upon the irradiance, area of the solar panel, and efficiency of the panel. While the manufacturer predefines area and efficiency for a given site, irradiance is an unpredictable and random variable. Therefore, on application of a direct formula to the raw irradiance data, errors get accumulated due to the uncertainty in the data. Hourly solar irradiance and its standard deviation for a location at a specified duration are the raw data available. This is a linear combination of two unimodal distributions, to which beta pdf is applicable individually [5–7]. The beta probability density curves for different values of α and β can be seen in Fig. 15.1.

The beta function can be defined by:

$$f_w(s) = I(s)(a, b) \tag{15.2}$$

where $a, b > 0$

Now the incomplete beta function ratio is given by:

$$I_y(a, b) = \frac{B_y(a, b)}{B(a, b)} \tag{15.3}$$

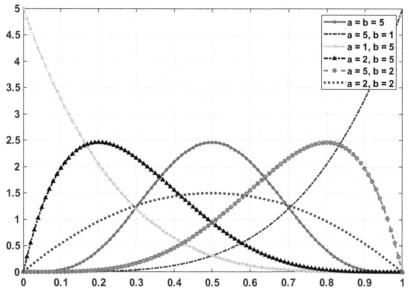

FIGURE 15.1   The beta probability density function for different values of α,β.

where $B_y(a,b)$ is the incomplete beta function and $B(a,b)$ is complete beta function.

$$B_y(a,b) = \int_0^y \omega^{\alpha-1}(1-\omega)^{\beta-1}d \tag{15.4}$$

Now, take $y=(\beta i)^{\alpha}$ where α is shape parameter and β is scale parameter. Therefore, the beta probability function becomes:

$$f_w(s) = I_{(\beta s)^{\alpha}}(s)(a,b) \tag{15.5}$$

$$f_w(s) = \frac{\int_0^{(\beta s)^{\alpha}} \omega^{\alpha-1}(1-\omega)^{\beta-1}d\omega}{B(a,b)} \tag{15.6}$$

for $0 < i < 1/\beta$

Now considering binomial distribution:

$$(1-z)^{b-1} = \sum_{i=0}^{\infty}(-1)^s \frac{\Gamma(b)}{\Gamma(b-s)s!}z^s \tag{15.7}$$

where $|z| < 1$, $b > 0$ and b is a real, nonintegral number.

On simplifying Eq. (15.6) using Eq. (15.7),

$$f_w(s) = \frac{\Gamma(\alpha + \beta)}{\Gamma(\alpha)} \sum_{k=0}^{\infty} 1 \frac{(-1)^k}{\Gamma(b-k)s!(\alpha+k)} (\beta s)^{\alpha(\alpha+k)} \tag{15.8}$$

Here, "si" is the solar irradiance, si_m is the mean, and σ is the standard deviation of the hourly irradiance data,

$$f(si) = \frac{\Gamma(\alpha + \beta)}{\Gamma(\alpha)} (si)^{(\alpha-1)} (1 - si)^{(\beta-1)} \quad 0 \le si \le 1; \alpha, \beta \ge 0$$

$$f(si) = 0 \qquad\qquad\qquad\qquad\qquad \text{otherwise}$$

and,

$$\alpha = \frac{si_m \beta}{1 - si_m} \tag{15.9}$$

$$\beta = (1 - si_m) \left(\frac{si_m(1 + si_m)}{\sigma^2} - 1 \right) \tag{15.10}$$

The power output for the solar module is calculated by:

$$P_{pv}(si) = \eta S si \tag{15.11}$$

where η is the efficiency of module (in %) and S is area of module (in m^2).

Example 1. For the hourly solar irradiance data given in Table 15.1, find out the hourly and daily power output from the solar farm using beta probability density model. Assume efficiency to be 24.4% and area to be 40 m^2.
  Solution:
  It is given that,

$\eta = 24.4\%$
$S = 40 \text{ m}^2$

The hourly values of α, β and power output for given data are calculated from Eqs. (15.9)–(15.11) respectively. All these values are shown in Table 15.2. This gives the daily power output equal to 46.72 MW.

2.2 Wind energy

Wind energy is a widely accepted replacement for the existing traditional energy generation sources. A setup of wind farm consists of multiple aeroturbines that generate electricity from wind. For a certain duration, the wind speed required to rotate the turbine blade is

TABLE 15.1   Hourly solar irradiance data.

Hour	si_m	σ	Hour	si_m	σ
1	0	0	13	0.65	0.282
2	0	0	14	0.58	0.265
3	0	0	15	0.46	0.237
4	0	0	16	0.33	0.204
5	0	0	17	0.21	0.163
6	0.03	0.035	18	0.08	0.098
7	0.11	0.11	19	0.01	0.032
8	0.24	0.182	20	0	0
9	0.41	0.217	21	0	0
10	0.52	0.253	22	0	0
11	0.62	0.273	23	0	0
12	0.67	0.284	24	0	0

constant. A synchronous generator converts the mechanical energy input of the wind into electrical energy output [8,9].

2.2.1 Wind turbine parameters

Each wind turbine has predefined rated power output. This output depends upon the following:

1. Location of the turbine
2. The height of the blades
3. The capacity of the synchronous generator
4. The wind velocity

The total power output of a wind farm is the sum of the power output of each turbine [10]. The turbine exist in binary condition i.e., "ON" and "OFF." In the "ON" condition, the turbine generates power output that is permissible within the set parameters. On the other hand, in the "OFF" condition, the turbine is out of service. Failure of any turbine within a wind farm does not depend upon other turbines. Therefore, the turbine works either at its maximum capacity or zero capacity [11].

The power output versus wind speed for any wind turbine is shown in the power curve in Fig. 15.2.

TABLE 15.2 Solar: power output using probability density function.

Hour	si_m	σ	B	α	si	P_{pv} (kW)
1	0	0	0	0	0	0
2	0	0	0	0	0	0
3	0	0	0	0	0	0
4	0	0	0	0	0	0
5	0	0	0	0	0	0
6	0.03	0.035	23.498	0.7267	31.53	30,775.4706
7	0.11	0.11	8.0909	1	3.94	3844.33712
8	0.24	0.182	6.0682	1.9163	0.24	229.918284
9	0.41	0.217	6.6533	4.6235	0.13	128.970263
10	0.52	0.253	5.4472	5.9011	0.72	699.235284
11	0.62	0.273	4.7411	7.7355	0.11	103.432342
12	0.67	0.284	4.2479	8.6246	0.07	69.1635718
13	0.65	0.282	4.3703	8.1162	0.67	654.771552
14	0.58	0.265	5.0608	6.9887	0.90	874.003416
15	0.46	0.237	5.9167	5.0401	0.09	85.9950534
16	0.33	0.204	6.3961	3.1503	0.89	870.685769
17	0.21	0.163	6.7654	1.7984	0.30	288.50619
18	0.08	0.098	7.3566	0.6397	1.46	1427.19009
19	0.01	0.032	8.7746	0.0886	6.83	6668.17295
20	0	0	0	0	0	0
21	0	0	0	0	0	0
22	0	0	0	0	0	0
23	0	0	0	0	0	0
24	0	0	0	0	0	0

The velocities lesser than "cut-in" velocity have an insufficient power output to start up the synchronous generator. Therefore, aeroturbine generates no electrical output. Generally, cut-in speed lies in the range of 3–5 m/s. Above the cut-in speed, the power output increases and reaches the rated power of the turbine. The velocity where the rated power output is achieved is "Rated" velocity. Beyond the rated velocity, the output is constant.

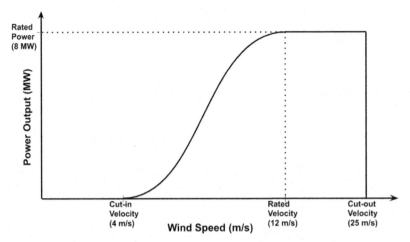

FIGURE 15.2   Power curve for wind energy.

At very high velocity, there are chances of mechanical damages to the turbine. Therefore, a "cut-out" velocity is specified for the turbine above which the control circuit shuts down the system to prevent harm. This velocity is usually between 20 and 25 m/s.

2.2.2 Power output using probability density functions

The power output of a wind turbine is very similar to a Rayleigh distribution \cite{orgis}. Rayleigh is a special case of a Weibull pdf, the equations are:

Let "$f_w(v)$" be the Rayleigh distribution function for wind velocity "v" and "c" be the scale index.

$$f_w(v) = \frac{2v}{c^2}e^{\frac{-v}{c^2}} \tag{15.12}$$

The mean wind velocity V_m is given by:

$$V_m = \int_0^\infty vf_w(v)dv = \int_0^\infty 1\frac{2v^2}{c^2}e^{\frac{-v}{c^2}}dv \tag{15.13}$$

On solving Eq. (15.13):

$$V_m = \frac{c\sqrt{\pi}}{2} \tag{15.14}$$

Or,

$$c = 1.128V_m \tag{15.15}$$

The power output is given as:

$$P_w = 0, \quad\quad\quad\quad \text{for } 0 \leq v \leq v_{ci}$$

$$P_w = P_r \times \frac{v - v_{ci}}{v_r - v_{ci}} \quad \text{for } v_{ci} \leq v \leq v_r$$

$$P_w = P_r \quad\quad\quad\quad \text{for } v_r \leq v \leq v_{co}$$ (15.16)

$$P_w = 0 \quad\quad\quad\quad \text{for } v \geq v_{co}$$

where v, v_{ci}, v_r and v_{co} are instantaneous, cut-in, rated, and cut-out velocities respectively.

Example 2. For the hourly wind velocity data given in Table 15.3, find out the hourly and daily power output of the turbine using pdf model. Assume cut-in, rated, and cut-out velocity as 4, 12 and 25 m/s respectively. The wind turbine has a rated power of 8 MW.

Solution:

It is given that, $v_{ci} = 4$ m/s

$v_r = 12$ m/s
$v_{co} = 25$ m/s
$P_r = 8$ MW

TABLE 15.3 Hourly wind speed data (m/s).

Time	Speed (m/s)	Time	Speed (m/s)
1	5	13	10
2	3	14	13
3	4	15	10
4	5	16	8
5	7	17	8
6	9	18	6
7	11	19	8
8	12	20	11
9	12	21	10
10	9	22	13
11	10	23	12
12	9	24	8

From Eq. (15.16),

$$P_w = 0 \qquad\qquad 0 \le v \le 4m/s$$

$$P_w = \frac{(v-4)8}{8} \quad 4 \le v \le 12m/s$$

$$P_w = 8 \qquad\qquad 12 \le v \le 25m/s$$

$$P_w = 0 \qquad\qquad v \ge 25m/s$$

From the aforementioned equations, the hourly power output is calculated as shown in Table 15.4.

2.2.3 Power output using wind speed

As seen in the previous section, the power output for wind turbine gets calculated using pdf.

This section analyzes the dependency of power output upon wind speed [11].
The power output is given by the formula:

$$P = \frac{C_p A \rho v^3}{2} \tag{15.17}$$

TABLE 15.4   Power output using probability density function.

Time	Speed (m/s)	Power (MW)	Time	Speed (m/s)	Power (MW)
1	5	1	13	10	6
2	3	0	14	13	8
3	4	0	15	10	6
4	5	1	16	8	4
5	7	3	17	8	4
6	9	5	18	6	2
7	11	7	19	8	4
8	12	8	20	11	7
9	12	8	21	10	6
10	9	5	22	13	8
11	10	6	23	12	8
12	9	5	24	8	4

The daily power output is the sum of the hourly power outputs i.e., 116 MW.

where A is the area of the cylindrical transverse section, ρ is the density of air, v is the velocity of wind, and C_p is the power coefficient of the turbine. It indicates the turbine conversion capacity from mechanical energy to electrical.

The power output is given as:

$$P_w = 0, \qquad \text{for } 0 \le v \le v_{ci}$$

$$P_w = \frac{C_p A \rho v^3}{2} \qquad \text{for } v_{ci} \le v \le v_r$$

(15.18)

$$P_w = P_r \qquad \text{for } v_r \le v \le v_{co}$$

$$P_w = 0 \qquad \text{for } v \ge v_{co}$$

Example 3. For the data in Table 15.3, find out the hourly and daily power output of the turbine using wind speed model. Take the density of air as 1.23 kg/m^3, power coefficient as 0.4, and length of blade as 52 m.

Solution:

Given:

Radius of circle formed by blade, $l = r = 52$ m; $v_{ci} = 4$ m/s; $v_r = 12$ m/s; $v_{co} = 25$ m/s; $P_r = 8$ MW; $= 1.23 \text{ kg/m}^3$; $C_p = 0.4$

The swept area A,

$A = \pi r^2$
$A = 8494.866 \text{ m}^2$

From Eq. (15.18),

$$P_w = 0 \qquad 0 \le v \le 4m/s$$

$$P_w = \frac{C_p A \rho v^3}{2} \qquad 4 \le v \le 12m/s$$

$$P_w = 8 \qquad 12 \le v \le 25m/s$$

$$P_w = 0 \qquad v \ge 25m/s$$

From aforementioned equations, the hourly output power is calculated as shown in Table 15.5.

The daily power output is the sum of the hourly power outputs i.e., 64.452 MW.

2.2.4 Power output using turbine output equation

The drawback of the wind speed model is that it deviates from the original probabilistic model. The turbine output equation is a fusion of probabilistic and wind speed model [12].

TABLE 15.5   Power output using wind speed model.

Time	Speed (m/s)	Power (MW)	Time	Speed (m/s)	Power (MW)
1	5	0.261	13	10	2.089
2	3	0	14	13	8
3	4	0	15	10	2.089
4	5	0.261	16	8	1.069
5	7	0.716	17	8	1.069
6	9	1.523	18	6	0.451
7	11	2.781	19	8	1.069
8	12	8	20	11	2.781
9	12	8	21	10	2.089
10	9	1.523	22	13	8
11	10	2.089	23	12	8
12	9	1.523	24	8	1.069

The power output is given as:

$$
\begin{aligned}
P_w &= 0, & \text{for } 0 \leq v \leq v_{ci} \\
P_w &= (A + Bv + Cv^2)P_r & \text{for } v_{ci} \leq v \leq v_r \\
P_w &= P_r & \text{for } v_r \leq v \leq v_{co} \\
P_w &= 0 & \text{for } v \geq v_{co}
\end{aligned}
\tag{15.19}
$$

where A, B, and C are

$$
A = \frac{1}{(v_{ci} - v_r)^2} \left[v_{ci}(v_{ci} + v_r) - 4(v_{ci}v_r)\left[\frac{v_{ci} + v_r}{2v_r}\right]^3 \right]
\tag{15.20}
$$

$$
B = \frac{1}{(v_{ci} - v_r)^2} \left[4(v_{ci} + v_r)\left[\frac{v_{ci} + v_r}{2v_r}\right]^3 - (3v_{ci} + v_r) \right]
\tag{15.21}
$$

$$
C = \frac{1}{(v_{ci} - v_r)^2} \left[2 - 4\left[\frac{v_{ci} + v_r}{2v_r}\right]^3 \right]
\tag{15.22}
$$

Example 4. For the data in Table 15.3, find out the hourly and daily power output of the turbine using turbine output equation.

Solution

It is given that,

$v_{ci} = 4\ \text{m/s}$
$v_r = 12\ \text{m/s}$
$v_{co} = 25\ \text{m/s}$
$P_r = 8\ \text{MW}$

From Eq. (15.20)

$$A = \frac{1}{\left(v_{ci} - v_r\right)^2}\left[v_{ci}(v_{ci} + v_r) - 4(v_{civ_r})\left[\frac{v_{ci} + v_r}{2v_r}\right]^3\right]$$

$$A = \frac{1}{(4 - 12)^2}\left[4(4 + 12) - 4(4 \times 12)\left[\frac{4 + 12}{2 \times 12}\right]^3\right]$$

$$A = 0.111$$

From Eq. (15.21)

$$B = \frac{1}{\left(v_{ci} - v_r\right)^2}\left[4(v_{ci} + v_r)\left[\frac{v_{ci} + v_r}{2v_r}\right]^3 - (3v_{ci} + v_r)\right]$$

$$B = \frac{1}{(4 - 12)^2}\left[4(4 + 12)\left[\frac{4 + 12}{2 \times 12}\right]^3 - (3 \times 4 + 12)\right]$$

$$B = -0.0787$$

From Eq. (15.22)

$$C = \frac{1}{\left(v_{ci} - v_r\right)^2}\left[2 - 4\left[\frac{v_{ci} + v_r}{2v_r}\right]^3\right]$$

$$C = \frac{1}{(4 - 12)^2}\left[2 - 4\left[\frac{4 + 12}{2 \times 12}\right]^3\right]$$

$$C = 0.0127$$

From Eq. (15.19),

$$P_w = 0 \qquad\qquad\qquad\qquad\qquad\qquad 0 \le v \le 4m/s$$
$$P_w = (0.111 + (-0.0787) \times v + (0.0127) \times v^2) \times 8 \quad 4 \le v \le 12m/s$$
$$P_w = 8 \qquad\qquad\qquad\qquad\qquad\qquad 12 \le v \le 25m/s$$
$$P_w = 0 \qquad\qquad\qquad\qquad\qquad\qquad v \ge 25m/s$$

The hourly power output is calculated from the aforementioned equations as shown in Table 15.6.

3. Multistate modeling of RES and ESD

Although RES is an excellent replacement for the existing conventional sources of energy, they do raise some problems. Firstly, renewable energy is variable as it depends upon natural factors. It is diffused and not concentrated on a location. Therefore, there is unpredictability in energy generation based on location as well as time. Also, the interaction of RES with conventional generation systems is complex [8].

At the mega-watt levels, the technology for RES differs from the conventional sources. To incorporate the errors generated by the aforementioned factors, multistate modeling methods are applied. The system gets divided into several states of partial capacity. Each capacity state corresponds to a level of partial energy [8].

TABLE 15.6 Power output using turbine output equations.

Time	Speed (m/s)	Power (MW)	Time	Speed (m/s)	Power (MW)
1	5	0.596	13	10	5.384
2	3	0	14	13	8
3	4	0	15	10	5.384
4	5	0.596	16	8	2.859
5	7	1.901	17	8	2.859
6	9	4.02	18	6	1.147
7	11	6.951	19	8	2.859
8	12	8	20	11	6.951
9	12	8	21	10	5.384
10	9	4.02	22	13	8
11	10	5.384	23	12	8
12	9	4.02	24	8	2.859

The daily power output is the sum of the hourly power outputs i.e., 103.176 MW.

The renewable energy data in consideration for this chapter is hourly values of wind speed (m/s) and solar irradiation kW/m^2. The data is physical phenomena that vary continuously and randomly with time and space. The data gets divided into predefined "N" states depending upon the values of instantaneous wind speed and solar irradiation. The "state 1" corresponds to the state of lowest partial capacity and "state N" corresponds to the highest [10]. The model is statistically stationary and does not depend upon the initial conditions at the beginning.

The following examples explain multistate modeling for a system.

Example 5. Table 15.7 shows hourly wind speed data for 5 days.

- **Plot wind speed** versus **time. What do you conclude?**
- **For multistate modeling, divide the system into four, six, and eight states. Predefine the limits for each state.**

Solution
The wind speed (m/s) versus time graph is shown in Fig. 15.3.

As seen in the curve, the hourly wind speed data is random and unpredictable. No 2 days have the same trend of hourly variation of wind speed. The maximum speed is 16 m/s, whereas the minimum speed is 0 m/s.

The multistate modeling can be performed for any duration of time. In the given example the data is corresponding to 5 days but modeling can be done for a single day, month, or year, as per the requirement.

To divide the system into four states, the following is defined.

Minimum speed = 0 m/s
Maximum speed = 16 m/s

To divide the system into four states, each state has wind speed "v" range as follows:

- State 1: $0 \leq v < 4$ m/s
- State 2: $4 \leq v < 8$ m/s
- State 3: $8 \leq v < 12$ m/s
- State 4: $12 \leq v < 16$ m/s

The day-wise duration of data in each state is tabulated in Table 15.8. Say wind speed is 6 m/s, it corresponds to "State 2."

To divide the system into six states, each state has a wind speed range as follows:

- State 1: $0 \leq v < 4$ m/s
- State 2: $4 \leq v < 6$ m/s
- State 3: $6 \leq v < 8$ m/s
- State 4: $8 \leq v < 10$ m/s
- State 5: $10 \leq v < 12$ m/s
- State 6: $12 \leq v < 16$ m/s

The day-wise duration of data and its overall duration in each state are tabulated in Table 15.9.

15. Reliability modeling of renewable energy sources with energy storage devices

TABLE 15.7 Hourly wind speed (m/s) data for 5 days.

Time	Day 1	Day 2	Day 3	Day 4	Day 5
1	4	1	5	3	6
2	3	0	3	4	5
3	2	1	5	13	6
4	2	1	6	6	4
5	3	4	8	6	2
6	2	3	10	10	1
7	1	2	6	10	2
8	2	4	3	13	8
9	4	15	2	7	3
10	3	14	9	10	2
11	4	11	4	10	3
12	6	10	3	9	3
13	10	7	7	13	4
14	8	4	6	13	7
15	4	6	4	8	4
16	3	4	2	10	4
17	5	8	1	7	14
18	4	5	3	6	11
19	5	4	4	11	5
20	4	3	3	16	7
21	5	7	3	5	9
22	9	8	2	4	9
23	6	7	5	3	8
24	6	9	2	6	9

To divide the system into eight states, each state has a wind speed range as follows:

- State 1: $0 \leq v < 6$ m/s
- State 2: $6 \leq v < 7$ m/s
- State 3: $7 \leq v < 8$ m/s
- State 4: $8 \leq v < 9$ m/s
- State 5: $9 \leq v < 10$ m/s
- State 6: $10 \leq v < 11$ m/s

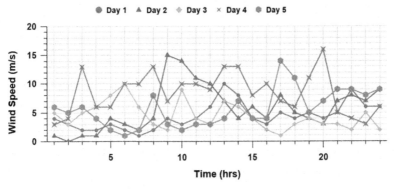

FIGURE 15.3   Wind speed versus time curve.

TABLE 15.8   Wind: day-wise multistate modeling for four states.

State	Duration for day 1 (h)	Duration for day 2 (h)	Duration for day 3 (h)	Duration for day 4 (h)	Duration for day 5 (h)	Total duration (h)
State 1	9	7	11	2	7	36
State 2	12	10	10	9	10	51
State 3	3	5	3	8	6	25
State 4	0	2	0	5	1	8

TABLE 15.9   Wind: day-wise multistate modeling for six states.

State	Duration for day 1 (h)	Duration for day 2 (h)	Duration for day 3 (h)	Duration for day 4 (h)	Duration for day 5 (h)	Total duration (h)
State 1	9	7	11	2	7	36
State 2	9	6	6	3	6	30
State 3	3	4	4	6	4	21
State 4	2	3	2	2	5	14
State 5	1	2	1	6	1	11
State 6	0	2	0	5	1	8

- State 7: $11 \leq v < 12$ m/s
- State 8: $12 \leq v < 16$ m/s

The day-wise duration of data and its overall duration in each state are tabulated in Table 15.10.

Example 6. Table 15.11 shows hourly solar irradiance data for 5 days.

TABLE 15.10 Wind: day-wise multistate modeling for eight states.

State	Duration for day 1 (h)	Duration for day 2 (h)	Duration for day 3 (h)	Duration for day 4 (h)	Duration for day 5 (h)	Total duration (h)
State 1	18	13	17	5	13	66
State 2	3	1	3	4	2	13
State 3	0	3	1	2	2	8
State 4	1	2	1	1	2	7
State 5	1	1	1	1	0	7
State 6	1	1	1	5	0	8
State 7	0	1	0	1	1	3
State 8	0	2	0	5	1	8

TABLE 15.11 Hourly solar irradiance (kW/m^2) data for 5 days.

Time	Day 1	Day 2	Day 3	Day 4	Day 5	Time	Day 1	Day 2	Day 3	Day 4	Day 5
1	0	0	0	0	0	13	0.92	0.83	0.42	0.74	0.31
2	0	0	0	0	0	14	0.63	0.64	0.30	0.60	0.14
3	0	0	0	0	0	15	0.30	0.41	0.15	0.42	0.21
4	0	0	0	0	0	16	0.19	0.27	0.01	0.23	0.08
5	0	0	0	0	0	17	0.12	0.05	0.01	0.04	0.01
6	0	0	0	0	0	18	0	0	0	0	0
7	0	0	0	0	0	19	0	0	0	0	0
8	0	0	0	0	0	20	0	0	0	0	0
9	0.71	0.55	0.27	0.55	0.26	21	0	0	0	0	0
10	0.74	0.70	0.40	0.66	0.36	22	0	0	0	0	0
11	0.73	0.71	0.47	0.72	0.28	23	0	0	0	0	0
12	0.10	0.81	0.48	0.74	0.66	24	0	0	0	0	0

- Plot irradiance versus time. What do you conclude?
- For multistate modeling, divide the system into 4 and 10 states. Predefine the limits for each state.

Solution

The solar irradiance (kW/m²) versus time graph is shown in Fig. 15.4.

As seen in the curve, the hourly solar irradiance data is random and unpredictable. No 2 days have the same trend of hourly variation of data. During the "night hours," the irradiance is nil. Therefore, the irradiance varies between 0 kW/m²and kW/m².

To divide the system into four states, each state has a solar irradiation range as follows:

- State 1: 0–0.25 kW/m²
- State 2: 0.25–0.5 kW/m²
- State 3: 0.5–0.75 kW/m²
- State 4: 0.75–1 kW/m²

The day-wise duration of data and overall duration in each state are tabulated in Table 15.12.

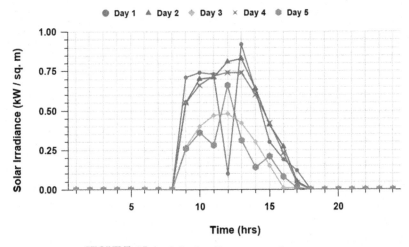

FIGURE 15.4 Solar irradiance versus time curve.

TABLE 15.12 Solar: day-wise multistate modeling for four states.

State	Duration for day 1 (h)	Duration for day 2 (h)	Duration for day 3 (h)	Duration for day 4 (h)	Duration for day 5 (h)	Total duration (h)
State 1	3	1	3	2	4	13
State 2	1	2	6	1	4	14
State 3	4	4	0	6	1	15
State 4	1	2	0	0	0	3

To divide the system into 10 states, each state has a solar irradiance range as follows:

- State 1: $0-0.1$ kW/m^2
- State 2: $0.1-0.2$ kW/m^2
- State 3: $0.2-0.3$ kW/m^2
- State 4: $0.3-0.4$ kW/m^2
- State 5: $0.4-0.5$ kW/m^2
- State 6: $0.5-0.6$ kW/m^2
- State 7: $0.6-0.7$ kW/m^2
- State 8: $0.7-0.8$ kW/m^2
- State 9: $0.8-0.9$ kW/m^2
- State 10: $0.9-1$ kW/m^2

The day-wise duration of data and overall duration in each state are tabulated in Table 15.13.

Example 7. For an ESD of capacity 100 MW, the power output data is given in Table 15.14. Perform multistate modeling for the system using five states.

Solution

For the data given in Table 15.14, the hourly power output for 3 days is shown. The data corresponding to negative sign signifies the discharging of the ESD. Thus, it acts as a source. Therefore, the positive sign signifies charging and ESD acts as a load. In the case when power output is 0 MW, the ESD is inactive.

TABLE 15.13 Solar: day-wise multistate modeling for 10 states.

State	Duration for day 1 (h)	Duration for day 2 (h)	Duration for day 3 (h)	Duration for day 4 (h)	Duration for day 5 (h)	Total duration (h)
State 1	0	1	2	1	2	6
State 2	3	0	1	0	1	5
State 3	1	1	2	1	3	8
State 4	0	0	1	0	2	3
State 5	0	1	3	1	0	5
State 6	0	1	0	1	0	2
State 7	1	2	0	2	1	6
State 8	3	1	0	3	0	7
State 9	0	2	0	0	0	2
State 10	1	0	0	0	0	1

TABLE 15.14 Energy storage device (ESD): power output data for 3 days.

Time (h)	Day 1 (MW)	Day 2 (MW)	Day 3 (MW)	Time (h)	Day 1 (MW)	Day 2 (MW)	Day 3 (MW)
1	3.06	81.4	21.83	13	0	41.41	0
2	1.58	1.62	1.51	14	11.12	21.59	5.83
3	−2.15	−2.29	−2.04	15	75.5	0.03	1.16
4	0	0	−0.31	16	13.39	0	−6.57
5	0	−13.96	−20.99	17	−20.34	−0.33	−0.43
6	−2.5	0	0	18	−79.66	−99.67	0
7	0	0	0	19	0	0	0
8	0	0	0	20	83.96	0	95
9	0	−66.77	0	21	0	83.75	0
10	0	0	0	22	−11.84	−0.47	−0.45
11	0.09	36.98	4.5	23	0	0	0
12	−0.09	0	−4.5	24	−52.12	−63.28	−74.55

For multistate modeling of ESD, the states are made based upon similarity in power outputs. To define a system with five states:

- State 1: $100 \leq P_{out} < -50$ MW
- State 2: $50 \leq P_{out} < 0$ MW
- State 3: $P_{out} = 0$ MW
- State 4: $0 < P_{out} \leq 50$ MW
- State 5: $50 < P_{out} \leq 100$ MW

The given Table 15.15 shows the multistate modeling for five states, day-wise and overall.

TABLE 15.15 Energy storage device (ESD): day-wise multistate modeling for five states.

State	Duration for day 1 (h)	Duration for day 2 (h)	Duration for day 3 (h)	Total duration (h)
State 1	2	3	1	6
State 2	5	4	7	16
State 3	10	10	10	30
State 4	5	5	5	15
State 5	2	2	1	5

4. Discrete Markov chains

As discussed in the previous sections, the RES data is diffused and discrete. It varies continuously in space and time. This variation in data is called stochastic process. In order to find the reliability of a system with RES, Markov chains are used for easy analysis of the data [8].

For the application of Markov chains, the following two things are mandatory.

- The system should lack any backup memory.

    There must be no dependency on the future states of the previous states. The state that is immediately preceding the current state is an exception here. Hence, the system should not possess any memory of the previous state's data.

    The data's behavior should be dependent on the instantaneous present location and not on how it landed at that location.
- The system should undergo homogeneous/stationary processes.

    Any system that has the same system parameters at all points in space and time is called a homogeneous/stationary system.

    If the system undergoes a process that is a function of time or has discrete steps of time, it becomes nonstationary. The Markov chains cannot be used for such a system.

In any system, space and time can be continuous or discrete. For reliability calculations of the system, time can be either continuous or discrete, but space needs to be discrete. The different states that each discrete step corresponds to contain distinct information about the component that is a part of that step.

4.1 Formation of Markov chains

To understand the formation of Markov chains, consider the system given in Fig. 15.5 [13]. Given is a system with two states, A and B. The probability of being in the state A and B is 0.25 and 0.5 respectively. Also, the probability of leaving state A and B is 0.75 and 0.5, respectively. All the probabilities are constant throughout the analysis. It should be noted that the movement between the states occurs in discrete steps.

If in the first-time interval, the system is assumed to be in state A, the probability to remain in state A is 0.25. The probability of transition to state B is 0.75. To determine the probability after the second interval when the system has entered state B, the probability to remain in state B is 0.5 and to transition to state A is 0.5. This behavior can be seen in the tree diagram in Fig. 15.6.

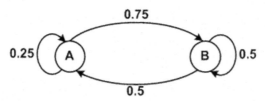

FIGURE 15.5   A system with two states for analysis of Markov chains.

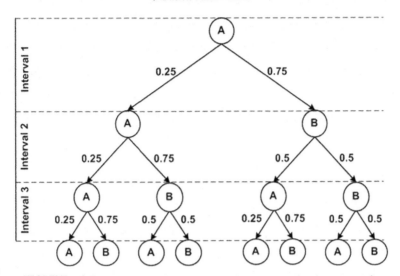

FIGURE 15.6 Tree diagram for system with two states for three intervals.

The probability of being in state A and B, after interval 1 is known. It should be noted that the sum of the probabilities of branches gives the probability of residing in a state. Therefore, to determine the probability after interval 2:
Probability for State A:

$$(0.25 \times 0.25) + (0.75 \times 0.5) = 0.4375$$

Probability for State B:

$$(0.25 \times 0.75) + (0.75 \times 0.5) = 0.5625$$

Similarly, after interval 3:
Probability for State A:

$$(0.25 \times 0.25 \times 0.25) + (0.25 \times 0.75 \times 0.5) + (0.75 \times 0.5 \times 0.25) + (0.75 \times 0.5 \times 0.5) = 0.390625$$

Probability for State B:

$$(0.25 \times 0.25 \times 0.75) + (0.25 \times 0.75 \times 0.5) + (0.75 \times 0.5 \times 0.75) + (0.75 \times 0.5 \times 0.5) = 0.609375$$

Table 15.16 summarizes the probabilities for different time intervals.

TABLE 15.16 State probabilities for the system with two states.

Time interval	Probability state A	Probability state B
1	0.25	0.75
2	0.4375	0.5625
3	0.390625	0.609375

4.2 Birth and death of Markov chains

For a predefined finite number of states (say "N"), birth and death of Markov chains are defined as in Fig. 15.7. The system passes from one state to another. These interstate transitions can be seen in Fig. 15.8, where λ and μ are failure and repair rates respectively.

The following assumptions need to be made for the application of Markov chains [10].

- In the multistate model explained in Section 3, the "N" data levels are predefined as per the input data.
- The transitions corresponding to only the immediately adjacent states are considered for the analysis.

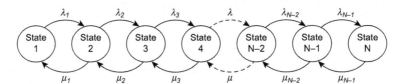

FIGURE 15.7 Birth and death of Markov chains.

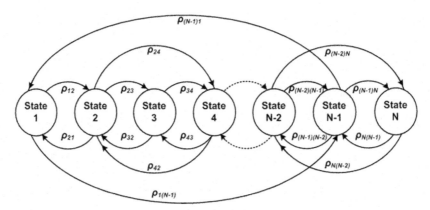

FIGURE 15.8 Transitions between "N" states of the system.

The failure rate, λ is analyzed by occurrences per year. If "N_f" is the number of forced outages, "T" is the total time duration (in hours) of the data set, and "h_s" is the time (in hours) that the turbine is in service, then:

$$\lambda = \frac{N_f T}{h_s} \tag{15.23}$$

The mean time to repair, r depends upon:

- The unit that is affected/damaged/out of service.
- The maintenance services available.
- Climatic conditions.

The repair rate, μ is:

$$\mu = \frac{1}{r} \tag{15.24}$$

Fig. 15.9 shows state modeling for an "N" state unit (turbine, pv module, etc.). The unit functions with binary conditions, "up" and "down." Whenever there is severe damage to a state, the "up" shifts to "down." The interchange between the two conditions depends on their failure and repair rates.

4.3 Calculation of transition rates and probabilities

The Markov approach follows probability distribution, such as exponential distribution [11]. The data may/may not follow the same trend but the long-term cumulated effects follow the distribution. Considering the states corresponding to one unit (turbine or pv module) and not the entire farm, the following is defined. $\rho_{i,j}$, the rate of transition from state "i" to "j," follows the exponential distribution formula:

$$\rho_{i,j} = \frac{N_{i,j}}{D_i} \tag{15.25}$$

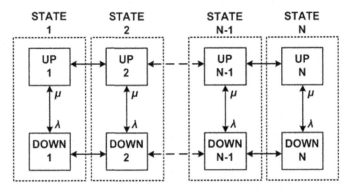

FIGURE 15.9 The state diagram for system with one unit.

where "$N_{i,j}$" is the number of transitions from "i" to "j" and "D_i" is the time duration of individual states:

$$D = \sum_{i=1}^{N} t_i \tag{15.26}$$

In a state "i," the probability "P_i" of occurrence of data is given by:

$$P_i = \frac{D_i}{T} \tag{15.27}$$

Hence,

$$P_{ci} = \frac{\sum_{j=1}^{N} N_{i,j}}{\sum_{k=1}^{N} \sum_{j=1}^{N} N_{k,j}} \tag{15.28}$$

Example 8. Table 15.7 shows hourly wind speed data for 5 days. Calculate the transition rates and state-wise probability for four, six, and eight states.

Solution

The system is divided into four states as explained in example 5. From this, the state-wise transition rates and probabilities are calculated and shown in Table 15.17.

The system is divided into six states as explained in example 5. From this, the state-wise transition rates and probabilities are calculated and shown in Table 15.18.

The system is divided into eight states as explained in example 5. From this, the state-wise transition rates and probabilities are calculated and shown in Table 15.19.

TABLE 15.17 Wind: rate of transitions and state-wise probability with four states.

	1	2	3	4	Probability
1	0.5833	0.3611	0.0556	0.0000	0.3000
2	0.2745	0.4706	0.1961	0.0588	0.4250
3	0.0400	0.4000	0.4000	0.1200	0.2000
4	0.0000	0.3750	0.3750	0.2500	0.0667

TABLE 15.18 Wind: rate of transitions and state-wise probability with six states.

	1	2	3	4	5	6	Probability
1	0.5833	0.2778	0.0833	0.0556	0.0000	0.0000	0.3000
2	0.4000	0.2333	0.2000	0.0667	0.0000	0.1000	0.2500
3	0.0952	0.2857	0.2381	0.1905	0.1905	0.0000	0.1750
4	0.0714	0.2857	0.1429	0.2143	0.1429	0.0714	0.1083
5	0.0000	0.0909	0.2727	0.1818	0.2727	0.1818	0.0917
6	0.0000	0.1250	0.2500	0.1250	0.2500	0.2500	0.0667

TABLE 15.19 Wind: rate of transitions and state-wise probability with eight states.

	1	2	3	4	5	6	7	8	Probability
1	0.7576	0.0758	0.0606	0.0303	0.0303	0.0000	0.0000	0.0455	0.5500
2	0.4615	0.2308	0.0000	0.0769	0.0000	0.1538	0.0769	0.0000	0.1083
3	0.2500	0.2500	0.0000	0.1250	0.2500	0.1250	0.0000	0.0000	0.0667
4	0.4286	0.0000	0.1429	0.0000	0.1429	0.2857	0.0000	0.0000	0.0583
5	0.2857	0.1429	0.0000	0.1429	0.1429	0.0000	0.0000	0.1429	0.0500
6	0.0000	0.1250	0.2500	0.1250	0.1250	0.2500	0.0000	0.1250	0.0667
7	0.3333	0.0000	0.0000	0.0000	0.0000	0.3333	0.0000	0.3333	0.0250
8	0.1250	0.1250	0.1250	0.1250	0.0000	0.0000	0.2500	0.2500	0.0667

Example 9. For the ESD given in Table 15.14, find out the transition rates and state-wise probability for a five-state system using Markov chains.

Solution

The system is divided into five states as explained in example 7. From this, the state-wise transition rates and probabilities are calculated and shown in Table 15.20.

4.4 Advantages of Markov chains

The following are the advantages:

- Using Markov chains, the analysis of one unit of RES or ESD can be easily extended for multiple units.
- As compared to single-state model, the multistate model using Markov chains gives more accurate results. The precision increases with increase in the number of states.

TABLE 15.20 Energy storage device (ESD): rate of transitions and state-wise probability with five states.

	1	2	3	4	5	Probability
1	0	0	0.5	0.1667	0.1667	0.0694
2	0.125	0.1875	0.6875	0	0	0.2222
3	0.1333	0.1667	0.4	0.2	0.1	0.4167
4	0	0.4667	0.1333	0.3333	0.0667	0.2083
5	0	0.2	0.4	0.4	0	0.0694

- Since the components with "similar," nonzero outputs are directly combined, the cumulative output is easily calculated.
- The existing systems/models can be easily combined with this model.

5. Reliability analysis

5.1 Capacity outage and probability table

The determination of capacity outage for any generating unit in power system network is an approach to determine the load losses in the system. A table consisting of capacity levels and its corresponding probabilities is called capacity outage and probability table (COPT) [14]. Consider the following example to understand the calculation.

Example 10. Consider a system of three units in which two units have capacity 50 MW with FOR 0.02 and one unit of 30 MW with FOR 0.03. Form a COPT for this system.
  Solution
  First of all, it is established that all the units exist in binary condition i.e., "ON" and "OFF." So, the units will work at their maximum capacity or zero capacity. Considering the unit with 50 MW, capacity outage can be 0 MW or 50 MW. Similarly, with 30 MW, it can be 0 MW or 30 MW. As per the given data, the probability of each unit for availability and unavailability is given in Table 15.21.
  The maximum capacity of the system is 130 MW (50 MW + 50 MW + 30 MW). The individual probability for each capacity level can be calculated. For a system to have "0 MW" capacity out of service, the probability will be calculated by $0.98 \times 0.98 \times 0.97 = 0.931588$. The sum of the individual probabilities of the system is 1. Therefore, cumulative probability will be the difference of individual probability from 1. Table 15.22 shows the capacity outage, individual probability, and cumulative probability for the system.
  Section 5.1.1 discusses the formation of COPT for multiple units of RES and ESD.

TABLE 15.21 Probability of availability and unavailability of three-unit system of example 10.

Unit (MW)	Capacity out of service (MW)	Probability out of service
30	0	0.97
	30	0.03
50	0	0.98
	50	0.02
50	0	0.98
	50	0.02

TABLE 15.22 Capacity outage and probability table for three-unit system.

Capacity out of service (MW)	U1+U2+U3 (30 MW, 50 MW, 50 MW)	Probability out of service	Cumulative probability
0	(0,0,0)	0.931588	1.000000
30	(30,0,0)	0.028812	0.068412
50	(0,50,0) + (0,0,50)	0.038024	0.039600
80	(30,50,0) + (30,0,50)	0.001176	0.001576
100	(0,50,50)	0.000388	0.000400
130	(30,50,50)	0.000012	0.000012

5.1.1 Markov chains for multiple units

In Section 4.3, one unit of RES is analyzed using Markov chains. To extend the discussion for multiple RES units, discrete convolution is used for calculating the capacity outages and probability of outage for the units [15]. The transitions between all the states of all the units are considered for the analysis, as seen in Fig. 15.10.

The entries of the matrix correspond to the outages of each state. The transitions to states with a lower capacity are considered as only these states are essential for the calculation of input power.

To calculate the capacity outage,

$$C_{i,j}^{out} = MG_N - (M - i + 1)G_j \tag{15.29}$$

where G_N is the state-wise power output, M is the number of RES units, and N is the number of states.

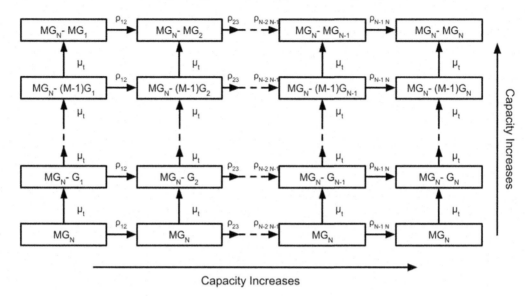

FIGURE 15.10 The state diagram for system with multiple units.

Forced outage is the condition of shutdown in a power system. The forced outage rate, FOR, is determined by:

$$FOR = \frac{\lambda}{\lambda + \mu} \qquad (15.30)$$

where λ is failure rate and μ is the repair rate.

The individual probability corresponding to outages is calculated as:

$$P_{in} = P_{ci} \left[C_r^q \prod_{i=1}^{r} (FOR) \prod_{r+1}^{q} (1-FOR) \right] \qquad (15.31)$$

where C_r^q is the combination of r units taken q at a time.

Example 11. For the wind speed data given in Table 15.7, find out the capacity if there are 10 wind turbines attached to the system. Model the system using eight states and take FOR = 0.04.

Solution

The system is divided into eight states, as shown in example 5. Using the results from Table 15.19 and the formulae given in Eqs. (15.29) and (15.31), the COPT is formulated and is shown in Table 15.23.

Example 12. For the ESD power output data given in Table 15.14, find out the capacity if there are 10 units attached to the system. Model the system using five states and take FOR = 0.04.

TABLE 15.23 Wind: capacity outage and probability table for eight states.

Capacity out of service (MW)	Individual probability	Cumulative probability	Capacity out of service (MW)	Individual probability	Cumulative probability
0	0.052656	1.000000	611	0.000028	0.786270
64	0.018468	0.947344	612	0.002597	0.786242
128	0.003463	0.928877	616	0.039072	0.783645
192	0.000385	0.925414	619	0.016180	0.744574
256	0.000028	0.925029	621	0.003030	0.728393
320	0.000001	0.925001	623	0.000338	0.725363
451	0.016621	0.925000	625	0.044322	0.725026
470	0.006925	0.908379	626	0.000025	0.680704
489	0.001299	0.901454	627	0.018468	0.680679
508	0.000144	0.900155	628	0.003464	0.662211
527	0.000011	0.900011	630	0.000385	0.658748
546	0.000001	0.900000	631	0.000028	0.658363
592	0.044322	0.900000	632	0.072024	0.658335
597	0.018468	0.855678	633	0.030011	0.586311
602	0.003463	0.837210	634	0.005627	0.556300
605	0.033242	0.833747	635	0.000671	0.550673
607	0.000385	0.800506	636	0.000002	0.550002
609	0.013851	0.800121	640	0.550000	0.550000

Solution

The system is divided into five states as shown in example 7. Using the results from Table 15.20 and the formulae given in Eqs. (15.29) and (15.31), the COPT is formulated and is shown in Table 15.24.

5.2 Loss of load expectation (LOLE)

In order to determine the reliability of any system, several indices are defined. For the study, a simple load model is considered representing the daily load peak values. The individual peaks for all the loads are arranged in descending order to form the "Daily Peak Load Duration Curve." The area under the curve represents the energy requirement for a given period.

LOLE is the loss of load expectation. A loss in load occurs when the load level of the system exceeds the generator capacity. It is different from the concept of capacity outage, which

TABLE 15.24 Energy storage device (ESD): capacity outage and probability table for five states.

Capacity out of service (MW)	Individual probability	Cumulative probability	Capacity out of service (MW)	Individual probability	Cumulative probability
0	0.060079	1.000000	840	0.416658	0.708318
84	0.019237	0.939921	876	0.000094	0.291660
168	0.003607	0.920684	882	0.001282	0.291567
252	0.000401	0.917077	888	0.011542	0.290284
336	0.000029	0.916676	894	0.061559	0.278742
730	0.138507	0.916647	900	0.147741	0.217184
741	0.057711	0.778140	1278	0.000029	0.069443
752	0.010821	0.720429	1351	0.000401	0.069414
763	0.001202	0.709608	1424	0.003607	0.069013
774	0.000088	0.708406	1497	0.019237	0.065406
			1570	0.046169	0.046169

means the generator's out-of-service condition. LOLE is a combination of COPT and the load characteristics of the system. It determines the risk of load loss and is calculated by the formula:

$$\text{LOLE} = \sum_{k=1}^{n} p_{ind,k} t_{p,k} \tag{15.32}$$

where $p_{ind,k}$ is the individual probability corresponding to capacity state C_k and $t_{p,k}$ is the time (%) corresponding to that capacity on the daily peak variation curve [14]. A detailed description is given in the following example.

Example 13. In the system considered in example 10, the COPT for same is given in Table 15.22. For the daily peak load variation curve given in Fig. 15.11, determine the LOLE for the system. Take the forecast peak load equal to 100 MW.

Solution

The first step is to form the COPT. All the probabilities that are less than the order 10^{-6} are neglected and their capacities are not considered in the table. Here, consider the COPT of Table 15.22. In this example, linear daily peak load variation is considered as shown in Fig. 15.11. However, the real-time model may or may not be linear. The next step is to determine the capacity in service, given by difference between maximum capacity and capacity out of service. Since maximum forecast peak load given is 100 MW, all the capacities in service (and capacity out of service) that are greater than 100 MW correspond to "0" time. Calculate all the other time (%) using line equation "$y = mx + c$." LOLE is the product of percentage time and corresponding probabilities and is shown in Table 15.25.

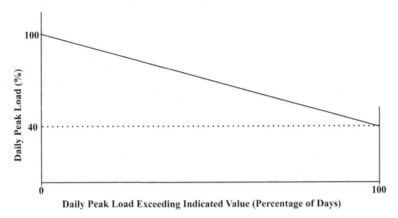

FIGURE 15.11 Daily peak load variation curve for loss of load expectations (LOLE).

TABLE 15.25 Loss of load expectations (LOLE).

Capacity out of service (MW)	Capacity in service (MW)	Individual probability	Total time (%)	LOLE
0	130	0.931588	0	0
30	100	0.028812	0	0
50	80	0.038024	33.33	1.267
80	50	0.001176	83.33	0.098
100	30	0.000388	–	0
130	0	0.000012	–	0
			Total LOLE	1.365

5.3 Expected energy not supplied (EENS)

The expected energy not supplied or EENS is a reliability index that works on the energy curtailment logic. It is a simple approach applicable for modeling of energy generated. The formula used is:

$$\text{EENS} = \sum_{k=1}^{n} p_{ind,k} E_{c_k} \tag{15.33}$$

where $p_{ind,k}$ is the individual probability corresponding to capacity C_k and E_{c_k} is the energy curtailed corresponding to C_k [14]. The curtailed energy is calculated by area under the load duration curve corresponding to C_k. A detailed description is given in the following example.

Example 14. In the system considered in example 10, the COPT for same is given in Table 15.22. For the load curve given in Fig. 15.12, determine the EENS for the system.

Solution

The first step is to form the COPT. All the probabilities that are less than the order 10^{-6} are neglected and their capacities are not considered in the table. Here, consider the COPT of Table 15.22 The system load model given in Fig. 15.12 is a linear representation taken for ease in calculations. The real time model may or may not be linear.

The next step is to determine the capacity in service, given by difference between maximum capacity and capacity out of service. Calculate all curtailed energy using the area under the curve for the corresponding load in service. EENS is the product of curtailed energy and corresponding probabilities and is shown in Table 15.26.

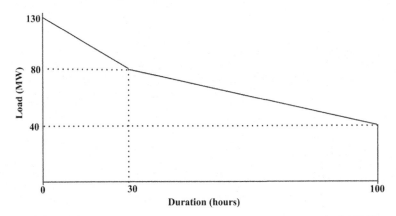

FIGURE 15.12   Load duration curve: expected energy not supplied (EENS).

TABLE 15.26   Expected energy not supplied (EENS).

Capacity out of service (MW)	Capacity in service (MW)	Individual probability	Energy curtailed (MWh)	EENS
0	130	0.931588	0	0
30	100	0.028812	270.000000	7.77924
50	80	0.038024	750.000000	28.518
80	50	0.001176	2437.500000	2.8665
100	30	0.000388	4350.000000	1.6878
130	0	0.000012	7350.000000	0.0882
			Total	40.93974

5.4 Expected demand not supplied (EDNS)

The calculations of EDNS are related to EENS as:

$$EDNS = \frac{EENS}{T} \tag{15.34}$$

where T is the duration of load curve corresponding to C_k [14].

---

6. Load flow analysis

Load flow is a crucial aspect of power system operation, planning, and management. Since load is a static quantity, there is a flow of power through the lines of the network. Thus the name power flow analysis is also applicable.

The question arises, why is load flow analysis done here? In order to analyze the effect of RES and ESD on any power system network, it is essential to find out the effect it has on the load flow.

At the steady state, the voltage, active-reactive powers, and line losses can be computed via load flow. If the voltage magnitude and angles are known, then using load flow analysis, the active and reactive power can be computed. The knowledge about sending and receiving end powers determines the losses in the lines.

Several nonlinear algebraic equations define the power flow of a network. Therefore, iterative methods are used for determination of power.

6.1 Types of buses

In any power system network, there are three types of buses.

- Slack bus
  The slack bus is the reference bus of the system. It's δ is zero.
- Load bus
  The load bus, or the PQ bus, has no connections of generator units onto them. Therefore, their P_G and $Q_G = 0$. The unknown quantities that need to be calculated for a PQ bus are $|V|$ and δ.
- Generator bus
  The generator bus, or the PV bus, has one or more connections of generator units onto them. The input power, P_i is constant. Values of P_G and $|V|$ are known. The unknown quantities that need to be calculated for a PQ bus are Q_G and δ.

6.2 Power flow equations

This section discusses the equations that determine the power flow. For the analysis, all the standard nomenclature is used. Also, the current/power entering a bus is considered positive, whereas that leaving a bus is negative.

For a given bus "x":

$$V_x = |V_x| \angle \delta_x = |V_x|(cos\delta_x + jsin\delta_x) \tag{15.35}$$

The self-admittance matrix, Y_{xx} is calculated by:

$$Y_{xx} = |Y_{xx}| \angle \theta_{xx} = |Y_{xx}|(cos\theta_{xx} + jsin\theta_{xx}) \tag{15.36}$$

Also,

$$Y_{xx} = G_{xx} + jB_{xx} \tag{15.37}$$

Similarly, the mutual admittance matrix, Y_{xy}:

$$Y_{xy} = |Y_{xy}|(cos\theta_{xy} + jsin\theta_{xy}) = G_{xy} + jB_{xy} \tag{15.38}$$

Given, total number of buses = "n,"

$$I_x = Y_{x1}V_1 + Y_{x2}V_2 + ... + Y_{xn}V_n \tag{15.39}$$

$$I_x = \sum_{k=1}^{n} Y_{xk}V_k \tag{15.40}$$

Complex power can be represented by

$$P_x - jQ_x = V_x^* I_x = V_x^* \sum_{k=1}^{n} Y_{xk}V_k \tag{15.41}$$

$$P_x - jQ_x = |V_x|(cos\delta_x - jsin\delta_x) \sum_{k=1}^{n} Y_{xk}V_k \tag{15.42}$$

$$P_x - jQ_x = \sum_{k=1}^{n} |V_x Y_{xk}V_k|(cos\delta_x - jsin\delta_x)(cos\theta_{xy} + jsin\theta_{xy})(cos\delta_k + jsin\delta_k) \tag{15.43}$$

$$P_x - jQ_x = \sum_{k=1}^{n} |V_x Y_{xk}V_k|(cos(\theta_{xk} + \delta_k - \delta_x) + jsin(\theta_{xk} + \delta_k - \delta_x)) \tag{15.44}$$

Hence,

$$P_x = \sum_{k=1}^{n} |V_x Y_{xk}V_k|cos(\theta_{xk} + \delta_k - \delta_x) \tag{15.45}$$

and

$$Q_x = \sum_{k=1}^{n} |V_x Y_{xk} V_k| sin(\theta_{xk} + \delta_k - \delta_x)$$ (15.46)

Let $P_{x,inj}$ be the power injected to the bus and $P_{x,calc}$ be the calculated power.

$$P_{x,inj} = P_{G,x} - P_{L,x}$$ (15.47)

$$\Delta P = P_{x,inj} - P_{x,calc} = P_{G,x} - P_{L,x} - P_{x,inj}$$ (15.48)

Similarly,

$$\Delta Q = Q_{x,inj} - Q_{x,calc} = Q_{G,x} - Q_{L,x} - Q_{x,inj}$$ (15.49)

The aim is to get the values of ΔP and ΔQ below the tolerance levels.

6.3 Load flow using Newton–Raphson method

There is a set of "n" nonlinear equations containing "n" variables.

$$f_1(X_1, X_2...X_n) = \eta_1$$
$$f_2(X_1, X_2...X_n) = \eta_2$$
$$.$$
$$.$$
$$.$$
$$f_n(X_1, X_2...X_n) = \eta_n$$

Define a function "g" as:

$$g_1(X_1, X_2...X_n) = f_1(X_1, X_2...X_n) - \eta_1 = 0$$
$$g_2(X_1, X_2...X_n) = f_2(X_1, X_2...X_n) - \eta_2 = 0$$
$$.$$
$$.$$
$$.$$
$$g_n(X_1, X_2...X_n) = f_n(X_1, X_2...X_n) - \eta_n = 0$$

Let the initial estimates be $X_1^{(0)}, X_2^{(0)}, \ldots, X_n^{(0)}$ and the corrections be $\Delta X_1^{(0)}, \Delta X_2^{(0)}, \ldots, \Delta X_n^{(0)}$. The new variables can be written as:

$$X_1^* = X_1^{(0)} + \Delta X_1^{(0)}$$
$$X_2^* = X_2^{(0)} + \Delta X_2^{(0)}$$
$$\cdot$$
$$\cdot$$
$$\cdot$$
$$X_n^* = X_n^{(0)} + \Delta X_n^{(0)}$$

Modify "g" with the new variables and use Taylor series expansion on it. All the partial derivatives of second order and higher are neglected.

Assuming that Bus 1 is slack bus, number of generator bus is n_g and number of load bus is n_p, then,

$$n = n_p + n_g + 1$$

Using Newton–Raphson algorithm, the modified equation can be written as:

$$J \begin{bmatrix} \Delta\delta_2 \\ \cdot \\ \cdot \\ \Delta\delta_n \\ \dfrac{\Delta|V_2|}{|V_2|} \\ \cdot \\ \cdot \\ \dfrac{\Delta|V_{1+n_0}|}{|V_{1+n_0}|} \end{bmatrix} = \begin{bmatrix} \Delta P_2 \\ \cdot \\ \cdot \\ \Delta P_n \\ \Delta Q_2 \\ \cdot \\ \cdot \\ \Delta Q_{1+n_0} \end{bmatrix} \tag{15.50}$$

This gets divided into Jacobian submatrices given as:

$$J = \begin{bmatrix} J_{11} & J_{12} \\ J_{21} & J_{22} \end{bmatrix} \tag{15.51}$$

Where,

$$J_{11} = \begin{bmatrix} \dfrac{\partial P_2}{\partial \delta_2} & \cdot & \cdot & \cdot & \dfrac{\partial P_2}{\partial \delta_n} \\ \cdot & \cdot & \cdot & \cdot & \cdot \\ \cdot & \cdot & \cdot & \cdot & \cdot \\ \cdot & \cdot & \cdot & \cdot & \cdot \\ \dfrac{\partial P_n}{\partial \delta_2} & \cdot & \cdot & \cdot & \dfrac{\partial P_n}{\partial \delta_n} \end{bmatrix} \tag{15.52}$$

$$J_{12} = \begin{bmatrix} |V_2|\dfrac{\partial P_2}{\partial |V_2|} & \cdot & \cdot & \cdot & |V_{1+n_0}|\dfrac{\partial P_2}{\partial |V_{1+n_0}|} \\ \cdot & \cdot & \cdot & \cdot & \cdot \\ \cdot & \cdot & \cdot & \cdot & \cdot \\ \cdot & \cdot & \cdot & \cdot & \cdot \\ |V_2|\dfrac{\partial P_n}{\partial |V_2|} & \cdot & \cdot & \cdot & |V_{1+n_0}|\dfrac{\partial P_n}{\partial |V_{1+n_0}|} \end{bmatrix} \tag{15.53}$$

$$J_{21} = \begin{bmatrix} \dfrac{\partial Q_2}{\partial \delta_2} & \cdot & \cdot & \cdot & \dfrac{\partial Q_2}{\partial \delta_n} \\ \cdot & \cdot & \cdot & \cdot & \cdot \\ \cdot & \cdot & \cdot & \cdot & \cdot \\ \cdot & \cdot & \cdot & \cdot & \cdot \\ \dfrac{\partial Q_{1+n_0}}{\partial \delta_2} & \cdot & \cdot & \cdot & \dfrac{\partial Q_{1+n_0}}{\partial \delta_n} \end{bmatrix} \tag{15.54}$$

$$J_{22} = \begin{bmatrix} |V_2|\dfrac{\partial Q_2}{\partial |V_2|} & \cdot & \cdot & \cdot & |V_{1+n_0}|\dfrac{\partial Q_2}{\partial |V_{1+n_0}|} \\ \cdot & \cdot & \cdot & \cdot & \cdot \\ \cdot & \cdot & \cdot & \cdot & \cdot \\ \cdot & \cdot & \cdot & \cdot & \cdot \\ |V_2|\dfrac{\partial Q_{1+n_0}}{\partial |V_2|} & \cdot & \cdot & \cdot & |V_{1+n_0}|\dfrac{\partial Q_{1+n_0}}{\partial |V_{1+n_0}|} \end{bmatrix} \tag{15.55}$$

6.4 Algorithm for Newton–Raphson method

The algorithm is as follows:

- Step 1
  Initialize the voltage magnitudes $|v|^{(0)}$ for all n_p load buses. Also initialize $\delta^{(0)}$ for n-1 buses (other than slack bus).
- Step 2
  Using these $|v|^{(0)}$ and $\delta^{(0)}$ values, determine all the "$n-1$," the injected power $\left(P^{(0)}_{calc}\right)$, and change in power $\Delta P^{(0)}$.
- Step 3
  Using these $|v|^{(0)}$ and $\delta^{(0)}$ values, determine all the n_p, the injected power $\left(Q^{(0)}_{calc}\right)$, and change in power $\Delta P^{(0)}$.
- Step 4
  Using the estimated $|v|^{(0)}$ and $\delta^{(0)}$ values to form $J^{(0)}$.
- Step 5
  Calculate $\delta^{(0)}$, $\frac{\Delta|v|^{(0)}}{|v|^{(0)}}$.
- Step 6
  Update:

$$\delta^{(1)} = \delta^{(0)} + \Delta\delta^{(0)}$$

$$|v|^{(1)} = |v|^{(0)} + \frac{\Delta|v|^{(0)}}{|v|^{(0)}}$$

- Step 7
  Check if mismatches are below tolerance levels. If yes, terminate the process, else go to step 1 for next iteration.

7. Results and discussions

The described methodology is modeled for real-time data collected from National Renewable Energy Laboratory (NREL) [16]. Hourly wind speed for 365 days is compiled and shown in Fig. 15.13.

The IEEE Reliability Test System 1979 (IEEE RTS-79) is used for simulation and implementation of the proposed methodology. This is a reference system set up to standardize database for testing reliability of bulk power systems [17,18]. The system has 24 buses having 32 generator units distributed at various bus locations. The total capacity of the system is 3405 MW. Bus 13 is the slack bus. The annual peak load is 2850 MW [17,18]. Original system of IEEE RTS-79 is referred as "System 1." The system is modified by replacing six generators of capacity of 860 MW by equivalent capacity wind farms. This modified system is referred as "System 2." The replaced generators are:

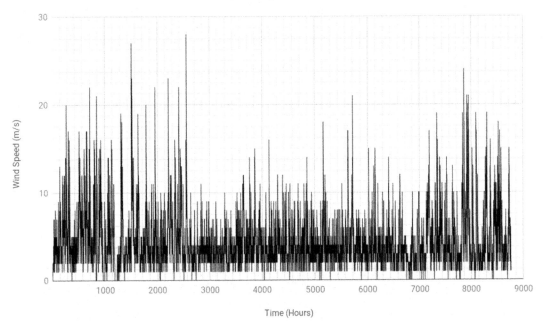

FIGURE 15.13 Hourly wind speed data for 365 days.

- Coal generator: four generators, capacity 76 MW
- Coal generator: 1 generator, capacity 155 MW
- Nuclear generator: 1 generator, capacity 400 MW

A wind power system consisting of several wind farms is considered. Each farm has 10 wind turbines of capacity 8 MW. The cut-in speed is 4 m/s, cut-out speed is 25 m/s, and rated speed is 12 m/s.

For the wind speed data considered, the output power is calculated using the turbine equations (Eq. 15.19). Fig. 15.14 shows the output power calculated for a duration of 100 h extracted from Fig. 15.13.

7.1 Results for discrete Markov chains

Based upon the wind speed data, multistate modeling is done as follows:

- State 1: $0 \leq v < 4$ m/s
- State 2: $4 \leq v < 6$ m/s
- State 3: $6 \leq v < 8$ m/s
- State 4: $8 \leq v < 10$ m/s
- State 5: $10 \leq v < 12$ m/s
- State 6: $12 \leq v < 25$ m/s

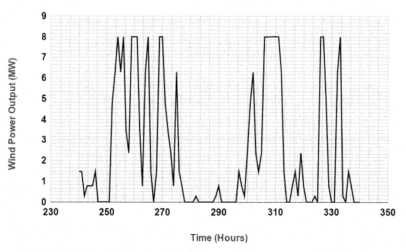

FIGURE 15.14   Wind speed output for 100 h duration using the given data.

The transition rates are calculated using Eq. (15.25) and are shown in Table 15.27. Notice, the transitions rates for each row sum up to approximately 1. The capacity outage, state-wise probability, and cumulative probability for one turbine unit are shown in Table 15.28.

TABLE 15.27   Wind energy data: transition rates for six states.

	1	2	3	4	5	6
1	0.8253	0.1357	0.0291	0.0075	0.0015	0.0007
2	0.3542	0.4305	0.1511	0.0481	0.0127	0.0033
3	0.1226	0.3085	0.3307	0.1799	0.0422	0.0161
4	0.0284	0.1246	0.3028	0.3517	0.1230	0.0694
5	0.0417	0.0545	0.1506	0.2372	0.2885	0.2276
6	0.0067	0.0156	0.0468	0.0824	0.1604	0.6882

TABLE 15.28   State-wise capacity for one wind turbine.

	Capacity State (MW)	Probability (State-wise)	Cumulative Probability		Capacity State (MW)	Probability (State-wise)	Cumulative Probability
1	0.00	0.520776	1.000000	4	3.41	0.072374	0.159247
2	0.28	0.206279	0.479110	5	6.23	0.035616	0.086872
3	1.42	0.113584	0.272831	6	7.88	0.051256	0.051256

TABLE 15.29 State-wise capacity for one wind farm.

Capacity out of service (MW)	Probability	Cumulative probability	Capacity out of service (MW)	Probability	Cumulative probability
0	0.034159	1.000000	52	0.003759	0.844847
8	0.014199	0.965809	55	0.000418	0.841088
16	0.026341	0.951610	58	0.000030	0.840671
23	0.009866	0.925269	62	0.000002	0.840640
24	0.000296	0.915403	65	0.075515	0.840639
29	0.001850	0.915107	66	0.031464	0.765124
32	0.000022	0.913257	67	0.005900	0.733660
35	0.000206	0.913235	69	0.000656	0.727760
39	0.000001	0.913030	70	0.000048	0.727104
41	0.000015	0.913029	72	0.000002	0.727057
45	0.048117	0.913014	76	0.204997	0.727054
48	0.020049	0.864897	77	0.001282	0.522057
			79	0.520776	0.520776

Using the formula given in Eqs. (15.29) and (15.31), this can be extended for one wind farm of 10 wind turbines as shown in Table 15.29.

7.2 Results for loss of load expectations (LOLE)

Reliability is a thinkable part of power system with newly incremental RES penetration in the power system network. As a serious concern about reliability, ESDs are used for satisfactory operation of the power system. The profit margin of wind farm owner could be also increased by using the ESD in the system. As the world is approaching toward deregulated power system, ESD will play an important role to improve the system performance without any load interruption in the system. ESD uses a different type of battery so the reliability parameter also changes with changing the characteristics of the storage devices. As some batteries have a greater discharging period, so those types of ESDs improve the system reliability for greater time durations. Transmission switching is performed with ESD devices to check the reliability indices values and compare with base case i.e., without ESD. Fig. 15.15 shows the flow chart to evaluate the reliability parameter LOLE, EDNS, and EENS in the RTS-79 system.

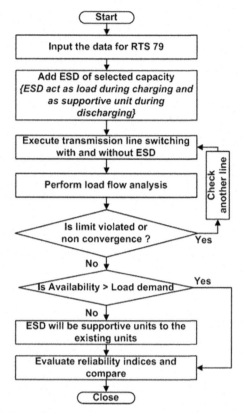

FIGURE 15.15   Flow chart for reliability analysis.

Power system reliability analysis has various indices to observe the risk level of the system. The risk level of power system is termed as a sensitive analysis of the system. Higher LOLE means more risk in the system. ESD has the capability to reduce the risk level of the system. ESD acts as a load during charging and as a supportive unit to the existing system during discharging.

LOLE is calculated at different system peak conditions. The RTS-79 system has installed capacity of 3405 MW and a load of 2850 MW. The system has ESD of 45 MW. The risk level of the system is checked for 3405 MW, 3000 MW, 2850 MW, and 2800 MW without and with using ESD and can be seen in Table 15.30. It can be inferred from the table that the risk level of the system gets reduced by using ESD.

If load demand in the system is reduced, then ESD can be charged to utilize the surplus power. Then the power stored in EDS can be used to support the system during peak load scenarios if required. When peak load in the system is 2850 MW, LOLE is improved by 54.35% (i.e., from 4.4101 to 2.0128) with ESD. The risk level will increase with increment in the system peak load.

TABLE 15.30 Loss of load expectations (LOLE) without and with energy storage device.

System LOLE	LOLE peak load	
(MW)	Without ESD (days/yr)	With ESD (days/yr)
3405	17.1038	16.7314
3000	9.0671	8.0461
2850	4.4101	2.0128
2800	2.3837	1.751

7.3 Results for expected demand not supplied (EDNS)

Transmission line switching is used to improve the system reliability according to load demand in the system. Load flow analysis is done by using MATPOWER. Optimal transmission switching (OTS) leads to the best reliable switching that exists in the system during a particular load demand [19–22]. ESD is used as supportive unit to the available generating units in the system. If there is access generation in the system or load demand gets reduced, then ESD will store the power. This stored power can be further utilized in peak load scenarios. EDNS is the expectation of power demand, which is not supplied to the system load. OTS is obtained in the system with using ESD [23–27].

Fig. 15.16 shows the variation in EDNS with each possible line switching using ESD. Table 15.31 shows the variation in EDNS with top five transmission switching among the existing line switching in the system. An outage of line 14 results in the best reliability so the switching of line 14 will be OTS for particular switching scenarios.

In base case scenarios, EDNS is reduced by 11.56% (i.e., 13.81275 to 12.2156 MW/yr) with incorporating ESD in the system. The storage device can be used to store power during the access generation or less load requirement at the end-user side. Effect on EDNS can be

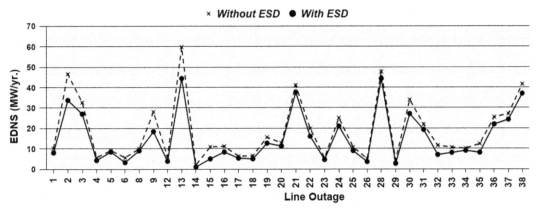

FIGURE 15.16 Variation in expected demand not supplied (EDNS) with every possible switching without and with energy storage device (ESD).

TABLE 15.31 Top five optimal transmission switching (OTS) with respect to expected demand not supplied (EDNS) without and with energy storage devices (ESD).

System peak load (MW)	EDNS Without ESD (MW/yr)	EDNS With ESD (MW/yr)
L 4	5.6495	4.321
L 6	5.6326	3.2721
L 14	2.0693	1.0657
L 26	5.065	3.596
L 29	4.3621	2.6984

observed through every possible transmission switching without and with using ESD in Fig. 15.16. As from results, it can be observed that ESDs have the capability to improve the reliability performance of the power system network.

7.4 Results for expected energy not supplied (EENS)

Expected energy not supplied is calculated at every possible switching. The outage of lines 7, 10, 11, and 27 results in the nonconvergence of load flow so these lines must be present in the system for the safe operation. Remaining outages of the lines will participate in OTS. The minimum values of reliability index give the best OTS. ESD, as a source, improves system reliability in different switching operations, if required. If there is no violation in equality and inequality constraints, then corresponding line outages are considered in switching operation. For the system simulated, the best OTS is obtained for switching of the line 14 and has better reliability with respect to the base case. Fig. 15.17 shows the variation in EENS with

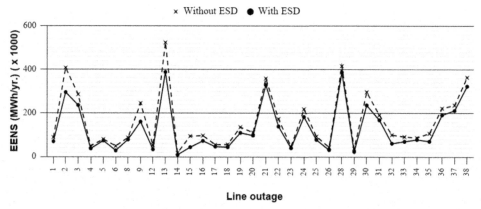

FIGURE 15.17 Variation in expected energy not supplied (EENS) with every possible switching without and with energy storage devices (ESD).

TABLE 15.32 Top five optimal transmission switching (OTS) and expected energy not supplied (EENS) details for system without and with energy storage device (ESD).

System peak load (MW)	EENS Without ESD (MWh/yr)	EENS With ESD (MWh/yr)
L 4	49,490.24	37,851.96
L 6	49,341.95	28,663.6
L 14	18,127.07	9,335.88
L 26	44,369.4	31,500.96
L 29	38,212	23,637.98

every possible line outages. Table 15.32 shows the five best switching results with their corresponding reliability index. The ESD improves EENS by 48.49% (i.e., from 18,127.07 to 9335.88).

Switching of lines 4, 6, 14, 26, and 29 gives the best reliability performance with respect to base case without and with ESD. This shows that EDS improves the reliability. The switching operation of lines that converges the load flow will participate in OTS operations.

7.5 Results for load flow analysis

Fig. 15.18 shows the hourly load variation for 1 day. The load flow results are shown in Table 15.33. The load power is equal for both the systems. However, there is a decrease in the power generated in case of system modified with RES. This is due to the decentralization of RES that has resulted in lower line losses. The comparison of line losses for 24 h is shown in Fig. 15.19.

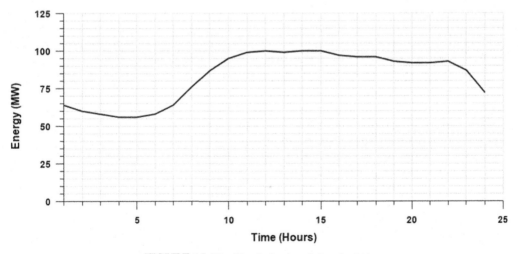

FIGURE 15.18 Hourly load variation for 24 h.

TABLE 15.33   Results for load flow analysis after 24 h.

	System 1	System 2
P_l (MW)	56,715	56,715
P_g (MW)	57,834.17	57,443.23
P_{loss} (MW)	1,119.186	728.224

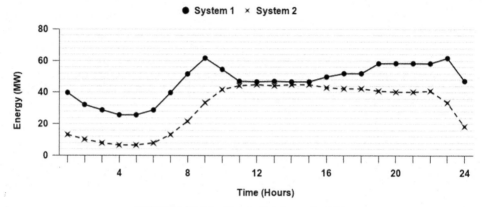

FIGURE 15.19   Hourly line losses for 24 h.

8. Conclusions

In the given chapter, modeling of RES and ESD using multistate modeling methods is discussed. Due to uncertainty in the initial data, conventional methods of power generation and calculation are not applicable for RES. Similarly, in ESD, the hourly data varies as per peak load value and demand. Hence, it is also modeled similar to RES. Multistate modeling methods increase the accuracy of the results. With the increased number of states, the error in calculations is reduced. This model uses discrete Markov chains to calculate the transitional probability and capacity outages for both the systems. The reliability of system 2 is calculated without and with ESD, and the results are tabulated. Due to the diffused data, reliability of system 2 is lesser than system 1. When ESD is incorporated into system 2, the reliability improves as there is a backup power supply to the system. On comparing the results for load flow, a reduction in the line losses in system 2 as compared to system 1 is observed. This is because RESs are decentralized.

Thus their location near the consumer site is advantageous. The consumer decides the priority between reliability indices and power flow. If the requirement is to minimize the generation cost, then RESs are preferred over conventional systems. On the other hand, if a reliable system is required, then either RES with ESD or conventional systems are preferred.

Acknowledgments

Prima facie, I would like to pay my sincere gratitude to God for blessing me with the well-being and good health, without which I wouldn't be standing on this platform. I extend this praise to my life coaches, my parents, who have been the constant pillars of support in my life, and my brother, the source of my enthusiasm.

I want to thank my guide and coauthor, Dr. Vasundhara Mahajan, Associate Professor, Electrical Engineering Department, SVNIT. Without her knowledge, insight, encouragement, and constant motivation, this manuscript would not be possible. This book chapter is dedicated to her for the confidence that she has showed in me.

I want to thank the head of the Electrical Engineering Department, Dr. R. Chudamani, for providing the necessary facilities and opportunities for this research. I thank my other coauthors, Mr. Atul Kumar Yadav and Dr. Vijay Prajapati, for their knowledge and help in bringing this draft together.

Lastly and most importantly, I thank the editors and publishers of "Elsevier Energy Storage in Energy Markets: Uncertainties, Modelling, Analysis and Optimization" and their teams for giving us this opportunity and selecting our chapter proposal for publication in their book.

References

[1] U. Shahzad, The need for renewable energy sources, Int. J. Inf. Technol. & Electr. Eng. (2015) 16—18.

[2] M. Rizwan, S. Waghmare, S. Labade, P. Fuke, A. Tekale, A review paper on electricity generation from solar energy, Int. J. Res. Appl. Sci. Eng. Technol. 887 (2017), https://doi.org/10.22214/ijraset.2017.9272.

[3] M. Gul, Y. Kotak, T. Muneer, Review on recent trend of solar photovoltaic technology, Energy Explor. Exploit. 34 (2016) 485—526, https://doi.org/10.1177/0144598716650552.

[4] P. Arjyadhara, A. Sm, J. Chitralekha, Analysis of solar pv cell performance with changing irradiance and temperature, Int. J. Eng. & Comput. Sci. 2 (2013). http://www.ijecs.in/index.php/ijecs/article/view/109.

[5] M. Mazidi, A. Zakariazadeh, S. Jadid, P. Siano, Integrated scheduling of renewable generation and demand response program in a microgrid, Energy Convers. Manag. 86 (2014) 1118—1127, https://doi.org/10.1016/j.enconman.2014.06.078.

[6] S. Mudgal, A.K. Yadav, V. Mahajan, Reliability evaluation of power system network with solar energy, in: 8th International Conference on Power Systems (ICPS), Malaviya National Institute of Technology Jaipur (MNIT Jaipur), Rajasthan, India, 2019.

[7] S. Mudgal, A.K. Yadav, P. Gupta, V. Mahajan, Impact of solar and wind energy on reliability of power system network, in: International Conference on 'Emerging Trends for Smart Grid Automation and Industry 4.0', Bit Mesra, Ranchi, Jharkhand, India, 2019.

[8] R. Billinton, A.A. Chowdhury, Incorporation of wind energy conversion systems in conven- tional generating capacity adequacy assessment, IEE Proc. Generat. Transm. Distrib. 139 (1992) 47—56, https://doi.org/10.1049/ip-c.1992.0008.

[9] S. Mudgal, V. Mahajan, Reliability and active power loss assessment of power sys- tem network with wind energy, in: 17th IEEE Student Conference on Research and Development (SCOReD), Universiti Teknologi PETRONAS, Perak, Malaysia, 2019, https://doi.org/10.1109/SCORED.2019.8896327.

[10] F. Castro Sayas, R.N. Allan, Generation availability assessment of wind farms, IEE Proc. Generat. Transm. Distrib. 143 (1996) 507—518, https://doi.org/10.1049/ip-gtd:19960488.

[11] A.P. Leite, C.L.T. Borges, D.M. Falcao, Probabilistic wind farms generation model for reliability studies applied to brazilian sites, IEEE Trans. Power Syst. 21 (2006) 1493—1501, https://doi.org/10.1109/TPWRS.2006.881160.

[12] P. Giorsetto, K.F. Utsurogi, Development of a new procedure for reliability modeling of wind turbine generators, IEEE Trans. Power Apparatus Syst. 102 (1983) 134—143, https://doi.org/10.1109/TPAS.1983.318006.

[13] R. Billinton, R.N. Allan, Reliability Evaluation of Engineering Systems, Springer science and business media, New York, 1983, https://doi.org/10.1007/978-1-4899-0685-4.

[14] R. Billinton, R.N. Allan, Reliability Evaluations of Power Systems, Springer science and business media, LLC, 1984, https://doi.org/10.1007/978-1-4899-1860-4.

[15] N. Nguyen, J. Mitra, Reliability of power system with high wind penetration under fre- quency stability constraint, IEEE Trans. Power Syst. 33 (2018) 985—994, https://doi.org/10.1109/TPWRS.2017.2707475.

[16] D. Jager, A. Andreas, NREL National Wind Technology Center (NWTC)s, 1996. https://midcdmz.nrel.gov/apps/daily.pl?site=NWTCstart=20010824yr=20 19mo=3dy=20.

[17] C. Grigg, P. Wong, P. Albrecht, R. Allan, M. Bhavaraju, R. Billinton, Q. Chen, C. Fong, S. Haddad, S. Kuruganty, W. Li, R. Mukerji, D. Patton, N. Rau, D. Reppen, A. Schneider, M. Shahidehpour, C. Singh, The IEEE reliability test system-1996. A report prepared by the reliability test system task force of the application of probability methods subcommittee, IEEE Trans. Power Syst. (1999), https://doi.org/10.1109/59.780914.

[18] Probability Methods Subcommittee, IEEE reliability test system, IEEE Trans. Power Apparatus Syst. PAS-98 (6) (1979).

[19] A.K. Yadav, V. Mahajan, Transmission system reliability evaluation by incorporating STATCOM in the system network, in: 17th IEEE Student Conference on Research and Development (SCOReD), Universiti Teknologi PETRONAS, Perak, Malaysia, 2019, https://doi.org/10.1109/SCORED.2019.8896263.

[20] A.K. Yadav, V. Mahajan, Transmission line switching for loss reduction and re- liability improvement, in: International Conference on Information and Commu- Nications Technology (ICOIACT), 24-25 July 2019, Yogyakarta, Indonesia, 2019, https://doi.org/10.1109/ICOIACT46704.2019.8938535.

[21] A.K. Yadav, V. Mahajan, Reliability improvement of power system network with optimal trans- mission switching, in: IEEE 1st International Conference on Energy, Systems and Information Processing (ICESIP), Chennai, India, 2019, https://doi.org/10.1109/ICESIP46348.2019.8938283.

[22] A.K. Yadav, S. Mudgal, V. Mahajan, Reliability test of restructured power system with capacity expansion and transmission switching, in: 8th International Conference on Power Systems (ICPS), Malaviya National Institute of Technology Jaipur (MNIT Jaipur), Rajasthan, India, 2019.

[23] V.K. Prajapati, V. Mahajan, Congestion management of power system with uncertain renewable resources and plug in electrical vehicle, in: IET Generation, Transmission and Distribution, 2019, pp. 927–938.

[24] V.K. Prajapati, V. Mahajan, Demand response based congestion management of power system with uncertain renewable resources, Int. J. Ambient Energy (2019) 1–20.

[25] V.K. Prajapati, V. Mahajan, Enhancement of ATC of transmission line using demand response programme for congestion management, in: 20th National Power Systems Conference (NPSC), NIT Trichy, 2018, https://doi.org/10.1109/NPSC.2018.8771800.

[26] V.K. Prajapati, V. Mahajan, Congestion management of power system with integration of renewable resources and energy storage system, in: Power India International Conference (PIICON), NIT Kurukshetra, 2018, https://doi.org/10.1109/POWERI.2018.8704420.

[27] V.K. Prajapati, V. Mahajan, Grey wolf optimization based energy management by generator rescheduling with renewable energy resources, in: 14th IEEE India Council International Conference (INDICON), IIT Roorkee, 2017.

Reliability and resiliency assessment in integrated gas and electricity systems in the presence of energy storage systems

Mohammad Taghi Ameli[1], Hossein Ameli[2], Goran Strbac[2], Vahid Shahbazbegian[1]

[1]Department of Electrical Engineering, Shahid Beheshti University, Tehran, Iran; [2]Department of Electrical and Electronic Engineering, Imperial College London, London, United Kingdom

1. Introduction

Energy systems (e.g., gas and electricity systems) include all components designed to supply and deliver energy to consumers. Gas and electricity systems deliver a major share of energy to consumers and play a vital role in people's lives all around the world [1]. On the other hand, these two sectors produce a considerable share of the total greenhouse gas emissions [2]. Therefore, there is a need to take corrective actions to curve this trend and decrease the potential consequences [3]. To meet this purpose, a high number of researches have been carried out in which investing more in renewable energies has been introduced as a solution to deal with the proposed problem.

The changeover to supply energy from renewable energy resources has started in the past few decades, especially in the electricity sector [4]. Renewable energy resources are expanded to replace the power plants with lower efficiency and characterized by high emissions (e.g., coal power plants). The intermittent nature of renewable energies in the power system affects the gas network demand through gas-fired power plants [5]. As a solution, energy storage is one of the flexibility measures that provides both downward and upward flexibility, discharging energy when there is a higher amount of demand and charging on the contrary case [6]. In the energy system, different types of storage systems, such as electricity storage

Energy Storage in Energy Markets
https://doi.org/10.1016/B978-0-12-820095-7.00003-0

systems, gas storage systems, and power to gas systems can play different roles depending on the timescale. Besides, in the gas network, due to the slow speed of gas within the pipelines, the amount of stored gas through the pipelines is another form of energy storage to meet rapid changes in gas demand (i.e., linepack) [7].

In energy systems, reliability is defined as "designing, running, and maintaining energy supply resources, transmission, and distribution systems to provide an adequate, safe, and stable flow of energy" [8]. In the integrated gas and electricity systems, flexibility options, such as energy storage systems, can take part in the reliable supply—demand balance, including mitigation of renewable unit output power fluctuation and peak load management.

On the other hand, resiliency is linked to the concept of reliability, which is defined as "the capacity of an energy system to tolerate disturbance and to continue to deliver affordable energy services to consumers. A resilient energy system can speedily recover from shocks and can provide alternative means of satisfying energy service needs in the event of changed external circumstances" [9]. In the energy system, resiliency is an important index that prevents widespread or extended interruptions due to potential threats. As a result, it is worthwhile taking resiliency into account in operation and planning of energy systems, although there is no standard to fit all various dimensions of resiliency [10]. In this regard, considering the feature of storage systems, their operation can either mitigate the curtailment of load during emergency operation or prevent serious consequences by discharging their stored energy, which boosts the resiliency of the energy system.

According to the proposed issues, including the role of storage systems in the enhancing reliability and resiliency of energy systems, this chapter is going to answer the following questions:

- What are the definitions of reliability and resiliency in these energy systems?
- How the reliability of the energy systems is analyzed? Which indices clarify a reliable energy system?
- How the resiliency of the energy system can be boosted?
- What is the role of storage systems in reliability and resiliency improvement of these energy systems?

The rest of this chapter is organized as follows. After the introduction, literature review is presented related to coordinated operation of gas and electricity networks in Section 2. In Section 3, model formulation and description for this problem are presented. In Section 4, solution methodology is proposed, including approaches to examine the resiliency and reliability. After introducing a case study to illustrate the applicability and performance of the proposed model and its solution in Section 5, simulation results are presented to assess the effectiveness of the proposed model in Section 6. Finally, the summary is presented in section 7.

2. Literature review

The most related stream of literature to this study is scheduling gas and electricity networks. The main focus of the studies is to address and examine: (a) optimization strategies,

(b) role of flexibility options, (c) reliability consideration, and (d) resiliency consideration in the cooptimization of these networks.

The coordinated operation of these networks, in literature, has been implemented through iterative and integrated strategies [11,12]. In the iterative strategy, first, the electricity network operation is optimized by electricity system operator, without accounting the gas network operation limits. According to the output power of gas-fired power plants, the required natural gas for electricity network is calculated and submitted to natural gas system operator. Then, the gas network operation is optimized and constraints violation and load shedding are examined [7]. On the other hand, in the integrated strategy, the operation of gas and electricity networks is optimized simultaneously in which objective function equates to the sum of objective functions of gas and electricity networks. Furthermore, a coupling constraint, which calculates the gas consumption of gas-fired power plant considering their output power, links these networks [13]. For instance, Ref. [14] present a security-constrained model for coordinated operation of gas and electricity networks in which the iterative strategy is devised to solve the problem. [15] developed a model for coordinated operation of gas and electricity networks. The model takes into account the uncertainty in output power of wind farms and the problem is optimized in an iterative manner. Alabdulwahab et al. [17] also proposed a model for coordinated operation of these networks that takes into uncertainty in outage of transmission lines and generating units, as well as the electricity demand. In the model, the nonlinear constraints of gas network are linearized using piecewise, and the problem is solved through the integrated strategy of operation. He et al. [18] present a model for these networks considering key uncertainties, including wind power and electrical load violations. They adopt a decomposition approach to solve the problem in the integrated manner. Ameli et al. [13] also proposed a model of gas and electricity networks and adopted a decomposition technique to optimize problem through integrated strategy. The results of these studies show that the gas network operation limits affect the operation of electricity network. Moreover, it is shown that integrated strategy of operation provides more accurate solution in comparison with iterative strategy of operation. However, solving the problem through the integrated strategy of operation considerably increases solving time due to the complexity of problem.

The potential of flexibility options, including gas storages, energy storages, power-to-gas systems, and flexible gas-fired power plants, is examined to mitigate supply—demand balance in some other studies. The storage systems are mostly charged during off-peak hours of operation period. These systems can be discharged during peak hours of demand, which can facilitate supply—demand balance and prevent shedding in the gas and electricity network. In addition, gas-fired power plants offer a flexible performance (e.g., providing short startup time and fast ramping rate), which can mitigate the variability and intermittency of the renewable energies. The power-to-gas systems also can utilize an amount of renewable energy curtailment to produce hydrogen by using electrolysis. A percentage of the hydrogen can be utilized in the gas network to boost the pressure or regenerate electricity in the electricity network [19]. Among the previous researches, Ameli et al. [3] studied the advantages of employing bidirectional compressors in operation gas and electricity networks. The study also investigates the impact of applying the integrated strategy to this problem in comparison with the iterative strategy. Qadrdan et al. [16] also studied the advantages of flexibility options, such as gas-fired power plants, electricity storage systems, and power-to-gas systems. Sun et al. [20] presented a probabilistic model for cooptimization of gas and electricity networks and investigated the ability of power-to-gas systems to avoid wind power curtailment. Antenucci and Sansavini [21] examined the possibility of overhead lines reenforcement and power-to-gas installation in operation of this networks. Gu et al. [22] propose a robust model for operation of the integrated energy systems to study the potential of power-to-gas systems to reduce wind curtailment considering the uncertainty in output power of wind farms. Thus, the roles of flexibility options have been addressed in the literature that shows their efficiency to mitigate the variability and intermittency in the coordinated operation of these networks.

The other relevant stream of literature to this study is the resiliency or reliability analysis in the coordinated operation of gas and electricity networks. Hao et al. [23], in this regard, proposed a three-stage robust model for operation of gas and electricity systems, which linearizes nonlinear constraints of the gas network. The robust model finds an optimal solution corresponding to the worst-case attack, which enhances resiliency of the integrated systems. Yan et al. [24] presented a two-stage robust model to improve the resiliency of these systems considering the outages in gas and electricity transmission systems, as well as power generation system. They also provide a decomposition technique to solve the presented model. Clegg and Mancarella [25] developed a model to include an integrated gas—electricity—heat optimization. The model takes the worst weather condition into account to enhance these networks' resiliency. On the other hand, Zhang et al. [26] introduce a stochastic model considering different scenarios on the random outages of electrical lines, gas pipelines, and power plants. In the model, Markov chain is utilized in reliability analysis to determine the available or unavailable state of components based on time to failure and time to repair. Yu et al. [27] proposed a methodology to examine the reliability of gas network based on Monte Carlo trials, reliability of gas wells, and reliability of gas storages. Zhang et al. [28] developed a multiobjective model for cooptimization of gas and electricity networks, which consists of the cost minimization and reliability maximization. The study also investigates the role of energy storage equipment and distributed generators. Therefore, the previous studies, which examine the reliability and resiliency of the gas and electricity systems, mostly determine the probable outage of components and the worst-case scenarios. Afterward, an approach provides to solve the cooptimization problem considering the determined scenarios, which the current study aims to do the same.

By reviewing previous studies, it reveals that, with the increasing share of the renewable energy systems, mitigating supply—demand is becoming more challenging. To cope with this problem, coordinated operation of gas and electricity networks could be helpful and beneficial. Furthermore, utilizing flexibility options, including gas-fired power plants, gas/electricity storage systems, and power-to-gas systems, is another solution to facilitate supplying gas and electricity demand. Besides, developing an approach to enhance the reliability or resiliency in optimizing the operation of these networks affects the frequency and magnitude of the consequences, which provide different economic and social benefits. Therefore, as depicted in Fig. 16.1, an overview of main steps in this chapter is declared as follows:

- Introducing the model for coordinated operation of gas and electricity networks considering flexibility options (first step);

First step

Coordinated operation of this networks

$Min\ Z_{total} = Z_{gas} + Z_{elec}$

Subject to

Electricty network constraints

Gas network constraints

Coupling constraints

Second step

Enhancing the resiliency of the systems

Improving the reliability of systems

Third step

Representing a case study to quantify the practicality as well as efficacy

- Examining the role of different strategies of operation
- Examining the role of storage system in enhancing resiliency
- Investigating the role of storage systems in the reliability improvement

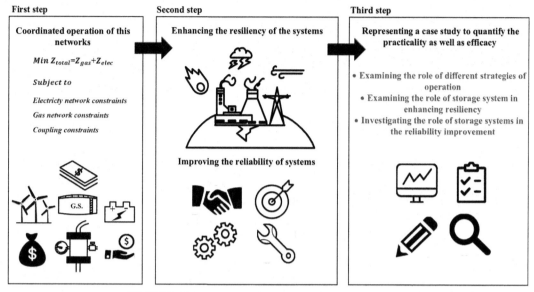

FIGURE 16.1   Overview of the main steps in this chapter.

- Presenting an approach to improve the reliability in the coordinated operation of gas and electricity networks (second step);
- Presenting an algorithm to enhance the resiliency in cooptimization of these networks (second step);

3. Model description and formulation

In this section, the main layers of gas and electricity networks are introduced and mathematical models for different components are presented as follows:

3.1 Gas network layers

Fig. 16.2 depicts the main layers of gas networks, including gas terminals, gas transmission systems, gas distribution systems. The natural gas is supplied through the terminals, such as liquefied natural gas carriers, onshore/offshore gas fields, and interconnectors (i.e., importing gas within the pipelines from neighboring countries). In transmission gas network, the natural gas is transmitted within the pipelines in high pressures (e.g., from 38 to 95 bars [33]). By using the high pressures, more compressed gas can be transmitted through long distances, which also reduces the cost of pipelines. The reason is that smaller diameter is necessary to transmit more compressed gas, which leads to the cost reduction. It should be noted that heavy industrial demands and power plants are directly supplied by the gas transmission systems. Finally, in order to deliver the natural gas to residential, commercial, and small industrial plants, the pressure is reduced through the regulator stations (e.g., to lower than

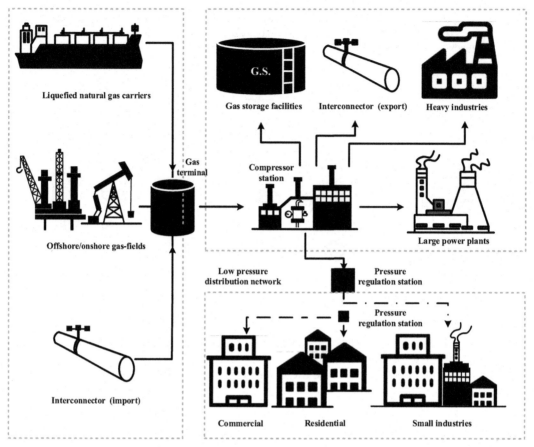

FIGURE 16.2   The main layers of supplying gas.

7 bars), and these consumers are supplied through the distribution systems. The performance and mathematical formulation to simulate the main components in this system, including the gas pipelines, compressors, storages, and regulators, are illustrated as follows.

3.1.1 Gas pipelines

The pipelines are in charge of transmitting natural gas from gas terminals to the final consumers. The main assumptions in the modeling gas flow through the pipeline include: (a) the pipelines are horizontal, (b) the temperature and speed are constant, (c) the proportion of the pipeline's length to its diameter is high, and (d) the change of pipelines' cross-sectional area is negligible [29]. The gas flow within the pipelines is calculated through the Panhandle, Continuity, and Momentum equations by applying the finite deference approach [30]. The gas flow within the pipelines for high-pressure system is indicted in (1), which applied to simulate the compressibility and loss pressure of gas flow within the pipelines.

$$\frac{\pi_{p,t}^{\text{out}} - \pi_{p,t}^{\text{in}}}{\text{Lenght}_p} = -\frac{2 \cdot Z \cdot R \cdot T \cdot \gamma \cdot (\rho_n)^2 \cdot \left(Q_{p,t}^{\text{pipe}}\right) \left| Q_{p,t}^{\text{pipe,}} \right|}{(A_p)^2 \cdot \text{Diameter}_p \cdot \pi_{p,t}^{av}} \quad \forall p, \forall t \qquad (16.1)$$

Where,

t Set of periods

p Set of pipelines

$\pi_{p,t}^{\text{out}}$ Output pressure of pipelines (bar)

$\pi_{p,t}^{\text{in}}$ Input pressure of pipelines (bar)

$\pi_{p,t}^{av}$ Average pressure within pipelines (bar)

Lenght$_p$ Length of pipelines (km)

Diameter$_p$ Diameter of pipelines (mm)

A_p Cross-sectional area of pipelines (m^2)

$Q_{p,t}^{\text{pipe}}$ Gas flow through pipelines (mm^3)

γ Friction factor in pipelines.

ρ_n Natural gas density in normal condition (kg/m^3)

Z Gas compressibility factor (0.95)

R Gas constant for natural gas (518 j/kg.K)

Moreover, the stored gas within the pipelines, which is called linepack, should be taken into consideration to simulate dynamic of gas flow [31]. The change of linepack equates to the difference between inlet and outlet flow of the pipeline (law of conservation of mass) that is simulated in (2). As the speed of gas flow within the pipelines is low, in the coordinated operation, the amount of linepack can be used to deal with the intermittency of renewable energies through gas-fired power plants [32].

$$LP_{p,t} = LP_{p,t-1} + \int_{t-1}^{t} \left(Q_{p,t}^{\text{pipe,in}} \cdot \eta p - Q_{p,t}^{\text{pipe,out}} / \eta p \right) dt \cdot \forall p, \forall t \qquad (16.2)$$

Where,

$LP_{p,t}$ Linepack (mm^3)

$Q_{p,t}^{\text{pipe,in}}$ Input gas flow (mm^3)

$Q_{p,t}^{\text{pipe,out}}$ Output gas flow (mm^3)

3.1.2 Gas compressors

A compressor is used for boosting the pressure between two points in the gas networks and compensating for pressure drop within the pipelines [7]. There are two main types of prime movers to derive the compressors, which are called (a) gas-driven prime movers and (b) electricity-driven prime movers [13]. Eqs. (16.3)–(16.5) determine the operation limits of compressors.

$$P_{c,t}^{\text{comp}} = \frac{\beta_{\text{comp}} \cdot \pi_{p,t}^{\text{in}} \cdot Q_{c,t}^{\text{comp}}}{\eta_{\text{comp}}} \cdot \left[\left(\frac{\pi_{p,t}^{\text{out}}}{\pi_{p,t}^{\text{in}}} \right)^{\frac{1}{\beta_{\text{comp}}}} - 1 \right] \cdot \forall c, \forall t \qquad (16.3)$$

$$1 \leq \frac{\pi_{p,t}^{\text{out}}}{\pi_{p,t}^{\text{in}}} \leq \text{PR}^{\text{max}} \cdot \forall c, \forall t \tag{16.4}$$

$$Q_{c,t}^{\text{comp}} \leq Q_{c,t}^{\text{comp,max}} \cdot \forall c, \forall t \tag{16.5}$$

$$P_{c,t}^{\text{comp}} \leq P_{c,t}^{\text{comp,max}} \cdot \forall c, \forall t \tag{16.6}$$

Where,

c Set of compressors

$P_{c,t}^{\text{comp}}$ Power consumption of compressors' prime mover (MW)

$Q_{c,t}^{\text{comp}}$ Gas flow through compressors (mm^3)

η_{comp} Efficiency factor of compressors

β_{comp} Polytrophic exponent of a gas compressors (4.7 MJ/cm)

PR^{max} Pressure ratio of compressor (1.5)

3.1.3 Gas storage facilities

Gas storages offer extra gas capacity by charging during off-peak hours of the gas demand and discharging during peak hours. These systems can work in different timeframe, such as seasonal or day-to-day time horizons [33]. The main technical limitations of the storage systems include the maximum capacity, the rate of charging and discharging, the efficiency of charging and discharging, and the number of times that these systems could be charged and discharged (Eqs. 16.7–16.11).

$$Gl_{q,t}^{\text{min}} \leq Gl_{q,t} \leq Gl_{q,t}^{\text{max}} \cdot \forall q, \forall t \tag{16.7}$$

$$Gl_{q,t} = Gl_{q,t-1} + \left(Q_{q,t}^{wd} \cdot \eta^{\text{ch}} - Q_{q,t}^{inj} / \eta^{\text{dch}} \right) \cdot \forall q, \forall t \tag{16.8}$$

$$0 \leq Q_{q,t}^{wd} \leq Q_{q,t}^{\text{wd,max}} \cdot I^{\text{ch}} \cdot \forall q, \forall t \tag{16.9}$$

$$0 \leq Q_{q,t}^{inj} \leq Q_{q,t}^{\text{inj,max}} \cdot I^{\text{dch}} \cdot \forall q, \forall t \tag{16.10}$$

$$I^{\text{ch}} + I^{\text{dch}} \leq N^{\text{max}} \cdot \forall q, \forall t \tag{16.11}$$

Where,

q Set of gas storages

$Gl_{q,t}$ Gas level in gas storage systems (mm^3)

$Gl_{q,t}^{\text{max/min}}$ Maximum/minimum gas level in gas storage systems (mm^3)

$Q_{q,t}^{wd/inj}$ Withdrawal/injected gas (mm^3)

$Q_{q,t}^{wd/inj,max}$ Maximum withdrawal/injected gas (mm^3)

$\eta^{ch/dch}$ Charging/discharging efficiency

$I^{ch/dch}$ State of charging/discharging

N^{max} Maximum number of charge/discharge during operation period

For each gas storage system, the maximum withdrawal and injected natural gas change in proportion to its gas level. Eqs.(16.12) and (16.13) indicate the maximum gas withdrawal and the maximum gas injection in which K_s, K_s^1, and K_s^2 depend on the type and dimension of these systems [30].

$$Q_{q,t}^{wd,max} = K_q \cdot \sqrt{Gl_{q,t}} \cdot \forall q, \forall t \tag{16.12}$$

$$Q_{q,t}^{wd,max} = -K_q^1 \cdot \sqrt{\frac{1}{Gl_{q,t} + Gl_{q,t}^{cush}} + K_q^2} \cdot \forall q, \forall t \tag{16.13}$$

Where,

K_q Coefficient of maximum withdrawal gas from gas storage systems

$K_q^{1/2}$ Coefficient of maximum injection gas into gas storage systems

$Gl_{q,t}^{cush}$ Cushion gas capacity of gas storage systems (mm^3)

3.1.4 Power-to-gas system

As stated earlier, power-to-gas systems use the curtailed wind or solar energy to produce hydrogen by implementing electrolysis process [34]. The produced amount of hydrogen is injected into the gas network to facilitate transmitting the natural gas. These systems facilitate supply—demand balance as well as produce low carbon gas. The operation limits of power-to-gas systems are demonstrated in (16.14)—(16.16), including the maximum allowance electrolyzer capacity for the amount of electricity power for hydrogen production and the maximum allowance hydrogen injection into the gas pipelines [35].

$$P_{d,t}^{p2g} = v^{H2} \cdot \eta^{p2g} \cdot Q_{d,t}^{p2g} \cdot \forall d, \forall t \tag{16.14}$$

$$P_{d,t}^{p2g} \leq P_{d,t}^{p2g,max} \cdot \forall d, \forall t \tag{16.15}$$

$$Q_{d,t}^{p2g} \leq \phi \cdot Q_{d,t}^{p2g,max} \cdot \forall d, \forall t \tag{16.16}$$

Where,

d Set of power-to-gas systems

$P_{d,t}^{p2g}$ Injected electric power to electrolyzers (MW)

$Q_{d,t}^{p2g}$ Injected hydrogen from electrolyzers to gas nodes (mm^3)

v^{H2} Constant for convert hydrogen energy to gas volume (90.9 cm/MWh)

η^{p2g} Efficiency of electrolyzers

ϕ Maximum allowance of hydrogen injection to natural gas system

3.1.5 Gas regulators

As low pressure is necessary for residential, commercial, and small-scaled industrial consumptions, the gas regulators are located between different pressure levels to control the pressure automatically [33]. When the pressure decreases in the low pressure system, it is indicated that gas demand increased and a more amount of natural gas is required. Therefore, the gas regulators must open their valves until the pressure increment shows the demand satisfaction. An ideal pressure regulator should supply the downstream demand without pressure drop in upstream network. However, in reality in gas regulator, pressure drop is always observable in upstream network, which should be controlled.

3.2 Electricity network layers

Fig. 16.3 depicts the main layers of electricity networks, including suppliers, electricity transmission systems, electricity distribution systems. The electricity power is supplied through the power plants, such as thermal power plants (e.g., coal power plants, gas-fired power plants, etc.), renewable energy power plants (e.g., wind farms, solar power plants, etc.), and interconnectors (i.e., supplying electricity through the transmission lines from neighboring regions). In transmission network, the electricity power is transmitted within the transmission lines in high-voltage level (e.g., 110 kV and above). Using a high voltage provides the ability to deliver a large amount of electricity power through long distances. This is due to the fact that using a higher voltage can reduce the amount of power loss along the transmission (i.e., the higher voltage, the lower current, which reduces power loss). Using a higher voltage level (i.e., a lower current level) reduces resistance in the conductors that means thin cables can be economically used to transmit power in long distance. It should

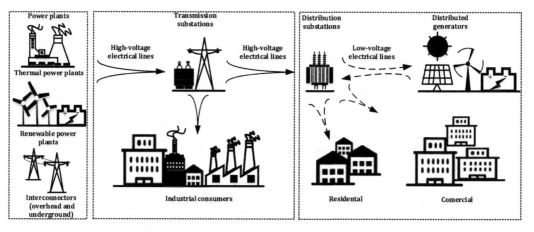

FIGURE 16.3   The main layers of supplying electricity.

be noted that heavy industrials are directly supplied by electricity transmission systems. Finally, in order to deliver the electricity to residential, commercial, and small industrial buildings, the voltage is reduced through the distribution substations (e.g., to 35 kV or less) and these consumers are supplied through the distribution systems.

Energy storage systems are also devised in the electricity network to provide both downward and upward flexibility. These systems are charged when there is a lower amount of demand and charged in contrary case that facilitates supply–demand balance. The storage systems can be used to deal with the intermittency and variability of renewable energy systems. Therefore, the operation of gas network is less affected by the renewable energies, which makes the linepack management less challenging. The operation limits of energy storage system are indicated in (16.17)–(16.20).

$$SOC_{r,t} = SOC_{r,t} + \left(P_{r,t}^{ch}.\eta_{ch} - P_{r,t}^{dch}/\eta_{dch}\right).\forall r, \forall t \tag{16.17}$$

$$SOC_r^{min} \leq SOC_{r,t} \leq .SOC_r^{max}.\forall r, \forall t \tag{16.18}$$

$$P_r^{ch,min} \leq P_{r,t}^{ch} \leq P_r^{ch,max}.\forall r, \forall t \tag{16.19}$$

$$P_{r,t}^{dch,min} \leq P_{r,t}^{dch} \leq P_r^{dch,max}.\forall r, \forall t \tag{16.20}$$

Where,

r Set of energy storage systems

$SOC_{r,t}$ State of charge of energy storage systems (MWh)

$P_{r,t}^{ch/dch}$ Charging/discharging power of energy storage systems (MW)

$P_r^{ch,max/min}$ Maximum/minimum charging power (MW)

$P_r^{dch,max/min}$ Maximum/minimum discharging power (MW)

$\eta_{ch/dch}$ Efficiency of charge/discharge

3.3 Coordinated operation of gas and electricity networks

A mixed-integer nonlinear model is formulated for optimizing coordinated operation of high-pressure gas network and transmission electricity networks. The problem consists of two subproblems, including (1) optimizing gas network operation, which is nonlinear programming due to gas flow and compressors equations [30], and (2) optimizing operation of electricity network, which mixed-integer linear programming due to binary variables in generation unit commitment [6]. In the gas network, constraints such as gas flow balance, gas supply limits for the terminals, linepack limits, pressure limits, gas compressors operation limits, gas storage operation limits, and power-to-gas systems operation limits are considered. In the electricity network, a network-constrained unit commitment (NCUC) is take into consideration, which takes into account power flow balance, spinning reserve requirements, energy storage systems operation limits, wind generators, and characteristics of thermal generating units, such as ramp up/down, minimum uptime/downtime, and minimum/maximum generation of thermal units. The coordinated operation of gas and electricity networks is optimized through iterative and integrated strategies of operation, which is presented as follows:

3.3.1 Iterative strategy of operation

In subsection, the iterative strategy has been introduced to solve the coordinated operation of natural gas and electricity networks [7]. As depicted in Fig. 16.4, in the first stage, the electricity network operation is optimized by electricity system operator, without accounting the gas network operation limits. The objective function of optimizing operation of electricity network consists of the startup/shutdown cost, cost of power generation, cost of wind curtailment, and cost of load shedding. This objective function is optimized subject to the proposed constraints in Subsection 3.3. According to the output power of gas-fired power plants, the required natural gas for electricity network is calculated using (16.21) and submitted to natural gas system operator. Then, the gas network operation is optimized subject to the constraints in Subsection 3.3 and constraints violation and load shedding are examined. The objective function of optimizing operation of gas network consists of three terms, including the cost of gas injection, cost of linepack management, and cost of gas shedding. If there is either constraint violation or gas shedding due to the excess of gas requirements of the gas-fired power plants, the power outputs of those generators are limited. It should be noticed that this algorithm is mainly used to solve the coordinated operation of these networks in a two-level manner (i.e., reducing complexity of the problem).

$$Pg_{g,t} = \psi \cdot Hv \cdot Q_{g,t}^{gen} \cdot \forall g, \forall t \tag{16.21}$$

Where,

$Pg_{g,t}$ Electrical power of thermal units (MW)

$Q_{g,t}^{gen}$ Required gas for gas-fired power plants (mm^3)

ψ Thermal efficiency of gas generator

Hv Gas heating value

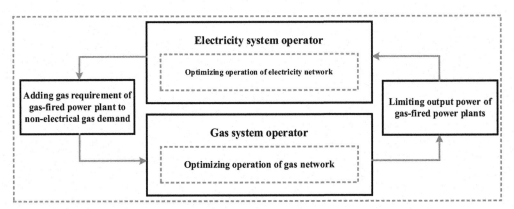

FIGURE 16.4   Iterative strategy for coordinated operation of gas and electricity networks.

3.3.2 *Integrated strategy of operation*

In the integrated strategy, the objective function is equal to sum of gas and electricity networks' objective functions (16.22). This problem is optimized subject to the proposed constraints in Subsection 3.3. Furthermore, a coupling constraint is also taken into account, which couples gas and electricity networks (i.e., taking (16.21) into account as constraint). The objective function and constraints of coordinated operation of gas and electricity networks are depicted in Fig. 16.5.

$$Z_{total} = Z_{gas} + Z_{elec} \tag{16.22}$$

Where,

Z_{total} Total objective function for coordinated operation of gas and electricity networks ($)
Z_{elec} Objective function for optimal operation of electricity network ($)
Z_{gas} Objective function for optimal operation of gas network ($)

4. Solution methodology

In this section, resiliency measures and reliability indices in the related literature are represented and categorized. Finally, an algorithm is proposed to enhance the resiliency and an approach is developed to make these networks more reliable.

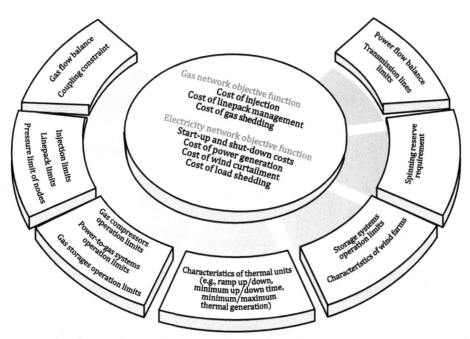

FIGURE 16.5 Objective function and constraints of coordinated operation of gas and electricity networks.

4.1 Reliability analysis

The reliability assessment is an important research area in the power system modeling, which has been advanced, continuously. Billinton and Hossain (1984) introduced the foundation of reliability analysis in the power system by evaluating power system components' outages. As stated in Subsection 3.2, generation, transmission, and distribution systems are three main layers of the power system. As depicted in Fig. 16.6, the reliability assessment is also conducted in three hierarchical level, including generation reliability analysis (first level), generation and transmission reliability analysis (second level), and generation, transmission, and distribution reliability analysis (third layer) [36,37].

It should be noted that the study of reliability in the power system is highly dependent on determining indicators and its calculation process. Different indicator has been used in the previous studies to evaluate the reliability in the power system, which can be mainly classified into power system generation indicators and distribution indicators [38]. Table 16.1 shows the indicators and its classification for analyzing the reliability in power system.

There are growing number of studies in which the role of energy storage systems in the reliability of power network is investigated [38, 43]. The reason is that these systems can efficiently mitigate the variability and intermittency in renewable energies' output power and facilitate the supply—demand balance. Therefore, operating energy storage systems in electricity network prevents transmitting the intermittency in renewable energy's output power into the gas network through the gas-fired power plants and makes the gas system operation less challenging. Therefore, in this section, an approach is presented to enhance the reliability in coordinated operation of gas and electricity networks by taking into account the role of energy storage systems. The main steps of this algorithm are presented as follows:

Step 1: Determining distribution function for repair and failure of each component, such as transmission lines and generating units. For component i, the probability distribution function is represented as follows:

$$P(\text{available} \rightarrow \text{available}) = \frac{\mu_i}{\mu_i + \lambda_i} + \frac{\lambda_i}{\mu_i + \lambda_i} \cdot e^{-(\mu_i + \lambda_i) \cdot (t - t_0)} \tag{16.23}$$

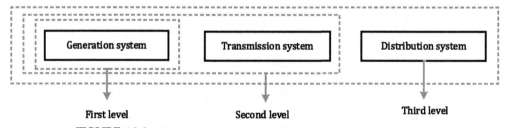

FIGURE 16.6   Hierarchical level of reliability assessment in power system.

TABLE 16.1 Reliability indicators [38].

Power system	Reliability indicators
Power generation	Frequency (F) [36,39]
	Average duration (D) [39]
	Unavailability (U) [36,39]
	Time to failure (λ) [39]
	Time to repair (μ) [36,39]
Distribution	Failure rate (λ) [40]
	Annual outage time (UOT) [40,41]
	Outage time (r) [40,42]
	System average interruption frequency index ($SAIFI$) [40,41, 42]
	System average interruption duration index ($SAIDI$) [40,42, 43]
	Average service availability index ($ASAI$) [40,41, 42]

$$P(\text{available} \rightarrow \text{unavailable}) = \frac{\lambda_i}{\mu_i + \lambda_i} \cdot \left(1 - e^{-(\mu_i + \lambda_i) \cdot (t - t_0)}\right) \qquad (16.24)$$

$$P(\text{available} \rightarrow \text{unavailable}) = \frac{\lambda_i}{\mu_i + \lambda_i} \cdot \left(1 - e^{-(\mu_i + \lambda_i) \cdot (t - t_0)}\right) \qquad (16.25)$$

$$P(\text{unavailable} \rightarrow \text{unavailable}) = \frac{\lambda_i}{\mu_i + \lambda_i} + \frac{\mu_i}{\mu_i + \lambda_i} \, e^{-(\mu_i + \lambda_i) \cdot (t - t_0)} \qquad (16.26)$$

Step 2: Using the probability distribution functions to generate scenario for each component during operation period by applying Monte Carlo simulation. For each component, a vector is generated in which the availability or unavailability for each hour of operation is indicated (i.e., when it is equal to 1/0, the component is available/unavailable).
Step 3: A two-stage stochastic model for coordinated operation of this networks is developed by considering the generated scenarios in the last step by (16.27), which is subject to other gas and electricity networks' constraints [15,44]. In this step, multiplying the generated scenarios (16.28)–(16.29) by maximum/minimum generating power of generating units and maximum/minimum power through the transmission line, the state of these components is indicated (i.e., availability or unavailability) (16.30)-(16.31).

$$Z_{\text{total}} = \sum_s (Z_{\text{gas}} + Z_{\text{elec}}) \qquad (16.27)$$

$$\beta_{g,t} = \begin{cases} 1 \text{if generating unit is available} \\ 0 \quad \text{O.W.} \end{cases} \tag{16.28}$$

$$\beta'_{l,t} = \begin{cases} 1 \quad \text{if transmission line is available} \\ 0 \quad \text{O.W.} \end{cases} \tag{16.29}$$

$$\beta_{g,t}.P_g^{\text{gen,min}} \leq P_{g,t}^{\text{gen}} \leq \beta_{g,t}.P_g^{\text{gen,max}}.\forall g, \forall t \tag{16.30}$$

$$\beta'_{l,t}.P_l^{\text{line,min}} \leq P_{l,t}^{\text{line}} \leq \beta'_{r,t}.P_l^{\text{line,max}}.\forall l, \forall t \tag{16.31}$$

Where,

    s Set of scenarios

    g Set of generating units

    l Set of transmission lines

    $P_{g,t}^{\text{gen}}$ Power output of generating units (MW)

    $P_g^{\text{gen,max/min}}$ Maximum/minimum power output of generating units (MW)

    $P_{l,t}^{\text{line}}$ Power through the transmission lines (MW)

    $P_l^{\text{line,max/min}}$ Maximum/minimum power through the transmission lines (MW)

Step 4: Comparing the cost of operation and the amount of load shedding for the stochastic and original model considering the occurrence of each generated scenario.

4.2 Resiliency analysis

Natural disasters as well as harsh climate may result in wide outages in the energy systems, such as natural gas and electricity networks. In the energy system, resiliency is defined as an ability of the system to supply affordable amount of demand and reduce wide outage in the case of high-impact and low-probability events [9]. These events, such as weather-related events, are commonly neglected in the reliability analysis that could lead to considerable number of outages or even blackout in the energy system [45]. Therefore, taking into account resiliency besides the reliability consideration makes the energy system modeling more realistic.

Investigating resiliency can be categorized according to the time event, including (a) resilience planning, (b) resilience response, and (c) resilience restoration [46]. The resilience planning is conducted, including the long-term measures to enhance the resiliency of the system, such resilience generation and transmission planning. However, resilience based response and restoration, such as emergency load curtailment include short-term activities. The classification of these measures in terms of occurrence time is depicted in Fig. 16.7.

The hardening and resilience investment (i.e., resilience planning) is implemented by modeling the energy system's damages, outages, and restoration durations. For this purpose, a set of parameters and data are required that are commonly obtained through the statistical data, data fitting models, or simulation-based models. The required data and parameters can be categorized into the energy system data (e.g., customer's data, system topology, and

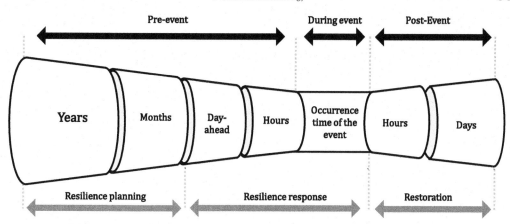

FIGURE 16.7 Classification of resiliency measures according to occurrence time.

transmission system availability) and the environmental data (e.g., topological data, geological data, and climatic data). Furthermore, the data fitting methods are commonly applied to estimate the damages to the energy systems (e.g., generalized linear model, generalized additive model [47]), duration of energy outage (e.g., accelerated failure time [48,53]), and the natural disaster impacts on the energy systems (e.g., fuzzy interface system [14], multivariate adaptive regression splines method [47]). Besides, with the lack of data, the simulation-based approaches, such as Monte Carlo simulation, are devised to simulate the outages and weather conditions [49].

The resilience response includes the corrective actions and resiliency activities applied by utilities to cope with severity of the disasters, such as advanced protection schemes [50], islanding schemes [51], preparation of sufficient emergency generation units [51], and black start capability [52].

The energy system restoration techniques can be categorized into conventional restoration strategies and using distributed generators or microgrids for restoration. The conventional restoration consists of three main steps, including preparation, system restoration, and demand restoration. It should be noticed that the load restoration needs an appropriate strategy to initially restore critical loads and minimize the unserved load. Different approaches for load restoration can be applied, such as fuzzy logic [56], heuristic approaches [54], and mathematical programming [57]. On the other hand, during and after these contingencies, microgrid can be efficiently used to manage distributed generators (e.g., microturbines, combined heat and power systems, fuel cells, small scale photovoltaic and wind systems with storage). The distributed generators aid the conventional load restoration [58] and can be used in island mode for supplying demand during contingency [59].

During the harsh conditions, resiliency of an energy system, such as electricity network, cannot solely depend on its own infrastructure [60]. For instance, the operation of gas-fired power plants in the electricity network is related to natural gas network and makes these networks interdependence. On the other hand, the storage systems capability can be utilized in energy system restoration [55]. Therefore, a two-stage algorithm is presented in this section to enhance the resiliency of electricity and gas networks in the presence of energy storage systems. In the first stage of proposed algorithm, the worst-case scenarios for electricity transmission lines and gas pipelines are recognized and the role of energy storage systems to deal with the contingencies is analyzed in the second stage. The algorithm of this resiliency consideration algorithm consists of four main steps, which are summarized as follows:

Step 1: Determining binary variables to indicate outages in the electricity and gas networks' components (16.32)-(16.33). When the binary variable is equal to 1, it is shown that the component is out of service (e.g., due to natural distress).

$$\alpha_p = \begin{cases} 1 & \text{if gas pipeline } p \text{ is out of service} \\ 0 & \text{O.W.} \end{cases} \tag{16.32}$$

$$\alpha'_l = \begin{cases} 1 & \text{if electricity line } l \text{ is out of service} \\ 0 & \text{O.W.} \end{cases} \tag{16.33}$$

Step 2: Modifying and substituting some constraints in the coordinated operation of gas and electricity networks, including gas and electricity flow limitations through the pipelines and electrical lines (16.34)−(16.35).

$$-(1 - \alpha_{p,t}).Q_{p,t}^{\text{pipe,max}} \le Q_{p,t}^{\text{pipe}} \le (1 - \alpha_{p,t}).Q_{p,t}^{\text{pipe,max}}. \forall p, \forall t \tag{16.34}$$

$$-\left(1 - \alpha'_{l,t}\right).Pl_l^{\max} \le Pl_{l,t} \le \left(1 - \alpha'_{l,t}\right).Pl_l^{\max}. \forall l, \forall t \tag{16.35}$$

Where,

l Set of transmission lines

$Q_{p,t}^{\text{pipe}}$ Gas through pipelines (mm^3)

$Q_{p,t}^{\text{pipe,max}}$ Maximum gas through pipelines (mm^3)

$Pl_{l,t}$ Power through transmission lines (MW)

$Pl_l^{\max}$ Maximum power through transmission lines (MW)

Step 3: Optimizing the objective function (16.36), subject to other gas and electricity networks' constraints, provides the worst-case scenarios for outage of gas pipelines and electricity transmission lines. This measure maximizes unfulfilled demand during operation period. In this problem, the number of outages should be limited, due to the fact that it is not probable for all the components to be out of service simultaneously (16.37). This step is repeated for analyzing (i.e., $N^{\max}$) and obtaining different scenarios.

$$\max_t \min \left(\omega. \left(\sum_n D_{n,t}^{\text{gas}} - \sum_n Q_{n,t}^{\text{sup}} + \sum_n Q_{p \in (n,n'),t}^{\text{pipe}} \right) + (1 - \omega). \left(\sum_b D_{b,t}^{elec} - \sum_b Pt_{b,t} + \sum_b Pw_{b,t} - \sum_b Pl_{l \in (b,b'),t} \right) \right) \tag{16.36}$$

$$\sum_p \alpha_p + \sum_l \alpha'_l \leq N^{\max} \tag{16.37}$$

Where,

$N^{\max}$ Maximum number of outages

ω Weight coefficient ($\omega \in [0,1]$)

Step 4: A two-stage stochastic model for coordinated operation of these networks is developed considering the worst-case scenarios in the last step by (16.38) subject to other constraints of gas and electricity networks [15,44]. Furthermore, in this step, two binary variables are also defined to investigate the role of storage systems in improving the resiliency of the system and mitigating the outages (16.39)–(16.40). When the binary variables are equal to 1, the storage systems are assumed to be installed in the potential location. The operation limits of energy and gas storage systems are rewritten and substituted as indicated in (16.41)–(16.44) and (16.45)–(16.48) respectively. The maximum number of storage systems that can be assumed is also limited in (16.49)–(16.50).

$$Z_{\text{total}} = \sum_s (Z_{\text{gas}} + Z_{\text{elec}}) \tag{16.38}$$

$$\alpha''_r = \begin{cases} 1 \text{ if electricity storage system is assumed in candidate location } r \\ 0 \text{ O.W.} \end{cases} \tag{16.39}$$

$$\alpha'''_q = \begin{cases} 1 \text{ if gas storage system is assumed in candidate location } q \\ 0 \text{ O.W.} \end{cases} \tag{16.40}$$

$$SOC_{r,t} = SOC_{r,t} + \left(P^{\text{ch}}_{r,t} \cdot \eta_{\text{ch}} - P^{\text{dch}}_{r,t} / \eta_{\text{dch}}\right) \cdot \forall r, \forall t \tag{16.41}$$

$$\alpha''_r \cdot SOC^{\min}_r \leq SOC_{r,t} \leq \alpha''_r \cdot SOC^{\max}_r \cdot \forall r, \forall t \tag{16.42}$$

$$\alpha''_r \cdot P^{\text{ch,min}}_r \leq P^{\text{ch}}_{r,t} \leq \alpha''_r \cdot P^{\text{ch,max}}_r \cdot \forall r, \forall t \tag{16.43}$$

$$\alpha''_r \cdot P^{\text{dch,min}}_{r,t} \leq P^{\text{dch}}_{r,t} \leq \alpha''_r \cdot P^{\text{dch,max}}_{r,t} \cdot \forall r, \forall t \tag{16.44}$$

$$Gl_{q,t} = Gl_{q,t-1} + \left(Q^{wd}_{q,t} - Q^{inj}_{q,t}\right) \cdot \forall q, \forall t \tag{16.45}$$

$$\alpha'''_q \cdot Gl^{\max}_{q,t} \leq Gl_{q,t} \leq \alpha'''_q \cdot Gl^{\max}_{q,t} \cdot \forall q, \forall t \tag{16.46}$$

$$0 \leq Q_{q,t}^{wd} \leq \alpha_q''' \cdot Q_{q,t}^{wd,max} \cdot \forall q, \forall t \tag{16.47}$$

$$0 \leq Q_{q,t}^{inj} \leq \alpha_q''' \cdot Q_{q,t}^{inj,max} \cdot \forall q, \forall t \tag{16.48}$$

$$\sum_r \alpha_{r,t}'' \leq N^{max} \tag{16.49}$$

$$\sum_q \alpha_{r,t}''' \leq N^{max} \tag{16.50}$$

Where,

    s Set of scenarios

    q Set of gas storage systems

    $SOC_{r,t}$ State of charge of energy storage systems (MWh)

    $P_{r,t}^{ch/dch}$ Charging/discharging power of energy storage systems (MW)

    $P_r^{ch,max/min}$ Maximum/minimum charging power (MW)

    $P_r^{dch,max/min}$ Maximum/minimum discharging power (MW)

    $\eta_{ch/dch}$ Efficiency of charge/discharge

    $Gl_{q,t}$ Gas level in gas storage systems (mm^3)

    $Gl_{q,t}^{max}$ Maximum gas level in gas storage systems (mm^3)

    $Q_{q,t}^{wd/inj}$ Withdrawal/injected gas (mm^3)

    N^{max} Maximum number of candidate gas and electricity storage systems

Step 5: Comparing the cost of operation and the amount of load shedding for the stochastic and original model considering the occurrence of the worst-case scenarios.

5. Case study

In this section, a modified six-bus and six-node system is introduced to optimize the coordinated operation of gas and electricity networks, and the role of gas and electricity storage systems is investigated (see Ref. [62]). The electricity system consists of six buses, seven transmission lines, three thermal power plants, a wind farm, and two energy storage systems. The installed capacity of wind farm is equal to 20 MW, and the characteristics of other generator units are indicated in Table 16.2. The characteristics of the energy storage systems are also demonstrated in Table 16.3.

TABLE 16.2 Characteristics of generating units.

No.	Cost coefficient ($/MW)	Minimum stable output power (MW)	Maximum stable output power (MW)	Ramp-up/ down (MW/ hour)	Minimum uptime/ downtime (hour)	Startup cost ($)	Type
1	13.51	100	220	26	4	100	Gas
2	17.63	10	80	47	1	10	Coal
3	27.7	10	20	7	2	50	Biomass

TABLE 16.3 Characteristics of energy storage system.

Parameter	Value	Parameter	Value
Minimum state of charge (MWh)	1	Maximum state of charge (MWh)	5
Minimum charging/discharging power (MW)	0	Maximum charging/discharging power (MW)	1
Charging efficiency (%)	95	Discharging efficiency (%)	90

TABLE 16.4 Nodes data.

Node	Minimum injection (mcm)	Maximum injection (mcm)	Minimum pressure (bar)	Maximum pressure (bar)	Gas demand (mcm)
1	–	–	17.23	25	2.75
2	–	–	18.27	25	2.75
3	0	2	18.62	25	–
4	–	–	18.96	25	2.75
5	0	2	19.65	25	–
6	–	–	10.34	25	2.75

TABLE 16.5 Characteristics of gas storage system.

Parameter	Value	Parameter	Value
Minimum gas level (mcm)	0.01	Maximum gas level (mcm)	0.1
Minimum charging/discharging gas (mcm)	0	Maximum charging/discharging gas (mcm)	0.01
Charging efficiency (%)	0.9	Discharging efficiency (%)	0.9

On the other hand, the gas network consists of six nodes, two gas terminals, six pipelines, two gas storage systems. Table 16.4 indicates the gas demand, limitation of pressure, and injection through each node. The characteristics of gas storage systems are also represented in Table 16.5.

The gas and electricity networks topology is depicted in Fig. 16.8, in which there is a flexible gas-fired power plant that is supplied through the gas network. The costs of gas shedding and load shedding are assumed as 11.10 m$/mcm and 10,000 $/MW, respectively [30]. The required data about the availability and unavailability of components are based on data from North American Electric Reliability Corporation's website [61].

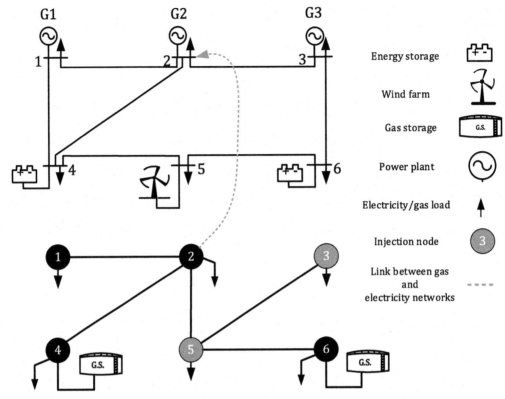

FIGURE 16.8   Six-bus and six-node gas and electricity networks.

6. Simulation results

In this section, the results of coordinated operation of gas and electricity networks are presented. For example, in Fig. 16.9, the output power of different generating units is presented. As mentioned previously, due to the characteristics of gas-fired power plant (G 2), such as fast ramping rate and short startup time, this power plant is mostly operated to deal with the intermittency in the output power of wind farm and the variability of demand. As depicted in the figure, the changes in the output power of the gas-fired power plant to deal with the variability are evident (from 11:00 to 13:00 and from 18:00 to 24:00). However, the generating power of other power plants is almost constant during operation period.

On the other hand, as the gas fired power plant is utilized to deal with the changes in the electricity network, the intermittency and variability in the electricity network transfer into the gas network through the gas-fired power plant. In the gas network, due to the low speed of natural gas through the pipelines, the linepack is commonly used to response to the rapid changes in demand. Therefore, the changes affect the linepack through the pipelines, although the gas and electricity storage systems can be employed to deal with the mentioned challenges. In Fig. 16.10, the changes in the sum of linepack through the pipelines with and without considering gas and electricity storage systems are presented.

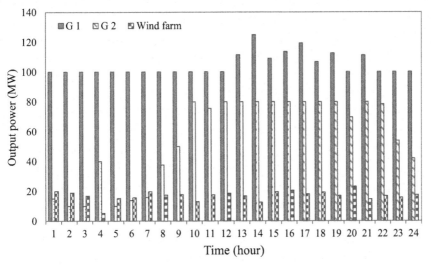

FIGURE 16.9 Generating power through the power generating units.

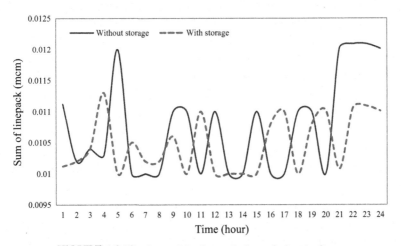

FIGURE 16.10 Sum of the linepack through the pipelines.

As demonstrated, there is an increment in the sum of the linepack through the pipelines from 4:00 to 6:00. This is due to the fact that the gas demand starts increasing during this period. Therefore, as it takes time to transport the gas from injection nodes to other nodes, the increment in the linepack initially occurred. On the other hand, the sum of linepack increment from 20:00 to 00:00 is due to the reason that the gas demand starts decreasing during this period. Therefore, it takes time to reduce the linepack due to the low speed of natural gas through the pipelines. The oscillations from 8:00 to 20:00 are also due to changes during this period. Besides, it is evident that when the gas and electricity storage systems are employed, the changes in the sum of linepack are reduced, which leads to operation cost reduction by preventing unnecessary injection through the terminals. This is due to the charging during off-peak hours of operation and discharging during peak hour of operation that facilitate linepack management through reducing the demand peak and oscillations.

6.1 Implementing reliability analysis

In order to analyze the reliability of coordinated operation of gas and electricity networks, initially, Monte Carlo simulation is applied to generate power generation unit outages (subsection 4.1, reliability analysis, and step 2). Afterward, the stochastic model for coordinated operation of these networks is optimized taking into account the generated scenarios (subsection 4.1, reliability analysis, and step 3). Finally, the total cost of operation and the amount of load and gas shadings are compared in the stochastic and deterministic models under different generated scenarios (subsection 4.1, reliability analysis, and step 4). In this regard, in Table 16.6, the generated scenarios of generating unit outages are shown.

In Table 16.7, the applied algorithm is quantified in which the decrement of total cost of operation and energy, not supply level, is demonstrated in the case of implementing the reliability analysis algorithm (i.e., the reliability increment). As depicted, in the normal operation, the operation cost of gas and electricity networks is less than when there is an outage. However, implementing reliability analysis reduces this cost in the case of probable outages through the stochastic model in which the problem is optimized by taking into account outage scenarios. Moreover, enhancing the energy system flexibility through the employing gas and electricity storage systems is illustrated in dealing with the probable outages. This is due to charging and discharging capacity of gas and electricity storage systems that facilitates supply—demand balance through reducing demand peak and oscillations.

6.2 Implementing resiliency analysis

In this subsection, the proposed resiliency analysis algorithm is implemented (subsection 4.2, resiliency analysis). After determining binary variables in the first step and modifying

TABLE 16.6   Outage scenario of generating units during operation period.

	Availability/unavailability of generating units											
Period	t = 1	t = 2	t = 3	t = 4	t = 5	t = 6	t = 7	t = 8	t = 9	t = 10	t = 11	t = 12
G 1	1	0	0	0	1	1	1	1	1	1	1	1
G 2	1	1	1	1	1	1	1	1	1	1	1	1
G 3	1	1	1	1	1	1	1	1	1	0	0	0
Period	t = 13	t = 14	t = 15	t = 16	t = 17	t = 18	t = 19	t = 20	t = 21	t = 22	t = 23	t = 24
G 1	1	1	1	1	1	1	1	1	1	1	1	1
G 2	1	1	0	0	0	0	1	1	1	1	1	1
G 3	1	1	1	1	1	1	1	1	1	1	0	0

TABLE 16.7 Total cost of operation and energy, not supply level, enhancement through the reliability analysis.

Outage scenario	Normal operation			Reliability consideration			Reliability consideration with storage		
	—	Outage of G 1	Outage of G 2	—	Outage of G 1	Outage of G 2	—	Outage of G 1	Outage of G 2
Energy not supply (MWh)	—	78.42	32.44	—	12.44	3.21	—	3.23	—
Cost of gas network (m$)	0.34	1.12	0.66	0.33	0.45	0.36	0.36	0.37	0.37
Cost of electricity network (m$)	1.43	1.45	1.41	1.43	1.44	1.42	1.45	1.46	1.45
Total cost (m$)	1.47	2.57	2.07	1.46	1.89	1.78	1.81	1.83	1.82

equations in the second step, in the third step, the worst-case scenarios for outage of electrical lines and gas pipelines are determined. After that, in the fourth step, the stochastic model is optimized through taking into account the worst-case scenarios, and the optimal locations of gas and electricity storage systems are specified. Finally, in the last step, the results of deterministic and stochastic models are compared under different scenarios.

In Table 16.8, the total cost of operation and energy, not supply level, are indicated with and without implementing resiliency analysis algorithm. The results show that implementing the resiliency analysis algorithm leads to reducing the operation cost as well as energy, not supply level, in the case of outage occurrence. For example, in the normal operation, since

TABLE 16.8 Total cost of operation and energy, not supply level, enhancement through the resiliency analysis.

	Normal operation		Resiliency consideration		Resiliency consideration with storage-I[a]		Resiliency consideration with storage-II[a]	
Outage scenario	—	Line 1–2 and pipeline 1–3	—	Line 1–2 and pipeline 1–3	—	Line 1–2 and pipeline 1–3	—	Line 1–2 and pipeline 1–3
Energy not supply (MWh)	—	221.11	—	54.22	—	49.72	—	41.86
Cost of gas network (m$)	0.34	11.45	0.52	6.34	0.51	5.68	0.46	4.72
Cost of electricity network (m$)	1.43	12.63	2.86	6.72	2.79	6.37	2.83	6.12
Total cost (m$)	1.77	24.08	3.38	13.06	3.30	12.05	3.29	10.84

[a]I: The locations of energy and gas storage system are based on the case study; II: The locations of these storage systems are determined based on the resiliency analysis algorithm.

there is no outage, the operation cost of gas and electricity networks is lower compared to when the resiliency algorithm is implemented. However, when the transmission line 1-2 and pipeline 1-3 are out of service (the worst-case scenarios), the resiliency analysis algorithm provides the solution with lower cost and energy not supply level. The reason is that, in this algorithm, the stochastic model of the gas and electricity networks is optimized by taking into account the worst-case scenarios. Therefore, in the case of outage occurrence, applying the resiliency approach provides better solutions. Moreover, according to the results of the resiliency analysis algorithm, the optimal location of gas storage systems is similar to the case study; however, the electricity storage systems are allocated in the buses 4 and 5. As indicted in the table, considering the gas and electricity storage systems either based on the case study or based on resiliency analysis algorithm reduces the total cost and energy not supply level in the case of the worst-case scenarios occurrences.

7. Summary

This chapter presents a study of coordinated operation strategy to optimize the operation of gas and electricity networks. To meet this purpose, firstly, a mathematical model of these networks was developed in which the characteristics of the natural gas within the pipelines are simulated through the Panhandle Equation. Furthermore, the mathematical model of gas and electricity storage systems was introduced to investigate the role of these systems in coordinated operation of gas and electricity networks. Besides, the reliability and resiliency analysis was explained and reviewed, and the algorithms were consequently presented to increase the reliability and resiliency in the coordinated operation of gas and electricity networks. Finally, in order to examine the role of gas and electricity storage systems in the reliability and resiliency enhancement of gas and electricity networks, a simple six-bus and six-node case study was introduced. The results show that taking into account the probable outage scenarios and the worst-case scenarios in the proposed algorithms boosts the reliability and resiliency in the coordinated operation of gas and electricity networks through the energy, not supply level, and cost reduction.

References

[1] M.S. Zantye, A. Arora, M.F. Hasan, Operational power plant scheduling with flexible carbon capture: a multi-stage stochastic optimization approach, Comput. Chem. Eng. 130 (2019) 106544.

[2] P. Agreement, United Nations Framework Convention on Climate Change, Paris Agreement, 2015 (In).

[3] H. Ameli, M.T. Ameli, S.H. Hosseinian, Multi-stage frequency control of a microgrid in the presence of renewable energy units, Elec. Power Compon. Syst. 45 (2) (2017a) 159–170.

[4] T. Adefarati, R. Bansal, Reliability, economic and environmental analysis of a microgrid system in the presence of renewable energy resources, Appl. Energy 236 (2019) 1089–1114.

[5] L. Hörnlein, The value of gas-fired power plants in markets with high shares of renewable energy: a real options application, Energy Econ. 81 (2019) 1078–1098.

[6] A. Ahmadi, A.E. Nezhad, B. Hredzak, Security-constrained unit commitment in presence of lithium-ion battery storage units using information-gap decision theory, IEEE Trans. Ind. Informat. 15 (1) (2019) 148–157.

[7] H. Ameli, M. Qadrdan, G. Strbac, Value of gas network infrastructure flexibility in supporting cost effective operation of power systems, Appl. Energy 202 (2017b) 571–580.

[8] G. Kovalev, L. Lebedeva, Reliability of Power Systems, Springer, 2019.

[9] N.M. Tabatabaei, S.N. Ravadanegh, N. Bizon, Power Systems Resilience: Modeling, Analysis and Practice, Springer, 2018.

[10] H.T. Nguyen, J.W. Muhs, M. Parvania, Assessing impacts of energy storage on resilience of distribution systems against hurricanes, J. Modern Power Syst. Clean Energy 7 (4) (2019) 731—740.

[11] C. He, X. Zhang, T. Liu, M. Shahidehpour, Coordination of interdependent electricity grid and natural gas network—a review, Curr. Sust.Renew. Energy Rep. 5 (1) (2018) 23—36.

[12] I.G. Sardou, M.E. Khodayar, M.T. Ameli, Coordinated operation of natural gas and electricity networks with microgrid aggregators, IEEE Trans. Smart Grid 9 (1) (2016) 199—210.

[13] H. Ameli, M. Qadrdan, G. Strbac, Coordinated operation strategies for natural gas and power systems in presence of gas-related flexibilities, Energy Syst. Integrat. 1 (1) (2019) 3—13.

[14] C. Liu, M. Shahidehpour, J. Wang, Coordinated scheduling of electricity and natural gas infrastructures with a transient model for natural gas flow, Chaos: An Interdiscip. J. Nonlinear Sci. 21 (2) (2011), 025102.

[15] M. Qadrdan, J. Wu, N. Jenkins, J. Ekanayake, Operating strategies for a GB integrated gas and electricity network considering the uncertainty in wind power forecasts, IEEE Trans. Sust. Energy 5 (1) (2014) 128—138.

[16] M. Qadrdan, H. Ameli, G. Strbac, N. Jenkins, Efficacy of options to address balancing challenges: integrated gas and electricity perspectives, Appl. Energy 190 (2017) 181—190.

[17] A. Alabdulwahab, A. Abusorrah, X. Zhang, M. Shahidehpour, Stochastic security-constrained scheduling of coordinated electricity and natural gas infrastructures, IEEE Sys. J. 11 (3) (2017) 1674—1683.

[18] C. He, L. Wu, T. Liu, M. Shahidehpour, Robust co-optimization scheduling of electricity and natural gas systems via ADMM, IEEE Trans. Sustain. Energy 8 (2) (2016) 658—670.

[19] S. Bolwiig, G. Bazbauers, A. Klitkou, P.D. Lund, A. Blumberga, D. Gravelsins, D. Blumberga, Review of modelling energy transitions pathways with application to energy system flexibility, Renew. Sustain. Energy Rev. 101 (2019) 440—452.

[20] G. Sun, S. Chen, Z. Wei, S. Chen, Multi-period integrated natural gas and electric power system probabilistic optimal power flow incorporating power-to-gas units, J. Modern Power Syst. Clean Energy 5 (3) (2017) 412—423.

[21] A. Antenucci, G. Sansavini, Extensive CO2 recycling in power systems via Power-to-Gas and network storage, Renew. Sustain. Energy Rev. 100 (2019) 33—43.

[22] C. Gu, C. Tang, Y. Xiang, D. Xie, Power-to-gas management using robust optimisation in integrated energy systems, Applied Energy 236 (2019) 681—689.

[23] C. Hao, H. Yang, W. Xu, C. Jiang, Robust optimization for improving resilience of integrated energy systems with electricity and natural gas infrastructures, J. Modern Power Syst. Clean Energy 6 (5) (2018) 1066—1078.

[24] M. Yan, Y. He, M. Shahidehpour, X. Ai, Z. Li, J. Wen, Coordinated regional-district operation of integrated energy systems for resilience enhancement in natural disasters, IEEE Trans. Smart Grid 10 (5) (2018) 4881—4892.

[25] S. Clegg, P. Mancarella, Integrated electricity-heat-gas modelling and assessment, with applications to the Great Britain system. Part II: transmission network analysis and low carbon technology and resilience case studies, Energy 184 (2019) 191—203.

[26] X. Zhang, M. Shahidehpour, A. Alabdulwahab, A. Abusorrah, Hourly electricity demand response in the stochastic day-ahead scheduling of coordinated electricity and natural gas networks, IEEE Trans. Power Syst. 31 (1) (2016) 592—601.

[27] W. Yu, J. Gong, S. Song, W. Huang, Y. Li, J. Zhang, X. Duan, Gas supply reliability analysis of a natural gas pipeline system considering the effects of underground gas storages, Appl. Energy 252 (2019) 113418.

[28] X. Zhang, J. Zhu, D. Yang, Y. Chen, W. Du, Research on operational optimization technology of regional integrated energy system considering operating cost and reliability, in: Paper Presented at the Proceedings of Purple Mountain Forum 2019-International Forum on Smart Grid Protection and Control, 2020.

[29] A.J. Osiadacz, C.s. centre, Simulation and Analysis of Gas Networks, 1987.

[30] M. Chaudry, N. Jenkins, G. Strbac, Multi-time period combined gas and electricity network optimisation, Elec. Power Syst. Res. 78 (7) (2008) 1265—1279.

[31] V. Shahbazbegian, H. Ameli, M.T. Ameli, G. Strbac, Stochastic optimization model for coordinated operation of natural gas and electricity networks, Comput. Chem. Eng. 142 (2020) 107060.

[32] H. Ameli, M. Qadrdan, G. Strbac, Coordinated operation of gas and electricity systems for flexibility study, Front. Energy Res. 8 (2020a) 120.

[33] M. Qadrdan, M. Abeysekera, J. Wu, N. Jenkins, B. Winter, The future of gas networks, in: The Future of Gas Networks, Springer, 2020, pp. 49—68.

[34] C. Gu, C. Tang, Y. Xiang, D. Xie, Power-to-gas management using robust optimisation in integrated energy systems, Appl. Energy 236 (2017) 681–689.

[35] H. Ameli, M. Qadrdan, G. Strbac, M.T. Ameli, Investing in flexibility in an integrated planning of natural gas and power systems, IET Energy Syst. Integrat. 2 (2) (2020b) 101–111.

[36] Y.-Y. Hong, L.-H. Lee, Reliability assessment of generation and transmission systems using fault-tree analysis, Energy Convers. Manag. 50 (11) (2009) 2810–2817.

[37] M. Shivaie, M.T. Ameli, M.S. Sepasian, P.D. Weinsier, V. Vahidinasab, A multistage framework for reliability-based distribution expansion planning considering distributed generations by a self-adaptive global-based harmony search algorithm, Reliab. Eng. Syst. Saf. 139 (2015) 68–81.

[38] Z. Zhou, P. Liu, Z. Li, E.N. Pistikopoulos, M.C. Georgiadis, Impacts of equipment off-design characteristics on the optimal design and operation of combined cooling, heating and power systems, Comput. Chem. Eng. 48 (2013) 40–47.

[39] K. Chaiamarit, S. Nuchprayoon, Modeling of renewable energy resources for generation reliability evaluation, Renew. Sustain. Energy Rev. 26 (2013) 34–41.

[40] C.L.T. Borges, An overview of reliability models and methods for distribution systems with renewable energy distributed generation, Renew. Sustain. Energy Rev. 16 (6) (2012) 4008–4015.

[41] L. Goel, Monte Carlo simulation-based reliability studies of a distribution test system, Elec. Power Syst. Res. 54 (1) (2000) 55–65.

[42] R. Billinton, R.N. Allan, Reliability evaluation of engineering systems, Plenum press, New York, 1992.

[43] F. Mohamad, J. Teh, C.-M. Lai, L.-R. Chen, Development of energy storage systems for power network reliability: a review, Energies 11 (9) (2018) 2278.

[44] P. Kall, S.W. Wallace, P. Kall, Stochastic Programming, Springer, 1994.

[45] M. Panteli, C. Pickering, S. Wilkinson, R. Dawson, P. Mancarella, Power system resilience to extreme weather: fragility modeling, probabilistic impact assessment, and adaptation measures, IEEE Trans. Power Syst. 32 (5) (2016) 3747–3757.

[46] M. Mahzarnia, M.P. Moghaddam, P.T. Baboli, P. Siano, A review of the measures to enhance power systems resilience, IEEE Syst. J. 14 (3) (2020) 4059–4070.

[47] S.D. Guikema, S.M. Quiring, S.R. Han, Prestorm estimation of hurricane damage to electric power distribution systems, Risk Anal. 30 (12) (2010) 1744–1752.

[48] R. Nateghi, S.D. Guikema, S.M. Quiring, Comparison and validation of statistical methods for predicting power outage durations in the event of hurricanes, Risk Anal. 31 (12) (2011) 1897–1906.

[49] Y. Wang, C. Chen, J. Wang, R. Baldick, Research on resilience of power systems under natural disasters—A review, IEEE Trans. Power Syst. 31 (2) (2015) 1604–1613.

[50] E. Yamangil, R. Bent, S. Backhaus, Resilient upgrade of electrical distribution grids, in: Paper Presented at the Twenty-Ninth AAAI Conference on Artificial Intelligence, 2015.

[51] Y. Fang, G. Sansavini, Optimizing power system investments and resilience against attacks, Reliab. Eng. Syst. Saf 159 (2017) 161–173.

[52] W. Sun, C.C. Liu, L. Zhang, Optimal generator start-up strategy for bulk power system restoration, IEEE Trans. Power Syst. 26 (3) (2010) 1357–1366.

[53] H. Liu, R.A. Davidson, T.V. Apanasovich, Statistical forecasting of electric power restoration times in hurricanes and ice storms, IEEE Trans. Power Syst. 22 (4) (2007) 2270–2279.

[54] A. Morelato, C. Monticelli, Heuristic search approach to distributionsystem restoration, IEEE Trans. Power Deliv. 4 (4) (1989) 2235–2241.

[55] C. Gouveia, C.L. Moreira, J.A.P. Lopes, D. Varajao, R.E. Araujo, Microgrid service restoration: the role of plugged-in electric vehicles, IEEE Indust. Elect. Magazine 7 (4) (2013) 26–41.

[56] S.-J. Lee, S.-I. Lim, B.-S. Ahn, Service restoration of primary distribution systems based on fuzzy evaluation of multi-criteria, IEEE Trans. Power Syst. 13 (3) (1998) 1156–1163.

[57] R. Pérez-Guerrero, G.T. Heydt, N.J. Jack, B.K. Keel, A.R. Castelhano, Optimal restoration of distribution systems using dynamic programming, IEEE Trans. Power Deliv. 23 (3) (2008) 1589–1596.

[58] H. Yang, Y. Zhang, Y. Ma, M. Zhou, X. Yang, Reliability evaluation of power systems in the presence of energy storage system as demand management resource, Int. J. Electr. Power Energy Syst 110 (2019) 1–10.

[59] N. Smith, R. McCann, Analysis of distributed generation sources and load shedding schemes on isolated grids case study: the Bahamas, in: Paper Presented at the 2014 International Conference on Renewable Energy Research and Application (ICRERA), 2014.

[60] A. Kwasinski, Quantitative model and metrics of electrical grids/quote resilience evaluated at a power distribution level, Energies 9 (2) (2016) 93.

[61] NERC Generator Availability Data System (GADS) [Online]. Available: http://www.nerc.com/pa.

[62] Electrical and Computer Engineering Department Illinois Institute of Technology Data [online]. Available: http://motor.ece.iit.edu/data.

Reliability analysis and role of energy storage in resiliency of energy systems

Mohammad Taghi Ameli, Kamran Jalilpoor,
Mohammad Mehdi Amiri, Sasan Azad

Department of Electrical Engineering, Shahid Beheshti University, Tehran, Iran

1. Introduction

Given the drastic changes in weather conditions and the occurrence of natural disasters over the past 10 years, and the connection between vital and daily human activities and the continuation of electricity supply, there is an urgent need to strengthen national energy infrastructure and increase its resiliency. Achieving this need will reduce the catastrophic effects of climate change and related events on life, economic activity, and national security [1].

In fact, in the vital artery of electricity, readiness for emergency condition is important in two ways. Firstly, the power outage and its lengthening can be a crisis in itself, and secondly, power outage has a negative impact on other arteries and thus exacerbates the crisis. For these reasons, the level of readiness of power grids against disruptions and types of events is very important, and it is necessary to take obligatory actions to improve the resiliency of power systems [2].

This chapter provides a model for examining the role of energy storage system (ESS) in the resiliency of an energy system and network reliability analysis. This method uses a broad programming problem, in which the objective function is considered to be the performance cost of sample microgrid and expected energy curtailment cost (EECC). The resiliency criterion is expected energy curtailment cost at times of a severe event, which is presented as a module of the objective function. The reliability criterion is loss of load expectation (LOLE), which is described as the expected fraction of the not provided load during the study period and is considered as a constraint on the problem. In this model, the resiliency criterion for large blackouts due to severe and rapid changes in the situation that have never been

Energy Storage in Energy Markets
https://doi.org/10.1016/B978-0-12-820095-7.00012-1

experienced before is considered, and the reliability criterion for small faults and transient failures due to technical reasons is examined.

Also, uncertainty is considered for the occurrence of natural disasters and production amount of renewable resources, where in each scenario, the conditions of the system elements and the production capacity of renewable energy sources are obtained. In the following, a mixed-integer programming (MIP) model is presented to effectively calculate the resiliency and reliability criteria in the optimization problem and evaluate the effects of ESS on the total cost of the microgrid. Numerical studies show that ESS can play an important role in improving the resiliency of power grids and also increasing the cost of network development to have higher reliability. The proposed goal is seen as a decision-making tool for providing information on long-term decisions that will help network designers make the right decisions on existing economic and technical issues.

2. Requirements and challenges of the future electricity network

The issue of blackouts in power systems has always been of interest to experts in the field. Although these blackouts can be caused by a variety of natural, technical, or human factors, the role of natural factors is more prominent.

According to the Executive Office of the President of the United States, 87% of power outages occurred in 2013 due to weather events [3]. In this regard, the study conducted in the Congress of this country is estimated to cost about 25−70 billion dollars annually due to weather conditions. In addition to natural disasters, the role of cyber issues has become increasingly important in recent years in the shutdown of power systems [4].

A recent report by the University of Cambridge estimated that the cost of the US electricity grid outages, which originated as a kind of cyber issue, was computed at 243 billion dollars to 1 trillion dollar [3]. This is due to the fact that power systems are largely intertwined with telecommunications and cyber systems, and this fusion has created new risks and threats for power systems.

As noted, the number of violent incidents has risen sharply in recent years around the world. Therefore, decisions must be made to prevent damage to such events in the future by increasing the resiliency of the electricity grid as well as other infrastructure.

3. The concept of resiliency in power systems

In the last few years, a new concept in the electricity industry called resiliency has come to the attention of researchers, which is the boundary between network strength and reliability. Today, the need to pay attention to the issue of increasing the resiliency of the network and the continuity of electricity supply is felt more and more due to the various consequences of blackouts at the social, political, and economic levels. Natural disasters and destructive actions are among the rare, unpredictable, and unlikely events that have severe effects and heavy damage on power and distribution systems.

Resiliency is the capacity of an energy system to withstand disruption and continue the process of delivering energy to consumers so that a resilient energy system can quickly recover from the disruption condition and provide alternative tools to provide energy services [5,6]. According to another definition, resiliency means the ability to be prepared and adapt to unpredictable conditions, sustain, and recover quickly. Thus, resiliency of a system is resilient to unpredictable disturbances [7].

In order to explain the concept of resiliency, the resiliency curve is shown in Fig. 17.1. As is clear, the level of resiliency to an unexpected event is a function of time. According to the figure before the accident at time t_e, the power system in normal operation must have sufficient strength and resistance to withstand the initial disturbances.

A well-designed and well-operated power system must have an acceptable resiliency to overcome various accidents, as shown here by $value_0$. Preventive flexibility ability in operation is essential at this stage as it enables operators to operate the network with resiliency capability.

Following the event, the system enters the postevent degraded state, where the resiliency of the system is significantly compromised ($value_{pe}$). Resourcefulness, redundancy, and adaptive self-organization are key resiliency features at this stage of the event, as they provide the corrective operational flexibility necessary to adapt to and deal with the evolving conditions (which possibly were never experienced before).

This helps minimize the impact of the event and the resiliency degradation (i.e., $value_0 - value_{pe}$) before the restoration procedure is initiated at tr. The system then enters the restorative state, where it should demonstrate the restorative capacity necessary for enabling the fast response and recovery to a resilient state as quickly as possible.

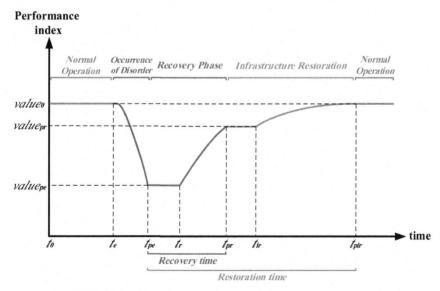

FIGURE 17.1 A conceptual resiliency curve associated with an event.

Once the recovery phase is completed, the system enters the postrestoration state. The postrestoration resiliency level $value_{pr}$ may or may not be as high as the preevent resiliency level $value_0$, i.e., $value_{pr} < value_0$. In particular, while the system may have recovered from the point of view of fully returning to its preevent operational state (thus showing a certain degree of operational resiliency), the infrastructure may take longer to fully recover (infrastructure resiliency), i.e., $(t_{pir}-t_{pr}) > (t_{pr}-t_{pe})$. This would depend on the severity of the event as well as on the resiliency features that the power system will demonstrate before, during, and after the external shock [8].

4. Comparison between reliability and resiliency

One of the main criteria in decision-making and planning of maintenance engineering is reliability. Reliability is a performance degree of power system in which electric power reaches consumers to the desired level and with certain standards.

On the other hand, the design and operation of power system infrastructure are usually based on key reliability criteria such as system security and adequacy. By considering these criteria in studies, we can be sure that the equipment will respond well to various faults and will create the least power outages for end users. But studies of reliability and relevant metrics cannot guarantee that an energy infrastructure can respond well to severe and sudden events, so designing a power system that is reliable and has necessary resiliency against certain threats and at the same time against unexpected events that have many consequences is a fundamental and major challenge [9]. In fact, a power network can be called efficient, which has a good quality in terms of reliability in the face of transient faults and acceptable resiliency to natural disasters and severe accidents [10,11].

Key features such as network reliability and security are sometimes used instead of resiliency concept. But the important thing to note is the difference between the concepts of resiliency and reliability. That means a network may be reliable, but the same network may not have the necessary resiliency. In fact, it can be said that resiliency concept is upstream and includes reliability, which, in addition to reliability, includes concepts such as durability, reaction, and rapid recovery. The events of the last decade show that electricity infrastructure must not only have good reliability against conventional and common accidents, but must also be resilient to very low probability accidents and with a wide range of effects [8].

5. The role of energy storage devices in improving resiliency

As storage technologies are advancing, the use of ESSs in future networks is attracting more and more attention from system operators so that they can be used economically in the power system [12].

According to Ref. [13], EES can have multiple attractive value propositions (functions) to power network operation and load balancing, such as: (i) helping in meeting peak electrical load demands, (ii) providing time-varying energy management, (iii) alleviating the intermittence of renewable source power generation, (iv) improving power quality/reliability,

(v) meeting remote and vehicle load needs, (vi) supporting the realization of smart grids, (vii) helping with the management of distributed/standby power generation, (viii) reducing electrical energy import during peak demand periods.

It is clear that resources such as distributed generation resources and energy storage devices lead to better energy management at lower levels of the network and better controllability in critical conditions and peak hours. Therefore, the way these resources are planned by the operator and the appropriate response in small and large disturbances at the network level are very important [14]. Operation of power system requires a complete balance between the supply and demand sectors. Establishing this balance is not easy, and to achieve it, the levels of supply and demand must be able to change quickly and unexpectedly. These changes can have reasons such as the forced departure of production units, the departure of transmission and distribution lines, and also sudden load changes. An ESS is one of the available resources for operating a system that can reduce the power error rate programmed by the network operator. These storage resources act as backups in power systems that can improve network performance and reduce network operating costs [15].

The storage device is usually charged during nonpeak hours when energy prices are low and discharged at peak hours when electricity prices are high or when the microgrid is experiencing a shortage of power due to disconnection from the upstream network. In fact, the stored power is discharged to meet the load requirement in the microgrid when a fault occurs or to sell power to the main grid and increase profitability. In other words, ESS improves network resiliency by storing energy during off-peak hours and delivering it when possible errors occur [16].

6. Problem modeling

In this chapter, we have tried to reduce the severity of network vulnerabilities and minimize the cost of loads shutdown by using an ESS as a way to increase resiliency. The expected shutdown cost is modeled as the value of the lost load (VOLL) and shutdown time.

It should be noted that the test network in this chapter is a microgrid consisting of a set of loads, distributed generation resources, and ESSs that can be operated in two states connected to the network and island mode. Fig. 17.2 shows a microgrid with its components.

The main objectives of this chapter are as follows:

- Investigating the importance of resiliency in power grids;
- Modeling of ESSs employed in microgrids;
- Energy generation and consumption planning in the presence of energy storage;
- Reducing operating costs and minimizing forced load cuts through the presence of energy storage devices; and
- Increasing network resiliency capability in the presence of distributed generation resources in microgrids.

The presented objectives are considered as a tool for providing information about long-term decisions that help network designers make the right decisions on technical, economic, and planning issues. Proposed model is a MIP that has been implemented in GAMS

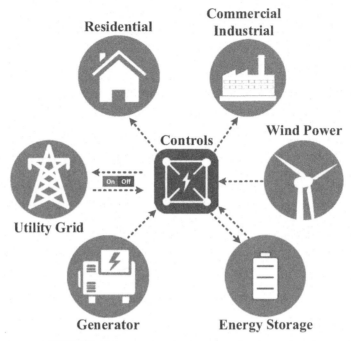

FIGURE 17.2   Studied microgrid and its component.

optimization software [17]. In this regard, MIP model has been used to formulate the problem, load consumption management, and ESS recharge and discharge program.

6.1 Uncertainty modeling of the problem

To address a more realistic issue and more accurately, it is necessary to consider uncertainties.

The uncertainty in the data of an optimization problem can be due to the forecasting, measurement, and implementation fault, or other factors.

Influential and important uncertainties in the planning of resiliency issues are the occurrence of natural disasters and the amount of renewable resources generation, which are discussed further.

❖ Uncertainty of heavy accidents

Generally, weather events are highly uncertain events, and it is difficult to predict them. This study is based on the probability of damage to the components of the power grid due to the severity of the weather event [18].

$$P(w_s) = \begin{cases} 0 & w < w_{critical} \\ P(w) & w_{critical} \leq w < w_{collapse} \\ 1 & w \geq w_{collapse} \end{cases} \tag{17.1}$$

Where $P(w_s)$ is the probability of failure as a function of weather parameter at scenario **s**.

Fig. 17.3 shows the failure probability of system component subject to the intensity of weather event.

❖ **Uncertainty in wind turbine power generation**

The generation schedule of a wind turbine mainly depends on the wind speed in the site. One of the most effective ways to consider the random nature of these resources is through probabilistic methods. In modeling the problem, wind speed can be calculated using statistical or probabilistic models. This chapter uses Rayleigh PDF to model wind speed changes. The following can be defined as the Rayleigh PDF:

$$PDF(v) = \left(\frac{2v}{c^2}\right) \exp\left[-\left(\frac{v}{c}\right)^2\right] \tag{17.2}$$

Where c is the scale parameter of the Rayleigh PDF and v is the wind speed in the region. The power generated and delivered to the grid when wind speed changes are specified can be calculated by Eq. (17.3) [19]:

$$P^w(v) = \begin{cases} 0 & \text{if } v \leq v_{in}^c \text{ or } v \geq v_{out}^c \\ \dfrac{v - v_{in}^c}{v_{rated}^c - v_{in}^c} P_r^w & \text{if } v_{in}^c \leq v \leq v_{rated} \\ P_r^w & \text{else} \end{cases} \tag{17.3}$$

Where P_r^w is the rated power of wind turbine. v_{in}^c and v_{out}^c are the connection and disconnection points of wind speed in the wind turbine, respectively.

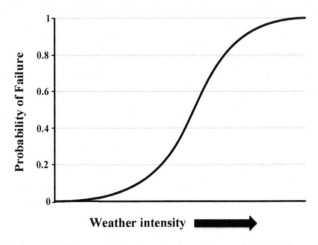

FIGURE 17.3 Failure probability curve of system components as a function of weather event intensity.

7. Problem formulation

In this section, the general optimization model for microgrid programming, the nature of the model, and the details of each component of the model are introduced. The objective function and ESS limitations are also reported. The following discusses microgrid limitations and generation units and operational constraints.

In operation of the microgrid, the objective function includes the expected operation cost of microgrid (OC) on a daily basis and expected energy curtailment cost (EECC). Therefore, the objective function (of) of the problem in this situation is formulated as Eq. (17.4):

$$\forall i \in N_i, s \in N_s, t \in N_t, h \in N_h$$

$$OF = \text{Min}\{OC + EECC\} \tag{17.4}$$

$$OC = \sum_{s=1}^{N_s} \pi_s \sum_{t=1}^{N_t} \sum_{h=1}^{N_h} \sum_{i=1}^{N_i} F_i\left(P_{s,t,h}^i\right) + \sum_{s=1}^{N_s} \pi_s \sum_{t=1}^{N_t} \sum_{h=1}^{N_h} \omega_{t,h} \cdot P_{s,t,h}^M \tag{17.5}$$

$$EECC = \sum_{s=1}^{N_s} \pi_s \sum_{t=1}^{N_t} \sum_{h=1}^{N_h} P_{s,t,h}^{lsh} \cdot VOLL_{t,h} \tag{17.6}$$

Where **i, r, s, t,** and **h** are index of thermal unit, renewable unit, scenario, day, and hour, respectively. Also N_i, N_r, N_s, N_t, and N_h are set of thermal units, renewable units, scenarios, days, and hours, respectively. π_s is probability of wind power generation scenarios, F_i is generation cost function of unit **i**, $P_{s,t,h}^i$ is the power generation thermal unit **i** at day **t** at hour **h** in scenario **s**, $\omega_{t,h}$ is electricity price at day **t** at hour h, $P_{s,t,h}^M$ is power exchanged between the microgrid and the main grid at day **t** at hour **h** in scenario s, $P_{s,t,h}^{lsh}$ is load shedding at day **t** at hour **h** in scenario s, and $VOLL_{t,h}$ is value of loss load at day **t** at hour **h**.

The fuel costs of electricity power generation, the cost of buying and selling electricity, and the exchange of energy with the main grid constitute the operation cost of microgrid, as shown in Eq. (17.5). In Eq. (17.6) expected energy curtailment cost in the microgrid is obtained as a resiliency criterion of the problem [20]. In general, ESS status is defined in three modes: charge, discharge, and no use. The EES charge and discharge equations and their limitations are modeled in Eqs. (17.7)–(17.12) [21]. The ESS discharge rate is neglected during noncharging and discharging mode.

$$\forall s \in N_s, t \in N_t, h \in N_h$$

$$SOC_{s,t,h} = SOC_{s,t,h-1} + \left(P_{s,t,h}^c \eta_c - P_{s,t,h}^d / \eta_d\right) \cdot \Delta h \tag{17.7}$$

$$0 \leq P_{s,t,h}^c \leq k \times SOC_{\text{max}} \tag{17.8}$$

$$0 \leq P^d_{s,t,h} \leq k \times SOC_{max} \tag{17.9}$$

$$SOC_{min} \leq SOC_{s,t,h} \leq SOC_{max} \tag{17.10}$$

$$SOC_{s,t,h(start)} \leq SOC^0 \tag{17.11}$$

$$SOC_{s,t,h(end)} = SOC^{end} \tag{17.12}$$

Where $SOC_{s,t,h}$ is state of charge at day **t** at hour **h** in scenario **s**, $P^c_{s,t,h}$ is ESS power charged at day **t** at hour **h** in scenario **s**, $P^d_{s,t,h}$ is ESS power discharged at day **t** at hour **h** in scenario **s**, η_c is ESS charge efficiency, η_d is ESS discharge efficiency, SOC_{max} is maximum state of charge capacity, SOC_{min} is minimum state of charge capacity, SOC^0 is state of charge at the start of every day, SOC^{end} is state of charge at the end of every day, and k is depth of discharge.

The amount of ESS energy is determined by Eq. (17.7). The energy stored per hour is determined by the energy stored in the previous hour plus the charge or discharge status at the same hour. Due to the time interval of 1 h, $\Delta h = 1$ is considered. ESS charging and discharging capabilities are limited by Eqs. (17.8) and (17.9). Eq. (17.10) restricts the amount of ESS energy. Eqs. (17.11) and (17.12) limit the amount of ESS energy at the beginning and end of each day.

The total generation capacity of local power generation units, the power exchanged between the microgrid and the main grid, and the battery power of charging or discharging must be equal to the microgrid load demand at any time [21]. Balance between generated power and demand is expressed using Eq. (17.13):

$$\forall i \in N_i, r \in N_r, s \in N_s, t \in N_t, h \in N_h, \left\{ UX^i_{s,t,h}, UY^M_{s,t,h} \right\} \in \{0, 1\}$$

$$\sum_{i=1}^{N_i} P^i_{s,t,h} + \sum_{r=1}^{N_r} P^r_{s,t,h} + P^d_{s,t,h} + P^M_{s,t,h} + P^{lsh}_{s,t,h} = P^L_{s,t,h} - P^d_{s,t,h} \tag{17.13}$$

$$P^i_{min} \cdot UX^i_{s,t,h} \leq P^i_{s,t,h} \leq P^i_{max} \cdot UX^i_{s,t,h} \tag{17.14}$$

$$P^i_{s,t,h} - P^i_{s,t,h-1} \leq RU_i \tag{17.15}$$

$$P^i_{s,t,h-1} - P^i_{s,t,h} \leq RD_i \tag{17.16}$$

$$\left| P^M_{s,t,h} \right| \leq P^M_{max} \cdot UY^M_{s,t,h} \tag{17.17}$$

$$0 \leq P^r_{s,t,h} \leq (0.01 \times \partial_s) \cdot P^r_{\text{max}} \tag{17.18}$$

$$0 \leq P^{lsh}_{s,t,h} \leq P^L_{s,t,h} \tag{17.19}$$

Where $P^L_{s,t,h}$ is microgrid load demand at day t at hour h in scenario s, $P^r_{s,t,h}$ is power generation renewable unit r at day t at hour h in scenario s, P^i_{min} is minimum power generation capacity of thermal unit i, P^i_{max} is maximum power generation capacity of thermal unit i, RU_i is ramp-up rate limit of thermal unit i, RD_i is ramp-down rate limit of thermal unit i, $UX^i_{s,t,h}$ is outage status of thermal unit i at day t at hour h in scenario s, $UY^M_{s,t,h}$ is outage status of the line connecting microgrid to the main grid at day t at hour h in scenario s, P^r_{max} is maximum power generation capacity of renewable unit r, P^M_{max} is maximum power capacity exchanged between the microgrid and the main grid, and ∂_s is percentage of existing wind power in scenario s.

The limitations of the maximum and minimum generated power values of unit i for every scenario s at day t at hour h are expressed by Eq. (17.14). Eqs. (17.15) and (17.16) control the increase or decrease of generation power between two consecutive hours. The power capacity exchanged between the microgrid and the main grid is limited by Eq. (17.17). $P^M_{s,t,h}$ is positive when the power is imported from the main grid, negative when the power is exported to the main grid, and zero when the microgrid operates in islanded mode [21].

Due to the stochastic nature of wind energy and the dependence of WT generation capacity on wind speed, the capacity of WTs cannot exceed the wind capacity of each scenario, which is limited by Eq. (17.18). [19] finds the times and scenarios in which the load is curtailed.

The reliability is defined in terms of LOLE. Eq. (17.20) finds the times and scenarios in which the load is curtailed. In case of load curtailment $w_{s,t,h}$ would be equal to 1. Using this curtailment indicator, the probability of curtailment scenarios is considered in LOLE Eq. (17.21). The obtained LOLE at each year should be less than its predefined targeted value Eq. (17.22) [22].

$$0 \leq P^{lsh}_{s,t,h} \leq M \cdot w_{s,t,h} \tag{17.20}$$

$$LOLE = \sum_{s=1}^{N_s} \pi_s \sum_{t=1}^{N_t} \sum_{h=1}^{N_h} w_{s,t,h} \tag{17.21}$$

$$LOLE \leq LOLE^{Target} \tag{17.22}$$

Where $w_{s,t,h}$ is a binary variable for status of load curtailment at day t at hour h in scenario s and $LOLE^{Target}$ is predefined targeted value for reliability index.

8. Case study and numerical results

A test microgrid was examined to express the performance of the proposed method.

This microgrid has three thermal units and one wind turbine that their specifications are presented in Table 17.1. The line power between the microgrid and main grid is 10 MW, which controls the transmission of power between them. According to the uncertainty of the wind, 12 scenarios have been created for it, which fully are given in Ref. [19].

The microgrid load is assumed to be constant in forward years, so all plans are for 1 year only. The reliability criterion is assumed to be 0.1 days per year (day/year) by the microgrid operator, which must be met. In modeling problem, it is assumed that resiliency criterion in the objective function is only for severe events and system reliability criterion is only for small and transient faults. The loads shutdown cost is $400/Mwh. The rated energy of ESS used is 19.731 MWh and its nominal power is 3.55 MW, with a charge and discharge efficiency of 95% and 90%, respectively.

Severe accidents, especially natural disasters, are widespread that destroy various parts of the network. These failures can be caused by trees falling on the lines, flooding of power sub-stations, and other network equipment failures. In fact, in addition to not having access to the upstream network, the adjacent microgrid may be detached from the upstream network due to the extent of the disturbance and damage caused by the severe accident and may not be able to help the microgrid studied and prefer to have an island mode performance of its own. As a result, the microgrid studied at the time of the power outage will act entirely as an island microgrid.

One of the best ways to model the outage state of an upstream network is the $T-\tau$ model. T is the planning time, which is 24 h in this study, and τ is the duration of the unavailability of the upstream network. Usually the time frame for most natural disasters such as tornadoes, hurricanes, heavy snowfalls, hail, and rain is predictable. Based on this, to model the upstream network outage, it is assumed that the accident will occur around 13:00. Due to the hard and destructive nature of the accident, its assumed 5 h for the upstream network to be disconnected [23].

To demonstrate the efficiency of the proposed model, different modes will be examined and analyzed to show the effects of the load consumption management program on the

TABLE 17.1 Characteristics of generating units.

Unite no.	Bus no.	Cost cofficient ($/MWh)	Minimum capacity (MW)	Maximum capacity (MW)	Ramp-up rate (MW/h)	Ramp-down rate (MW/h)
1	Gas	27.7	1	3	2.5	2.5
2	Gas	39.1	1	2	1.5	1.5
3	Gas	61.6	0.5	1	1	1
4	Wind	0	0	1	—	—

existence of ESS in a sample microgrid. Therefore, in this section, the microgrid is discussed in four cases including normal and critical operation conditions in the presence and absence of ESS.

Case 1: Network operation in normal mode without ESS.
Case 2: Network operation in normal mode with ESS.
Case 3: Network operation during severe accident occurrence without ESS.
Case 4: Network operation during severe accident occurrence with ESS.

❖ **Network operation in normal mode without ESS.**

Case 1 is assumed to be the basic state that the microgrid is programmed without ESS. As the planning of production units and energy exchange with upstream network on a typical day of the year in normal operation is shown in Fig. 17.4, at this stage all the power purchased from the upstream network or produced by thermal and renewable units to supply microgrid loads and no storage device is available on the network.

❖ **Operation of the network in normal mode with ESS**

The storage device can be charged in less than 6 h to reach the maximum SOC. The storage device is usually charged during nonpeak hours when the price of energy is cheap and discharged at peak hours when the price of electricity is expensive.

Discharged power stored during peak hours is used to meet the load requirement in the microgrid during overload or to sell power to the main grid and increase profitability. The ESS charge and discharge clock schedule is shown in Fig. 17.5.

Given that electricity prices are higher at 13 and 19.23 than other hours, the ESS helps make microgrid profitable by discharging power and preventing the microgrid from buying energy from the upstream network during the hours when electricity is expensive. It can be seen that even at 22 and 23 h, part of the power discharged by ESS is sold to the main

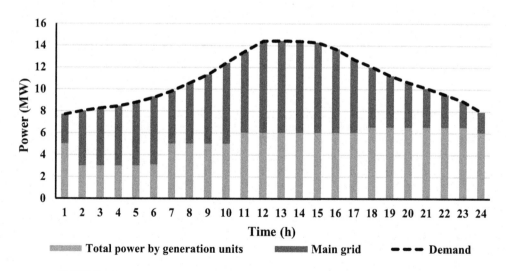

FIGURE 17.4 Operation planning of the microgrid and production units in case 1.

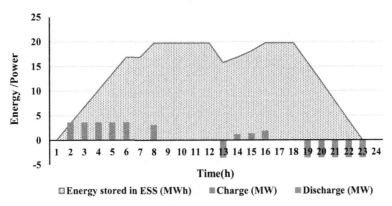

FIGURE 17.5 The hourly dispatch of ESS at the optimal size in case 2.

network. The operation planning of the dispatchable units and the power exchange between the microgrid and main grid in case 2 are shown in Fig. 17.6.

❖ **Network operation during severe accident occurrence without ESS**

In this operation, it is assumed that the microgrid is programmed without a storage system in confronting with a disturbance or a power outage. As shown in Fig. 17.7, an error occurred at 1 p.m., and the microgrid is operated as an island for 5 h from 1 to 6 p.m. During these hours, the generation of thermal units inside the microgrid increases to supply the existing load, thus trying to optimally plan the microgrid to minimize its shutdown costs; nevertheless, the microgrid is only able to provide part of the existing load. The final cost of curtailment loads is $13,756.

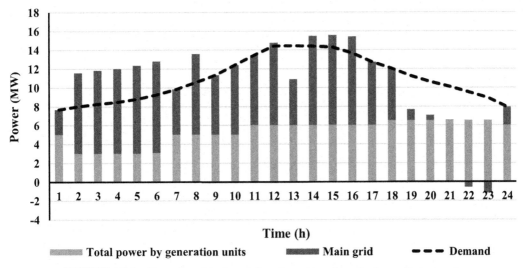

FIGURE 17.6 Operation planning of the microgrid and production units in case 2.

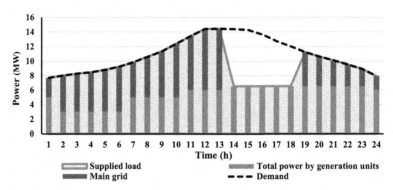

FIGURE 17.7 Operation planning of the microgrid and production units in case 3.

❖ Network operation during severe accident occurrence with ESS

The ESS charge and discharge clock schedule is shown in Fig. 17.8 in case 4. The storage device is usually charged during nonpeak hours when energy prices are low and discharged during peak hours when electricity prices are high or when the microgrid has a power shortage due to disconnection from the upstream network. In fact, the discharge of stored power is done to meet the load requirement in the microgrid when an error occurs or to prevent the purchase of power from the main grid and increase profitability.

During the hours when the microgrid is disconnected from the main grid, the power stored in the ESS is discharged and prevents forced outage of part of the loads. The planning of the dispatchable units and the power exchange between the microgrid and the main grid in case 4 are shown in Fig. 17.9. As shown in Fig. 17.8, the status of the fed loads is improved between 1 and 7p.m., and the role of ESS as a backup is well illustrated. Also, the final cost of curtailment loads will be reduced to $6,653, which is compared to case 3 because of the discharge of power stored in the ESS in state of island microgrid.

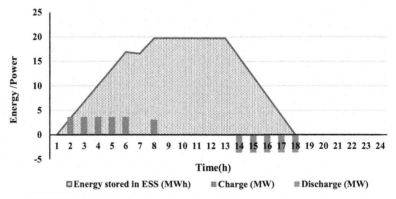

FIGURE 17.8 The hourly dispatch of ESS at the optimal size in case 4.

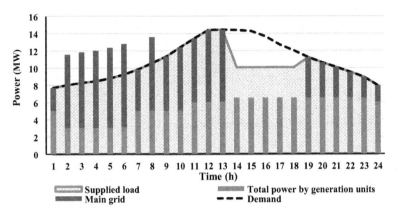

FIGURE 17.9 Operation planning of the microgrid and production units in case 4.

8.1 Resiliency criterion investigation

Table 17.2 summarizes the results from cases 1 to 4 for the sample day for comparison. The results show that ESS's participation in energy storage during the hours when electricity prices are low and its discharge at high electricity prices or the possible hours of disturbance occurrence improve grid economic profitability and increase network resiliency by reducing load curtailment. In case 4, a large part of the curtailment loads in case 3 is fed through ESS.

The results show that ESS reduces need for power imports from the main grid as well as reduces loads shutdown cost by energy storage during the hours when the price of electricity is cheap and by providing low cost power in case of disturbance occurrence and power outage. Fig. 17.10 shows the supplied loads and the intermittent loads during the disturbance period for cases 3 and 4. When an error occurs and the microgrid becomes an island, 34.3 MW of load is turned off, where ESS supplies 52% of them and decreases this amount to 16.6 MW.

TABLE 17.2 Summary of system costs for a sample day of the year.

Case No.	Case1	Case2	Case3	Case4
ESS rated power (MW)	0	3.55	0	3.55
ESS rated energy (MWh)	0	19.731	0	19.731
Generation cost ($)	4127	4130	4250	4253
Power import ($)	5794	5088	4034	4513
Power export ($)	0	−116	0	0
Load shedding cost ($)	0	0	13,756	6653
Total cost ($)	9921	9102	22,040	15,419

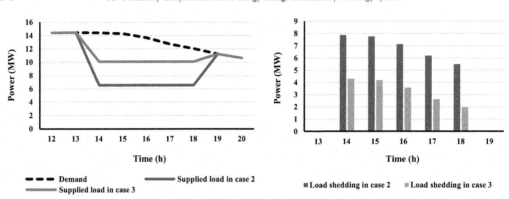

FIGURE 17.10   Supplied load and load shedding at the time of extreme fault.

Therefore, ESS can play an effective role in improving the resiliency of power grids in addition to reducing environmental pollution, compliance with load, power quality, and load demand management during peak hours.

It should be noted that improving the resiliency of a power network requires different strategies in fields of tolerance, adaptation, and recovery, whereas in this chapter only the role of ESSs is examined in improving network performance and resiliency.

8.2 Reliability criterion investigation

In this section, three different values are considered for the expected reliability index of microgrid, namely 0.1, 0.2, and 0.3 day/year. As they change, ESS programming changes to ensure the reliability of the microgrid. As shown in Table 17.3, in case 2, the ESS acts as a backup by discharging energy in emergencies and largely eliminates the network need to purchase energy in the event of a fault. Also, having a lower reliability index (less blackout) or increasing the reliability of the microgrid leads to an increase in the total cost of the

TABLE 17.3   Summary of system costs for a year.

Reliability criterion	0.0 day/year		0.1 day/year		0.2 day/year	
Case No.	Case1	Case2	Case1	Case2	Case1	Case2
ESS rated power (MW)	0	3.55	0	3.55	0	3.55
ESS rated energy (MWh)	0	19.731	0	19.731	0	19.731
Generation cost ($)	1,506,277	1,507,150	1,506,277	1,507,150	1,506,277	1,507,150
Power import ($)	2,115,012	1,856,406	2,097,131	1,830,335	2,075,069	1837,334
Power export ($)	0	−42,465	0	−43,732	0	−71,932
Total cost ($)	3,621,289	3,321,091	3,603,408	3,293,753	3,581,346	3,272,552

microgrid. The reason for this is that in order to have less curtailment load, the load of the microgrid especially during peak hours must be provided in any way from the expensive market price or from the expensive units inside the microgrid. Therefore, it will increase the total cost of the microgrid.

9. Conclusion

In this chapter, a MIP model is presented to examine the role of ESS in resiliency of an energy system and network reliability analysis. The resiliency criterion for reducing the impact of severe accidents on network was considered as a module of the objective function. The system reliability index, which is used to ensure reliable network performance against minor errors and transient failures, was cited as a constraint. The numerical studies presented indicate that using an ESS in power systems can reduce the power factor error programmed by the network operator.

These storage resources act as backup in power systems that can improve network performance and reduce operating costs. Finally, ESS can reduce cost of curtailment loads and improve network resiliency in the face of severe and sudden events by energy savings when electricity prices are low and providing low-cost power in time of disturbance occurrence or power outage.

References

[1] Z. Bie, Y. Lin, G. Li, F. Li, Battling the extreme: a study on the power system resiliency, Proc. IEEE 105 (2017) 1253–1266.

[2] E. Rosales-Asensio, M. de Simón-Martín, D. Borge-Diez, J.J. Blanes-Peiró, A. Colmenar-Santos, Microgrids with energy storage systems as a means to increase power resiliency: an application to office buildings, Energy 172 (2019) 1005–1015.

[3] A.B. Smith, R.W. Katz, US billion-dollar weather and climate disasters: data sources, trends, accuracy and biases, Nat. Hazards 67 (2013) 387–410.

[4] W. House, Economic Benefits of Increasing Electric Grid Resiliency to Weather Outages, Executive Office of the President, Washington, DC, 2013.

[5] M.S. Khomami, K. Jalilpoor, M.T. Kenari, M.S. Sepasian, Bi-level network reconfiguration model to improve the resiliency of distribution systems against extreme weather events, IET Gener. Transm. Distrib. 13 (2019) 3302–3310.

[6] A. Hossein, A. Mohammad Taghi, H. Seyed Hossein, Multi-stage frequency control of a microgrid in the presence of renewable energy units, Electric Power Components and Systems 45 (2) (2016) 1–12.

[7] A. Hussain, V.-H. Bui, H.-M. Kim, Microgrids as a resiliency resource and strategies used by microgrids for enhancing resiliency, Appl. Energy 240 (2019) 56–72.

[8] M. Panteli, P. Mancarella, The grid: stronger, bigger, smarter?: Presenting a conceptual framework of power system resiliency, IEEE Power Energy Mag. 13 (2015) 58–66.

[9] H. Xie, X. Teng, Y. Xu, Y. Wang, Optimal energy storage sizing for networked microgrids considering reliability and resiliency, IEEE Access 7 (2019) 86336–86348.

[10] T. Khalili, M.T. Hagh, S.G. Zadeh, S. Maleki, Optimal reliable and resilient construction of dynamic self-adequate multi-microgrids under large-scale events, IET Renew. Power Gener. 13 (2019) 1750–1760.

[11] M. Shivaie, et al., A multistage framework for reliability-based distribution expansion planning considering distributed generations by a self-adaptive global-based harmony search algorithm, Reliab. Eng. Syst. Saf. 139 (2015) 68–81.

[12] H. Ameli, et al., A fuzzy-logic–based control methodology for secure operation of a microgrid in interconnected and isolated modes, Int. Trans. Electr. Energy Syst. 27 (11) (2017) e2389.

[13] X. Luo, J. Wang, M. Dooner, J. Clarke, Overview of current development in electrical energy storage technologies and the application potential in power system operation, Appl. Energy 137 (2015) 511–536.

[14] A. Ingalalli, A. Luna, V. Durvasulu, T.M. Hansen, R. Tonkoski, D.A. Copp, et al., Energy storage systems in emerging electricity markets: frequency regulation and resiliency, in: 2019 IEEE Power & Energy Society General Meeting (PESGM), 2019, pp. 1–5.

[15] A.S. Sidhu, M.G. Pollitt, K.L. Anaya, A social cost benefit analysis of grid-scale electrical energy storage projects: a case study, Appl. Energy 212 (2018) 881–894.

[16] M. Nazemi, M. Moeini-Aghtaie, M. Fotuhi-Firuzabad, P. Dehghanian, Energy storage planning for enhanced resiliency of power distribution networks against earthquakes, IEEE Transactions on Sustainable Energy 11 (2) (2020) 795–806.

[17] A. Soroudi, Power System Optimization Modeling in GAMS, vol. 78, Springer, 2017.

[18] S. Nikkhah, K. Jalilpoor, E. Kianmehr, G.B. Gharehpetian, Optimal wind turbine allocation and network reconfiguration for enhancing resiliency of system after major faults caused by natural disaster considering uncertainty, IET Renew. Power Gener. 12 (2018) 1413–1423.

[19] A. Soroudi, Possibilistic-scenario model for DG impact assessment on distribution networks in an uncertain environment, IEEE Trans. Power Syst. 27 (2012) 1283–1293.

[20] H. Farzin, M. Fotuhi-Firuzabad, M. Moeini-Aghtaie, A stochastic multi-objective framework for optimal scheduling of energy storage systems in microgrids, IEEE Trans. Smart Grid 8 (2016) 117–127.

[21] K. Jalilpoor, R. Khezri, A. Mahmoudi, A. Oshnoei, Optimal sizing of energy storage system, Var. Scalability Stab. Microgrids 139 (2019) 263.

[22] S. Bahramirad, W. Reder, A. Khodaei, Reliability-constrained optimal sizing of energy storage system in a microgrid, IEEE Trans.Smart Grid 3 (2012) 2056–2062.

[23] A. Khodaei, Resiliency-oriented microgrid optimal scheduling, IEEE Trans. Smart Grid 5 (2014) 1584–1591.

Electric vehicles and electric storage systems participation in provision of flexible ramp service

Sajjad Fattaheian-Dehkordi[1,2], *Ali Abbaspour*[2], *Matti Lehtonen*[1]

[1]Aalto University, Espoo, Finland; [2]Sharif University of Technology, Tehran, Iran

Nomenclature

Variables

$P_{i,t}^{Gen}$ Power generation by generation unit i at time t.

$P_{i,t}^{Ch,ESS/EV}$, $P_{i,t}^{Dis,ESS/EV}$ Charging/discharging power of storage/electrical-vehicle unit i at time t.

$P_{i,t}^{RU,Gen}$, $P_{i,t}^{RD,Gen}$ Ramp-up/down power provided by generation unit i at time t.

$P_{i,t}^{RU,Ch,ESS/EV}$, $P_{i,t}^{RD,Ch,ESS/EV}$ Ramp-up/down power provided by charging state of storage/electrical vehicle unit i at time t.

$P_{i,t}^{RU,Dis,ESS/EV}$, $P_{i,t}^{RD,Dis,ESS/EV}$ Ramp-up/down power provided by discharging state of storage/electrical vehicle unit i at time t.

$E_{i,t}^{ESS/EV}$ Stored energy of storage/electrical vehicle unit i at time t.

$\alpha_{i,t}^{Ch,ESS/EV}$, $\alpha_{i,t}^{Dis,ESS/EV}$ Binary variables for impeding simultaneous charging/discharging power of storage/electrical vehicle unit i at time t.

$P_{i,t}^{Driving,EV}$ Power consumption for driving electrical vehicle unit i at time t.

Parameters

$Cost_i^{Gen}$ Operational cost of generation unit i.

393

λ_t^{RT} Real time energy price at time t.

λ_t^{RU}, λ_t^{RD} Ramp-up/down price at time t.

P_t^{RU}, P_t^{RD} Estimated ramp-up/down requirements at time t.

$P_i^{Max,Gen}$, $P_i^{Min,Gen}$ Maximum/minimum possible power generation by generation unit i.

$\rho_i^{Ch,ESS}/_{EV}$, $\rho_i^{Dis,ESS}/_{EV}$ Charging/discharging efficient of storage/electrical vehicle unit i.

$E_i^{Max,ESS}/_{EV}$, $E_i^{Min,ESS}/_{EV}$ Maximum/minimum energy of storage/electrical vehicle unit i.

$\alpha_{i,t}^{V2G,EV}$ Connected to the grid condition of the electrical vehicle unit i at time t.

Δt Time-period of activating flexible ramp product.

Sets

ω^{Gen} Set of all generation units in the system.

$\omega^{ESS}/_{EV}$ Set of all ESS/EV units in the system.

1. Introduction

The increasing trend of integrating renewable energy sources (RESs) has resulted in a paradigm shift of traditional procedures employed for operating power systems. In this context, while the installation of these resources is supported by governments based upon the environmental concerns associated with conventional generation units, the variable and uncertain nature of them impedes concise prediction and control of their power production in future time periods [1,39]. In this regard, the net load of the system, which equals the actual load minus nondispatchable power generation (i.e., RESs), would become more volatile (i.e., uncertain and variable) in systems with high penetration of RESs. This new operational environment would lead to new challenges in the operation and planning of power systems. In other words, the uncertainty and variability associated with RESs would be added to the current condition of the system; therefore, it would challenge the traditional ways employed by system operators to maintain the reliability and flexibility of the system. Among the arisen challenges, system operators would frequently confront with the shortage of ramping capacity to balance supply and demand, which could cause very high electricity prices during the real-time operation of the system [2,3].

As mentioned, increasing the installation rate of RESs, i.e., wind turbines and photovoltaic (PV) units, would result in high ramping in the net load of the system, which is not also perfectly predictable. Specifically, the net load of systems with high rate integration of PV units, such as the California transmission system, would have a duck curve shape, which is a common issue raised by high fluctuation in power generation of PV units in the morning and evening time periods [4]. While supply and demand should be continuously balanced in the power system, this condition could significantly increase the operational risk of the system. Consequently, the operational flexibility of power system to address the ramp concerns seems to be an essential characteristic that facilitates accommodating a significant share of

RESs replacing fossil-fuel-based power plants. In other words, sufficient operational flexibility in power systems plays a key role in the effective integration of large shares of RESs in the system. In this context, the lack of operational flexibility in power systems could result in load shedding, RESs curtailments, and high volatile market prices, which show the importance of improving system operational flexibility [5].

The operational flexibility of a system primarily relies on the ability of the system operators to procure flexibility services from the eligible flexible resources to address the net-load changes in real time. In this regard, system operators would require operational frameworks that enable them to activate flexibility services in order to improve the ramping capability of the system. Traditionally, utilities rely on bulk generation units in transmission systems, based on their speed and quantity, to provide flexibility services for balancing supply and demand in the power system. However, the declining operational and investment costs of RESs as well as significant investment costs, operational costs, and construction time associated with conventional resources would cause the lack of flexibility capability of power systems. In other words, commitment of bulk generation units, as well as investment in their expansion planning, would be decreased significantly, while the share of RESs in supplying demand would extensively be increased. That is why system operators could not completely rely on these resources to provide operational flexibility in power systems. Moreover, congestion in transmission systems would result in activating operational flexibility in local areas. In this regard, system operators would significantly strive to procure flexibility services from local flexible resources [3]. Consequently, flexible resources such as EVs and EESs could play a key role in improving the flexibility of the system by incorporating operational frameworks that enable system operators to exploit their operational scheduling in order to procure flexibility service to address operational issues such as high ramping in net load of the system [6,7].

Aligning with RESs, investments of private investors and governments in promoting the deployment of EVs and ESSs in power systems have widely been increased in recent years. In this context, it is considered that these flexible resources based on their fast responsiveness nature could facilitate the integration of RESs by providing flexibility services for reliable operation of the system. Moreover, EVs have also environmental benefits as they could decrease carbon emission in the transportation sector. It is noteworthy that EVs could be operated in grid-to-vehicle (G2V) or vehicle-to-grid (V2G) modes, when they are connected to the electricity grids. Putting all together, EVs and ESSs could improve reliability and flexibility of the power system by providing flexibility services, while their operational constraints such as capacity, charging/discharging rate, and operational costs are taken into account in their operational scheduling optimization.

2. Ramping issue in the system

In order to illustrate the effects of RESs on the system net load, Fig. 18.1 presents the ramping requirements in a test system considering different levels of integrating PV units. In this regard, the ramping requirements in the morning as well as evening periods, when power generation by PV units significantly changes, seem to be very high, which could cause operational issues in real-time operation of the system. It is noteworthy that the intermittent nature of RESs would exacerbate the operational condition of the system, which shows the importance of improving flexibility of the system by procuring flexibility ramp service

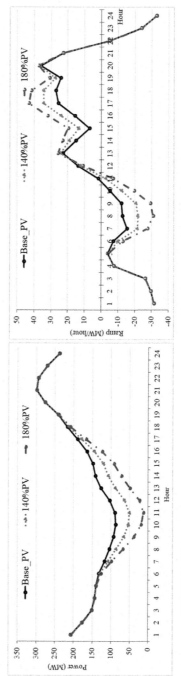

FIGURE 18.1   Net load of the test system and its associated ramping for different cases of integrating PV units.

from responsive resources such as EVs and ESSs. In this context, the system flexibility for operation and planning procedures could be analyzed by utilizing various analytic frameworks; i.e., visualizing diagrams [8,9], flexibility metrics [10–12], and power system operational margins obtained from simulation analysis [13].

3. Flexible ramp effects on operation of power systems

Ramping associated with the net load of the system could result in high market prices in case that state-of-the-art techniques for improving its operational flexibility are not taken into account in the operational management procedure. To elaborate on the condition of a power system without considering flexibility ramp service, based on [14], the energy price in a hypothetical system during a time period with significant ramping is investigated. In this regard, it is considered that ancillary services for addressing ramping requirements in the system are not deployed by the system operator. Based upon Fig. 18.2, it is considered that the basic load of the system is supplied by committing low-cost generators with the operational cost of 20 $/MWh. However, these low-cost generators have a ramping capability of 1 MW/min, which could not meet the 10 MW/min ramping associated with the net load of the system from 10:00 to 10:30. In other words, ramp-up (RU) of 300 MW in the 30 min period requires 10 MW/min ramping capability, which could not be supplied by low-cost generators. Noted that the ramping of the net load at 10:00–10:30 is originated from the fluctuation in power generation by RESs due to extreme changes in the weather conditions. In this context, expensive generators with the operational cost of 100 $/MWh have to be committed by the system operator to meet the ramping of the net load and supply the demand. Moreover, these expensive generators have to be operated for 5 h in order to enable low-cost generators to supply all the demand of the system. Consequently, the cleared energy price of the system would be equal to the operational cost of expensive generators (i.e., 100 $/MWh) in these time periods. This example shows the importance of developing flexibility ramp service in power systems to improve the ramping capability and finally prevent high market prices in the system.

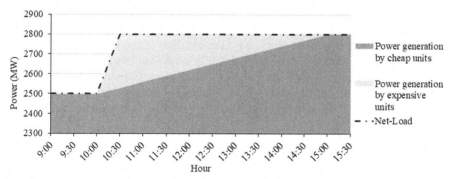

FIGURE 18.2   Net load and power production by committed units in the test system.

4. Flexible ramp service

Ancillary services have been deployed by utilities in order to efficiently operate the system by procuring required reserve products in a cost-effective manner. However, employing previously defined ancillary services seems not to be appropriate for addressing ramping requirements in the system. In this context, regulation services and contingency reserves are respectively defined for balancing the supply and demand in real-time operation as well as contingency conditions. However, the high rate of activating ramp service and the high price of regulation and contingency reserves impede their utilization for providing flexible ramp service. In this regard, a new ancillary service should be developed for addressing flexible ramping requirements in power systems with a high share of intermittent RESs.

As mentioned, the issues associated with the integration of RESs originate from the variability and uncertainty of RESs. Variability addresses the change in the operational condition of the system over a sequence of time periods, while uncertainty means that the net load of the system could not be forecasted with perfect accuracy. These conditions would be deteriorated by integrating a huge amount of RESs in the system. Conventional reserve services in the power system are mostly designated to address the uncertainty associated with the operational condition of the system. In this regard, a new reserve product seems to be essential for addressing the shortage in the ramping capacity of the system, which resulted from the uncertainty and variability of the net load. On the other hand, the limitation of flexibility ramp service that could be provided by conventional bulk generation units shows the importance of incurring ramp service from flexible resources such as EVs and ESSs to ensure the reliable and flexible operation of the system. It is noteworthy that the short time and frequent activation of flexible ramp service in real time are associated with operational constraints of EVs and ESSs in case they have capacity available for addressing ramp up/down requests by the system operator. In other words, the high-speed response time of these resources enables their contribution to the provision of flexible ramp service, while aiming to maximize their profit.

The new operating reserve known as flexible ramp product (FRP) or flexible ramp service is designated to ensure the ramping capability of the system beyond the operating time frame. In this regard, due to the uncertainty and variability of the resources in the system, the ramp service should be activated in a short time interval before the operating time of the system. Moreover, the developed frameworks for activating flexible ramp service from EVs and ESSs should be aligned with the operational characteristics of these resources, management structure of the system, and structures of energy/reserves markets [15]. Consequently, operational modeling as well as optimization procedure associated with the participation of EVs and ESSs in providing flexible ramp service could be addressed with different perspectives. In the following sections, the concepts behind frameworks devised for procuring FRP as well as the operational optimization modeling associated with the participation of EVs and ESSs in providing flexible ramp service are studied.

4.1 Ramp market modeling

Operational management of flexible resources in order to activate flexible ramp products could be conducted by utilizing a market that facilitates exchanging of flexible ramp service among the participant entities. In other words, system operators would be able to purchase flexible ramp service from flexible resources that participated in the ramp market. Ramp market is a new concept that has been developed by some utilities to be conducted in the power system to ensure its reliable operation. In this regard, flexible ramp-up (FRU) and flexible ramp-down (FRD) services could be procured by system operators from the ramp markets based on their anticipation of ramping capability requirements in the system. Noted that this new ancillary service market is now actively utilized by California independent system operator (CAISO) and Midcontinent independent system operator (MISO) to address the flexibility requirements in the system [16–18].

Ancillary service markets are generally aimed to support the reliable operation of power systems. In this regard, their structures are tied tightly with the energy market designs. However, FRP is the only ancillary service market that considers fluctuations in the net load of the system between two consecutive time intervals. Moreover, unlike other ancillary services, FRP aims to simultaneously address both variability and uncertainty in the real-time operation of the system; therefore, it would be deployed very frequently during the operation of the system. As a result, this new service would facilitate utilizing greater responsiveness from the EVs/ESSs in the system. In addition, deploying this ancillary service would result in the reduction of activating regulation service in the system by improving the system flexibility.

Reserve services are created to address the changes in the net load of the system and so maintain the balance between net generation and net demand. In this context, regulation and operating reserves are two well-known ancillary services in the power system that are developed to enable system operators to manage the system in different conditions. In this regard, regulation service is designed to be operated in a matter of seconds to deal with the difference between supply and demand in the system. On the other hand, operating reserve (i.e., contingency reserve) is developed in two categories (i.e., spinning reserve and nonspinning reserve) based on the potential contingency events in the system (i.e., loss of generators). Therefore, it is typically designed to be in an upward direction to cover the loss of generation units or a sudden increase in load demand. In general, the price of activating regulation and the operating reserve is higher than utilizing FRP to address the high ramping of the net load. Moreover, FRP is designed in two directions (i.e., upward ramping and downward ramping), while the operating reserve is typically considered as the upward reserve capacity. Consequently, FRP seems to be a better practical cost-efficient way of ensuring supply–demand balance when the system confronts with ramping shortages.

FRP is currently deployed by CAISO and MISO in their energy and ancillary services co-optimization models to address challenges raised as a result of the high installation of RESs. Specifically, due to the high rate of PV integration, meeting the power balance during morning and evening is challenging in CAISO. In this regard, the net load deviates significantly in these time periods from the actual load and so the system needs considerable flexibility to respond quickly to the changes in the net load. Consequently, by deploying a well-organized FRP, the system would be able to cope with the potential net-load movements. A study in MISO regarding the benefits of deploying FRP shows the possibility of annual savings of $3.8–5.4 million, which is aside from the fact that the system would become significantly more flexible in case of employing FRP [18]. In general, deploying FRP would benefit the electricity market by reserving ramp capacity for potential future ramp shortages in the system. The benefits of employing FRP procured from flexible resources such as EVs/ESSs could be categorized as follows:

- Better management of the ramping capacity procured from flexible resources to address changes in the system's net load.

- Decreasing the rate of energy imbalance and price spikes in the system due to ramping issues.
- Decreasing the rate of employing regulation services.
- Utilizing a market model that enables flexible resources to fairly exchange flexibility services.
- Increasing the efficiency of the energy and ancillary markets.
- Considerable cost savings in operation and investment procedures by applying a relatively small change in the operation of the system to efficiently balance the supply—demand, while using the independently operated flexible resources and delaying the required investments in conventional bulk generation power plants.

The idea of utilizing ramp markets to incorporate ramp capacity reserve for balancing the supply and demand in the California and Midcontinent grids firstly emerged in 2012, and during these years the modeling of FRP has been completed to address the raised operational issues. It is noteworthy that while the developed ramp markets have differences from implementation perspectives, both are aimed to meet the uncertainty and variability of the net load to ensure reliable and efficient operation of their corresponding power systems. In general, flexible resources would participate in the provision of flexible ramp market provided that their loss of opportunity in parallel markets such as energy markets would be compensated by system operators. In this regard, the system operators could procure FRP service in real-time operation of the system and pay the collaborated resources the loss of opportunity values. It is noteworthy that New York independent system operator (NYISO) has also planned to establish day-ahead/real-time ramp markets to procure FRP [19]. Based on the current planned market, the resources that participated in providing FRP would be compensated based on their loss of opportunity costs [5].

- 4.1.1 Estimation of required ramp capacity

In order to direct the flexible ramp market in the system, the ramping requirements of the system must be estimated by the system operators. In this regard, system operators could estimate the ramping requirements, i.e., FRD and FRU, based on the uncertainty associated with the forecasted net load of the future dispatch intervals. As a result, system operators consider the changes in the net load of the system in two sequential intervals and estimate the FRU/FRD based on the forecasted net load and its associated uncertainty. A simplified model of estimating FRU/FRD in the system is represented in Fig. 18.3 [20]. Based on this methodology, system operators strive to decrease the possibility of violation in supply—demand balance by considering a ramp margin that covers the sidelines of the forecasted ramping in each dispatch interval. In this respect, the margins of the system's net load are calculated considering the distribution probability of the error associated with the forecasted net load, which is developed based on the historical data. Correspondingly, FRD and FRU would be estimated based on the forecasted net load and its up and down margins considering a certain confidence level. Noted that machine learning methodologies could also be taken into account by system operators to improve the forecasting of the FRP for reliable operation of the system. Moreover, the Copula concept could be incorporated to model the correlation between the forecasted data and its associated uncertainty to increase the overall accuracy of estimating the ramping of the net load [21]. Finally, it is noteworthy that the effective implementation of the flexible ramp service depends on the accurate estimation of the FRU/FRD associated with the net load, which is the basis for conducting the ramp market in the system.

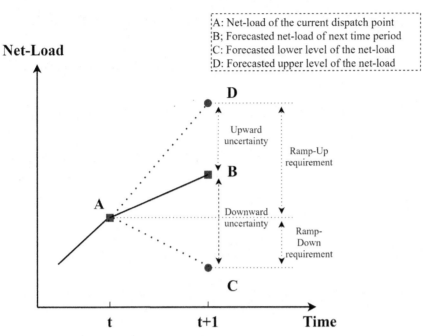

FIGURE 18.3 Estimation of ramp-up/down requirements in the system.

- 4.1.2. Participation of EVs and ESSs in ramp markets

Based on the development of the electricity markets, different perspectives could be employed with the aim of facilitating the participation of EVs and ESSs in the ramp market. In general, the resources could participate in the market based on their announced bids, which could be estimated by taking into account their marginal costs for providing FRP. On the other hand, system operators could procure FRP from flexible resources considering their loss of opportunity costs, which shows the profits that could have been obtained by resources in case of participating in energy markets or other ancillary service markets.

The centralized/decentralized management of flexible resources could also affect their participation in the ramp market. As a result, employing bid-in participation of flexible resources as well as the loss of opportunity cost concept depends on the centralized/decentralized operational scheduling of these resources. In this regard, loss of opportunity cost concept could be deployed by the system operators provided that the scheduling optimization of the flexible resources (i.e., ESSs and EVs) is conducted in a central manner. However, loss of opportunity cost concept could not be utilized as a general rule by the system operators to procure FRP from the flexible resources operated by an independent entity. In this regard, loss of opportunity cost could normally be employed by the system operator to activate ramp flexibility service from generation units. Nevertheless, procuring FRP from independently operated responsive resources such as EVs and ESSs could not merely be compensated by the loss of opportunity cost estimated based on the price of energy/ancillary service at the same time interval.

The operational scheduling of EVs and ESSs in each time interval depends on the condition of the system and scheduling of EVs/ESSs in the previous as well as the future time intervals. In this regard, the loss of opportunity cost associated with exploiting their operational scheduling could not merely be estimated based on the price of parallel markets in the same interval. In other words, any change in the operational scheduling of these resources results in rescheduling their energy exchange in future time intervals, which determines the loss of opportunity cost associated with providing FRP. In this regard, in the following sections, the optimization of EVs and ESSs participation in the ramp market is conducted based on the two previously discussed concepts.

- Centralized operational scheduling of flexible resources

Based on the previous explanations, scheduling of EVs and ESSs for participation in the ramp market could be conducted in a central manner aligning with the generation units. In this respect, based on the scheme used by CAISO and MISO, it is considered that activating FRP is determined in 15−10 min before the dispatch time [22]. However, the optimization could be updated based on the scheme deployed by the system operator of the respective power grid for ensuring the balance of supply and demand. It is noteworthy that the future hours are considered in the optimization to accurately estimate the loss of opportunity cost associated with the scheduling of EVs and ESSs. In this regard, the optimization model is developed as shown in (1)−(14) considering the model predictive control (MPC) concept in order to take into account the scheduling of the EVs/ESSs in future time intervals.

$$\text{Min } Cost_{t''} = \sum_{i \in \omega^{Gen}} \left[P_{i,t}^{Gen} \cdot \left(Cost_i^{Gen} - \lambda_t^{RT} \right) \right] \Big|_{t=t''} + \sum_{i \in \omega^{ESS}/_{EV}} \left[\sum_{t \geq t''} \left(P_{i,t}^{Ch,ESS}/_{EV} - P_{i,t}^{Dis,ESS}/_{EV} \right) \cdot \lambda_t^{RT} \right]$$

$$\tag{18.1}$$

$$\sum_{i \in \omega^{Gen}} P_{i,t}^{RU,Gen} + \sum_{i \in \omega^{ESS}/_{EV}} \left(P_{i,t}^{RU,Ch,ESS}/_{EV} + P_{i,t}^{RU,Dis,ESS}/_{EV} \right) \geq P_t^{RU} \tag{18.2}$$

$$\sum_{i \in \omega^{Gen}} P_{i,t}^{RD,Gen} + \sum_{i \in \omega^{ESS}/_{EV}} \left(P_{i,t}^{RD,Ch,ESS}/_{EV} + P_{i,t}^{RD,Dis,ESS}/_{EV} \right) \geq P_t^{RD} \tag{18.3}$$

$$P_{i,t}^{RU,Gen} + P_{i,t}^{Gen} \leq P_i^{Max,Gen} \tag{18.4}$$

$$P_{i,t}^{Gen} - P_{i,t}^{RD,Gen} \geq P_i^{Min,Gen} \tag{18.5}$$

$$P_{i,t}^{Ch,ESS}/_{EV} - P_{i,t}^{RU,Ch,ESS}/_{EV} \geq 0 \tag{18.6}$$

$$P_{i,t}^{Ch,ESS}/_{EV} + P_{i,t}^{RD,Ch,ESS}/_{EV} \leq P_i^{Max,Ch,ESS}/_{EV} \cdot \alpha_{i,t}^{Ch,ESS}/_{EV} \tag{18.7}$$

$$P_{i,t}^{Dis,ESS}/_{EV} + P_{i,t}^{RU,Dis,ESS}/_{EV} \leq P_i^{Max,Dis,ESS}/_{EV} \cdot \alpha_{i,t}^{Dis,ESS}/_{EV} \tag{18.8}$$

$$P_{i,t}^{Dis,ESS}/_{EV} - P_{i,t}^{RD,Dis,ESS}/_{EV} \geq 0 \tag{18.9}$$

$$\alpha_{i,t}^{Ch,ESS/EV} + \alpha_{i,t}^{Dis,ESS/EV} \leq 1 \cdot \alpha_{i,t}^{V2G,EV} \tag{18.10}$$

$$E_{i,t}^{ESS/EV} = E_{i,t-1}^{ESS/EV} - \left(P_{i,t}^{Ch,ESS/EV} \cdot \rho_i^{Ch,ESS/EV} - \frac{P_{i,t}^{Dis,ESS/EV}}{\rho_i^{Dis,ESS/EV}} \right) \cdot \Delta t - P_{i,t}^{Driving,EV} \cdot \Delta t \tag{18.11}$$

$$P_{i,t}^{RD,Ch,ESS/EV} \cdot \rho_i^{Ch,ESS/EV} + \frac{P_{i,t}^{RD,Dis,ESS/EV}}{\rho_i^{Dis,ESS/EV}} \leq \left(E_i^{Max,ESS/EV} - E_{i,t}^{ESS/EV} \right) \bigg/ \Delta t \tag{18.12}$$

$$P_{i,t}^{RU,Ch,ESS/EV} \cdot \rho_i^{Ch,ESS/EV} + \frac{P_{i,t}^{RU,Dis,ESS/EV}}{\rho_i^{Dis,ESS/EV}} \leq \left(E_{i,t}^{ESS/EV} - E_i^{Min,ESS/EV} \right) \bigg/ \Delta t \tag{18.13}$$

$$E_i^{Min,ESS/EV} \leq E_{i,t}^{ESS/EV} \leq E_i^{Max,ESS/EV} \tag{18.14}$$

Based on the mathematical optimization model, the ramp-up/down services provided by generation units and ESSs/EVs should cope with their operational characteristics. In this regard, the constraints of the optimization model are organized in a way that addresses the capacity as well as the charging and discharging rates of their batteries. Moreover, the objective function (18.1) aims to minimize the real-time operational cost of the system, while ensuring that the total required ramp-up/down capacities are addressed in (18.2)–(18.3) by dispatchable generation units and ESSs/EVs. Finally, the loss of opportunity cost associated with the flexible units that participated in providing FRP could be determined by comparing the profits of each unit while considering the FRP service and without considering the FRP service in the optimization model. In this regard, utilizing MPC concept for incorporating the optimization of ESSs/EVs in future time intervals, while scheduling the current dispatch period, would enable the system operator to efficiently estimate the loss of opportunity cost associated with each of the flexible resources. It is noteworthy that the developed model would result in maximizing the social welfare as the optimum operating point of the system would be determined by considering cooptimization of energy and ramp markets.

- Decentralized operation of flexible resources

Nowadays, it is typical that ESSs and EVs would be operated by independent agents, which strive to maximize their profits. In this regard, the system operator would not be able to directly exploit their scheduling to activate flexible ramp service from the flexible resources. That is why the loss of opportunity scheme for compensation of the costs of flexible resources for participation in the ramp market could not be deployed for operational scheduling of independently operated EVs/ESSs. In other words, the system operator could not determine the cost associated with the rescheduling of EVs and ESSs to procure FRP. Accordingly, a model of auction markets should be employed by the system operator, where the system operator determines the incorporation of EVs and ESSs in providing FRP based on the received offers/bids from their respective agents. In this regard, the ramp market could be organized in an auction form, where all the flexible resources would be able to participate in providing ramp service for the next time interval. On the other hand, EVs/ESSs could also determine their participation in providing ramp service by announcing FRP offers to the system operator. In the current form of ramp markets run by CAISO and MISO, system operators activate FRP from generation units based on their respective loss of opportunity costs; therefore, the price of activating FRP from generation units could be announced to responsive energy limited flexible units (i.e., EVs and ESSs). A simplified optimization model conducted by the independent agents responsible for the operational scheduling of EVs/ESSs is represented in (18.15)–(18.24). Consequently, the price of FRP could be forecasted by the agent or taken as a parameter based on the received prices from the system operator. Moreover, considering the current form of ramp markets, the optimization model is structured based on determining FRP for the next time interval by employing MPC concept to model the scheduling in the future time intervals. It is noteworthy that the optimization model could also be revised to be deployed for scheduling the resources for the ramp market in the day-ahead [7]. In case the ramp market is cleared in day-ahead, resources would provide FRP based on the deployment request by the system operator. In this respect, the system operator intends to procure FRP in the day-ahead market to ensure the ramping capability of the system in the real-time dispatch. Consequently, the performance of the resources in the real time would be monitored by the system operator and resources would have to compensate their shortages for providing ramp services in the prespecified limits based upon the designated penalty factor by the operator. In these conditions, the penalty factor would be proportionate with the real-time price of procuring FRP. It is noteworthy that the day-ahead market is typically conducted on an hourly basis, while ramp service activation would be deployed in 10–5 min before the dispatch time. In this regard, the resources that participated in the day-ahead market should be able to adapt to the perquisite deployment time period of the FRP. Furthermore, the FRU/FRD deployment possibilities could be modeled by the agents of EVs/ESSs using their associated probability models derived from the historical data. Finally, machine learning methodologies could be employed to increase the accuracy of the optimization model by improving the accuracy of ramp capability forecasting as well as FRP deployment probability and energy prices in future time periods.

$$\text{Max Profit}_{t''} = \sum_{i \in \omega^{ESS}/EV} \left(\left(P_{i,t}^{RU,Ch,ESS}/EV + P_{i,t}^{RU,Dis,ESS}/EV \right) \cdot \lambda_t^{RU} + \left(P_{i,t}^{RD,Ch,ESS}/EV + P_{i,t}^{RD,Dis,ESS}/EV \right) \cdot \lambda_t^{RD} \right)\Big|_{t=t''}$$

$$+ \sum_{i \in \omega^{ESS}/EV} \sum_{t \geq t''} \left(\left(P_{i,t}^{Dis,ESS}/EV - P_{i,t}^{Ch,ESS}/EV \right) \cdot \lambda_t^{RT} \right)$$

$$(18.15)$$

$$P_{i,t}^{Ch,ESS}/EV - P_{i,t}^{RU,Ch,ESS}/EV \geq 0 \tag{18.16}$$

$$P_{i,t}^{Ch,ESS}/EV + P_{i,t}^{RD,Ch,ESS}/EV \leq P_i^{Max,Ch,ESS}/EV \cdot \alpha_{i,t}^{Ch,ESS}/EV \tag{18.17}$$

$$P_{i,t}^{Dis,ESS/EV} + P_{i,t}^{RU,Dis,ESS/EV} \leq P_i^{Max,Dis,ESS/EV} \cdot \alpha_{i,t}^{Dis,ESS/EV} \tag{18.18}$$

$$P_{i,t}^{Dis,ESS/EV} - P_{i,t}^{RD,Dis,ESS/EV} \geq 0 \tag{18.19}$$

$$\alpha_{i,t}^{Ch,ESS/EV} + \alpha_{i,t}^{Dis,ESS/EV} \leq 1 \cdot \alpha_{i,t}^{V2G,EV} \tag{18.20}$$

$$E_{i,t}^{ESS/EV} = E_{i,t-1}^{ESS/EV} - \left(P_{i,t}^{Ch,ESS/EV} \cdot \rho_i^{Ch,ESS/EV} - \frac{P_{i,t}^{Dis,ESS/EV}}{\rho_i^{Dis,ESS/EV}} \right) \cdot \Delta t - P_{i,t}^{Driving,EV} \cdot \Delta t \tag{18.21}$$

$$P_{i,t}^{RD,Ch,ESS/EV} \cdot \rho_i^{Ch,ESS/EV} + \frac{P_{i,t}^{RD,Dis,ESS/EV}}{\rho_i^{Dis,ESS/EV}} \leq \left(E_i^{Max,ESS/EV} - E_{i,t}^{ESS/EV} \right) \Big/ \Delta t \tag{18.22}$$

$$P_{i,t}^{RU,Ch,ESS/EV} \cdot \rho_i^{Ch,ESS/EV} + \frac{P_{i,t}^{RU,Dis,ESS/EV}}{\rho_i^{Dis,ESS/EV}} \leq \left(E_{i,t}^{ESS/EV} - E_i^{Min,ESS/EV} \right) \Big/ \Delta t \tag{18.23}$$

$$E_i^{Min,ESS/EV} \leq E_{i,t}^{ESS/EV} \leq E_i^{Max,ESS/EV} \tag{18.24}$$

The optimization model is aimed to maximize profits of ESSs/EVs while participating in real-time energy and ramp markets. In this regard, the operational constraints of EVs/ESSs are organized in a way that proposed ramp-up/down services fulfill their operational characteristics, i.e., capacity and charging/discharging rates. It is considered that the price of ramp-up/down would be received from the operator; however, the agents could also estimate the market prices and employ a similar model to determine their scheduling. It is noteworthy that based on the current ramp-market modeling in CAISO/MISO, the system operator would determine the price based on the generation units. In this regard, the prices associated with providing ramp service are predetermined and the provided ramp service by EVs/ESSs could cover the ramping shortage or even replace the ramping capacity that could be procured from generation units. However, the ramp market could also be conducted in an auction form where all the units provide the system operator with their offered FRPs and prices to clear the market. Furthermore, in case that the offered FRU/FRD would not completely be accepted by the system operator, the chance of acceptance and deployment could be modeled utilizing a distribution probability model developed based on the historical data [7]. Generally, the ramp market regulations and constraints would be updated by increasing the role of independently operated flexible units in power systems. In this context, the optimization model associated with the participation of EVs/ESSs in the flexible ramp market could be revised to cope with the structure of the ramp market.

- 4.1.3. New concepts for activating flexible ramp service

- Ramping capacity of distribution systems

The new developments in flexibility ramp capability of the system as well as modern control concepts could be incorporated to devise new frameworks for addressing the ramp issues in the system. In this regard, potential congestions in transmission networks have resulted in the development of the zonal flexible ramp capacity analysis concept. Zonal flexibility ramp analysis is also in accordance with the concept of locational based ancillary service market design, which aims to ensure that each area would be able to address its local operational issues. Similarly, this new viewpoint intends to provide practical operational management schemes enabling the system operators to ensure the ramping capability of the system in each predefined zone of a network [23]. On the other hand, the integration of flexible resources such as EVs and ESSs in distribution systems has been increased in recent years. Consequently, providing FRP from local flexible resources in distribution systems could play a key role in order to ensure the supply—demand balance in each zone of the system. In this regard, new control management schemes are required in order to enable distribution system operators (DSOs) to activate flexibility services from local responsive resources connected to distribution grids [24].

Local flexible resources due to their small sizes are typically unable to participate in the electricity markets conducted by the ISOs. In this regard, the idea of managing flexible ramp services in distribution systems would also facilitate the collaboration of local flexible resources, which based on the ongoing procedure of the significant integrations of responsive resources in the distribution systems would considerably improve flexibility of the system [25,40]. One of the methodologies for procuring FRP would be conducting a local ramp market in the distribution system. It is noteworthy that the transition from the electricity markets run by the ISOs in transmission level to the markets conducted in distribution systems is one of the potential management choices for future systems with distributed structures. In this regard, the previous discussions about the participation of EVs and ESSs in the ramp market would be applicable for local ramp markets conducted by DSOs. Finally, the elevation of local flexibility ramp markets requires the coordination between DSO and ISO in order to define the flexibility zones and the required ramping capacity in each zone.

- FRP management in distribution systems using TE concept

Flexibility ramp service could also be procured at the distribution level without deploying conventional electricity markets. In this context, new control algorithms have recently been developed that facilitate the scheduling of independently operated flexible resources in the distribution system to ensure supply—demand balance. Noted that the independent operation of local flexible resources impeding the direct control of local flexible resources is the main issue in exploiting operational scheduling of these resources to procure flexibility services. In this regard, the control algorithm should enable system operators to incentivize the collaboration of flexible resources in providing flexibility ramp service in real time, while addressing the objective functions of the flexible resources as well as their privacy concerns. In this regard, along with conventional control algorithms, transactive energy (TE) concept, which is structured based on organizing a control scheme utilizing value-driven signals, seems to be a practical way of activating flexibility service from EVs/ESSs in distribution systems [26]. In other words, TE could be employed by DSOs in order to incentivize the independent entities to reschedule their respective local flexible resources with the aim of providing flexibility ramp services.

TE concept as a new energy management framework copes with the multi-agent structure of distribution systems [39]. In other words, TE facilitates the integration of distributed energy resources (DERs) scheduled by an independent entity into the distribution grids. In this regard, a set of economic-based mechanisms would be employed in energy management frameworks to enable the system operators to incentivize the contribution of independent agents in the operational management of power systems. In this regard, TE control signals could be deployed by system operators to procure flexibility service from local responsive resources. This would finally result in the improvement of the power systems' reliability and security by providing flexibility service required for dynamic supply—demand balance in the system. In this vein, field studies in Refs. [27,28] have also shown the practicality of employing TE control signals for procuring flexibility service for addressing operational issues (i.e., congestion) in the distribution grid. Consequently, utilizing TE control concept seems to be a practical option for activating local flexibility from consumers and independently operated resources in order to improve the ramping capacity of the system. This control method could be well organized by the transition of power systems toward smart grids, which requires the state-of-the-art structures for facilitating information exchange between independent agents of the system.

The simplified conceptual model of information exchange between the system operator and agents is represented in Fig. 18.4. Based upon the TE concept, the information exchange should be limited to flexibility capacities and economic signals (i.e., bonus) in order to address agents' concerns regarding the privacy of their respective resources. In this regard, DSOs would increase the bonus as a TE control signal with the aim of procuring more flexibility capacity from the agents scheduling EVs/ESSs in the respective time interval [29]. This iterative control scheme would be conducted until the step where DSO ensures that the flexibility ramping capacity of the system meets the forecasted ramping up/down in the net load. It is noteworthy that the simple structure and practicality of the TE-based

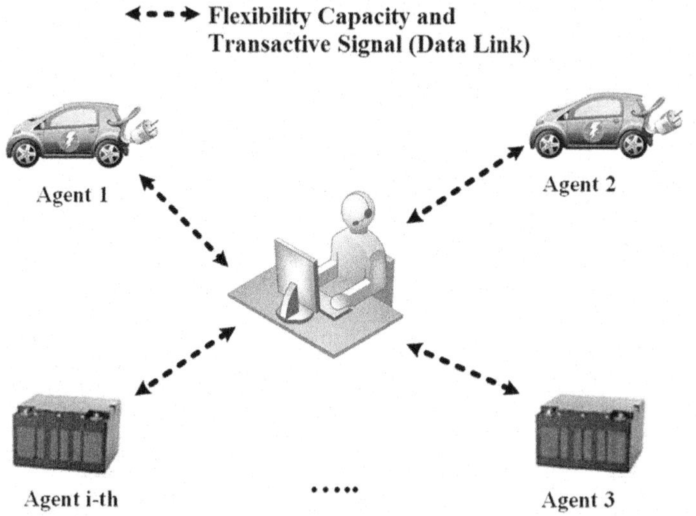

FIGURE 18.4 Simplified model of information exchange between system controller and EVs/ESSs agents.

control frameworks would result in the expansion of flexibility ramp management algorithms devised based on this new concept. Furthermore, it is noteworthy that the TE control signal could be directly sent to the customers, which would minimize the risks associated with agents responsible for the operational management of resources. As a result, TE-based flexibility ramp management in distribution systems seems to be a valuable asset based on the current increasing trend of integrating EVs/ESSs as well as the expansion of multi-agent structures in distribution systems, which would lead to promoting practical local electricity markets. Note that the participation of local flexible resources such as ESSs/EVs in ramp market would increase their profits, which would finally cause more investments in the expansion of these flexible resources in the system.

- Flexibility ramp service exchange between independent agents

Along with procuring FRP from responsive resources by DSO/ISO, flexibility ramp service in power systems could be exchanged between independent entities. In this context, independent entities would be able to procure flexibility service from the independently operated flexible resources with the aim of maximizing their profits or addressing the ramping constraints deployed by the system operators [6,30]. In other words, entities that participate in transmission-level electricity markets could procure flexibility ramp service in order to meet their offered energy/regulation services in a cost-efficient way. In addition, by the significant integration of variable energy resources in local systems, it would be conceivable that system operators limit the ramping possibility in net load of the local areas in order to meet the flexibility ramp service that could be provided by resources connected to transmission systems. Consequently, receiving flexibility ramp service from the flexible resources such as EVs and ESSs would be an efficient way for independent entities to avoid penalties determined by system operators as well as real-time prices in the upper-level system for compensation of the shortage in the ramping capacity.

- 4.1.4. Further operational points associated with the participation of EVs/ESSs in ramp markets

- As mentioned, ramp service would frequently be deployed by the system operator to address the shortage of ramping capacity in the system. However, frequent discharging and charging of EVs/ESSs would decrease their batteries' lifetime, which should be taken into account in case of participation in ramp markets. In this regard, the decrease in the lifetime of EVs/ESSs could be modeled as operational cost in the optimization of ramp service. Consequently, precise modeling of the cost associated with charging/discharging of the EVs/ESSs should be considered in the optimization associated with the operational scheduling of these resources.
- EVs could be operated in G2V or V2G modes, which should be taken into account in their operational optimization. In this respect, while V2G could increase their operational profits, it could result in decreasing their lifetime, which should be taken into account in the optimization modeling [6,31].
- The agents responsible for the scheduling of EVs/ESSs have to estimate the electricity prices, behaviors of drivers, system ramping capacity condition, etc., which could not be forecasted precisely. In this regard, efficient algorithms should be taken into account to address the uncertainty associated with these parameters.

✔ The stochastic programming method, which considers a set of operational scenarios as well as their respective probability distributions, could be deployed to model the operational optimization of EVs/ESSs in order to determine a more economical solution [32]. Moreover, risk analysis techniques such as conditional value at risk (CVaR) could be employed to consider a compromise between operational costs and operational risks in the optimization of EVs/ESSs [33].

✔ Robust optimization technique, which aims to provide an optimal solution that addresses all realizations of uncertain parameters, is an effective algorithm that could be taken into account for operational optimization of EVs/ESSs. This approach requires limited information in comparison with the stochastic optimization method, which necessitates the distribution probability of the uncertain parameters to develop operational scenarios.

✔ Besides stochastic and robust optimization algorithms, distributionally robust optimization [34], data-driven robust optimization [35], chance-constrained optimization [36], and information gap decision theory (IGDT) [37] could be deployed for addressing uncertainty in the optimization of flexible resources.

• While EVs could provide the system operator with the flexibility ramp service, their uncoordinated operational scheduling could deteriorate the ramping associated with the net load of the system. In other words, charging of EVs in the evening after the working time could significantly increase the load of the system, which could cause high ramps in the net load. In this regard, system operators as discussed in Ref. [38] could employ a tariff control system in order to impede the simultaneous charging of EVs in charging stations during the evening periods, which finally results in decreasing the ramp-up in the net load of the system.

5. Conclusion

One of the operational issues in power systems is the net-load ramping, which requires ancillary services to ensure the balance of supply and demand in real-time operation of the system. This condition would be deteriorated by the current increasing trend of integrating RESs in power system. In this regard, uncertainty and variability associated with power production by RESs would cause significant ramping in the net load of systems. That is why new frameworks have to be developed to improve ramping capability of the system. Consequently, incorporation of flexible resources such as EVs and ESSs could enable system operators to efficiently manage the shortage of ramping capacity in the system. In this regard, the operational optimization of EVs/ESSs with the aim of participation in the frameworks developed to provide flexible ramping services in power systems has been investigated in this chapter.

References

[1] H. Holttinen, P. Meibom, C. Ensslin, L. Hofmann, J. Mccann, J. Pierik, Design and operation of power systems with large amounts of wind power, in: VTT Research Notes, vol. 2493, Citeseer, 2009.

[2] T. Leiskamo, Definition of Flexibility Products for Multilateral Electricity Markets, 2019.

[3] J. Villar, R. Bessa, M. Matos, Flexibility products and markets: literature review, Elec. Power Syst. Res. 154 (2018/01/01) 329–340.

[4] P. Denholm, et al., Overgeneration from Solar Energy in California: A Field Guide to the Duck Chart, U.S. National Renewable Energy Laboratory NREL/TP-6A20-65023 (2015).

[5] IRENA, Innovative Ancillary Services, 2019. Available: https://www.irena.org/-/media/Files/IRENA/Agency/Publication/2019/Feb/IRENA_Innovative_ancillary_services_2019.pdf?la=en&hash=F3D83E86922DEED7AA3DE3091F3E49460C9EC1A0.

[6] B. Zhang, M. Kezunovic, Impact on power system flexibility by electric vehicle participation in ramp market, IEEE Trans. Smart Grid 7 (3) (2016) 1285–1294.

[7] J. Hu, M.R. Sarker, J. Wang, F. Wen, W. Liu, Provision of flexible ramping product by battery energy storage in day-ahead energy and reserve markets, IET Gener. Transm. Distrib. 12 (10) (2018) 2256–2264.

[8] Y. Yasuda, et al., Flexibility chart: evaluation on diversity of flexibility in various areas, in: 12th International Workshop on Large-Scale Integration of Wind Power into Power Systems as Well as on Transmission Networks for Offshore Wind Farms, WIW13, Energynautics GmbH, 2013.

[9] Q. Wang, H. Wu, A.R. Florita, C.B. Martinez-Anido, B.-M. Hodge, The value of improved wind power forecasting: grid flexibility quantification, ramp capability analysis, and impacts of electricity market operation timescales, Appl. Energy 184 (2016) 696–713.

[10] E. Lannoye, D. Flynn, M. O'Malley, Evaluation of power system flexibility, IEEE Trans. Power Syst. 27 (2) (2012) 922–931.

[11] M.A. Bucher, S. Chatzivasileiadis, G. Andersson, Managing flexibility in multi-area power systems, IEEE Trans. Power Syst. 31 (2) (2015) 1218–1226.

[12] A.A. Thatte, L. Xie, A metric and market construct of inter-temporal flexibility in time-coupled economic dispatch, IEEE Trans. Power Syst. 31 (5) (2015) 3437–3446.

[13] J. Zhao, T. Zheng, E. Litvinov, A unified framework for defining and measuring flexibility in power system, IEEE Trans. Power Syst. 31 (1) (2015) 339–347.

[14] B. Kirby, M. Milligan, Examination of Capacity and Ramping Impacts of Wind Energy on Power Systems, Technical Report 2008 National Renewable Energy Laboratory.

[15] D. Godoy-González, E. Gil, G. Gutiérrez-Alcaraz, Ramping ancillary service for cost-based electricity markets with high penetration of variable renewable energy, Energy Econ. 85 (2020/01/01) 104556.

[16] CAISO, Flexible Ramping Product Refinements, 2020. Available: http://www.caiso.com/InitiativeDocuments/DraftFinalProposal-FlexibleRampingProductRefinements.pdf.

[17] L. Xu, D. Tretheway, Flexible Ramping Products, Calif. ISO (2012). Available: http://www.caiso.com/Documents/RevisedDraftFinalProposal-FlexibleRampingProduct-2012.pdf.

[18] N. Navid, G. Rosenwald, Ramp Capability Product Design for MISO Markets, 2013. Available: https://cdn.misoenergy.org/Ramp%20Capability%20for%20Load%20Following%20in%20MISO%20Markets%20White%20Paper271169.pdf.

[19] NYISO, Master Plan: Wholesale Markets for the Grid of the Future, 2018. Available: https://www.nyiso.com/documents/20142/4347040/2018-Master-Plan.pdf/88225d15-082b-c07a-b8ef-ccac3619a1ce.

[20] CAISO, Flexible Ramping Product Refinements: Appendix B, 2020. Available: http://www.caiso.com/InitiativeDocuments/DraftTechnicalDescription-FlexibleRampingProduct-Procurement-Deployment-Scenarios%20.pdf.

[21] S. Sreekumar, K.C. Sharma, R. Bhakar, Gumbel copula based multi interval ramp product for power system flexibility enhancement, Int. J. Electr. Power Energy Syst. 112 (2019/11/01) 417–427.

[22] L. Xu, D. Tretheway, Flexible Ramping Products: Incorporating FMM and EIM, Calif. ISO (2014). Available: http://www.caiso.com/Documents/DraftFinalProposal_FlexibleRampingProduct_includingFMM-EIM.pdf.

[23] Q. Wang, B. Hodge, Enhancing power system operational flexibility with flexible ramping products: a review, IEEE Trans. Ind. Inform. 13 (4) (2017) 1652–1664.

[24] H.K. Nguyen, et al., Incentive mechanism design for integrated microgrids in peak ramp minimization problem, IEEE Trans. Smart Grid 9 (6) (2018) 5774–5785.

[25] A. Majzoobi, A. Khodaei, Application of microgrids in supporting distribution grid flexibility, IEEE Trans. Power Syst. 32 (5) (2017) 3660–3669.

[26] T. Council, Gridwise Transactive Energy Framework Version 1.0, The GridWise Architecture Council, Tech. Rep., 2015.

[27] A. Masood, A. Xin, J. Hu, S. Salman, A.R. Sayed, M.U. Jan, FLECH services to solve grid congestion, in: 2018 2nd IEEE Conference on Energy Internet and Energy System Integration (EI2), 2018, pp. 1−5.

[28] A.R. Soares, et al., Distributed optimization algorithm for residential flexibility activation − results from a field test, IEEE Trans. Power Syst. 35 (5) (2019) 4119−4127.

[29] S. Fattaheian-Dehkordi, M. Tavakkoli, A. Abbaspour, M. Fotuhi-Firuzabad, M. Lehtonen, Distribution grid flexibility-ramp minimization using local resources, in: 2019 IEEE PES Innovative Smart Grid Technologies Europe (ISGT-Europe), 2019, pp. 1−5.

[30] M. Tavakkoli, et al., Bonus-based demand response using stackelberg game approach for residential end-users equipped with HVAC system, IEEE Trans. Sustain. Energy (2020). In press.

[31] X. Zhang, et al., Electric vehicle participated electricity market model considering flexible ramping product provisions, IEEE Trans. Ind. Appl. 56 (5) (2020) 5868−5879.

[32] H. Gharibpour, F. Aminifar, Multi-stage equilibrium in electricity pool with flexible ramp market, Int. J. Electr. Power Energy Syst. 109 (2019/07/01) 661−671.

[33] A.J. Conejo, M. Carrión, J.M. Morales, Decision Making under Uncertainty in Electricity Markets, Springer, 2010.

[34] E. Guevara, F. Babonneau, T. Homem-de-Mello, S. Moret, A machine learning and distributionally robust optimization framework for strategic energy planning under uncertainty, Appl. Energy 271 (2020/08/01) 115005.

[35] F. Shen, L. Zhao, W. Du, W. Zhong, F. Qian, Large-scale industrial energy systems optimization under uncertainty: a data-driven robust optimization approach, Appl. Energy 259 (2020/02/01) 114199.

[36] J.D. Vergara-Dietrich, M.M. Morato, P.R.C. Mendes, A.A. Cani, J.E. Normey-Rico, C. Bordons, Advanced chance-constrained predictive control for the efficient energy management of renewable power systems, J. Process Contr. 74 (2019/02/01) 120−132.

[37] M.A. Mirzaei, A. Sadeghi-Yazdankhah, B. Mohammadi-Ivatloo, M. Marzband, M. Shafie-khah, J.P.S. Catalão, Integration of emerging resources in IGDT-based robust scheduling of combined power and natural gas systems considering flexible ramping products, Energy 189 (2019/12/15) 116195.

[38] Z. Moghaddam, I. Ahmad, D. Habibi, M.A.S. Masoum, A coordinated dynamic pricing model for electric vehicle charging stations, IEEE Trans. Transp. Electr. 5 (1) (2019) 226−238.

[39] S. Fattaheian-Dehkordi, et al., An incentive-based mechanism to alleviate active power congestion in a multi-agent distribution system, IEEE Trans. Smart Grid (2020). In press.

[40] S. Fattaheian-Dehkordi, et al., Incentive-Based Ramp-Up Minimization in Multi-Microgrid Distribution Systems, 2020 IEEE PES Innovative Smart Grid Technologies Europe (ISGT-Europe) (2020) 1−5.

Author Index

Subject Index

Printed in the United States
by Baker & Taylor Publisher Services